MECHANISMS OF OXIDIZING ENZYMES

INTERNATIONAL SYMPOSIUM ON MECHANISMS OF OXIDIZING ENZYMES

La Paz, Baja California Sur, Mexico
December 5-7, 1977

ORGANIZING COMMITTEE

RAUL N. ONDARZA, *Chairman*
FELIX CORDOBA, DALE E. EDMONDSON,
VINCENT MASSEY, and THOMAS P. SINGER, *Co-Chairmen*

SPONSORS

Government of the State of Baja California Sur
National Council of Science and Technology
Department of Health and Welfare of Mexico
International Union of Biochemistry
Organization of American States
Panamerican Association of Biochemical Societies

National Autonomous University of Mexico
National Polytechnical Institute
Metropolitan Autonomous University of Mexico, Iztapalapa Branch
Center of Research and Advanced Studies, Polytechnic Institute of Mexico
University of California, San Francisco

Miles Institute of Experimental Therapeutics, Mexico
Hoffman-La Roche, Inc.
Burroughs-Wellcome, Inc.

Developments in Biochemistry, volume 1

International Symposium on

MECHANISMS
OF
OXIDIZING ENZYMES

Edited by

Thomas P. Singer

Department of Biochemistry and Biophysics,
University of California, San Francisco,
and Molecular Biology Division,
Veterans Hospital, San Francisco

Raul N. Ondarza

Faculty of Medicine, National Autonomous
University of Mexico, and National Council
of Science and Technology, Mexico City

ELSEVIER/NORTH-HOLLAND

AMSTERDAM · NEW YORK · OXFORD

ELSEVIER NORTH-HOLLAND, INC.
52 Vanderbilt Avenue, New York, New York 10017

ELSEVIER/NORTH-HOLLAND BIOMEDICAL PRESS
Jan Van Galenstraat 335
Amsterdam, The Netherlands

Library of Congress Cataloging in Publication Data

International Symposium on Mechanisms of Oxidizing
    Enzymes, La Paz, Mexico, 1977.
    Mechanisms of oxidizing enzymes.

    (Developments in biochemistry ; v. 1)
    Proceedings of the symposium held Dec. 5-7, 1977.
    Includes bibliographical references and index.
    1. Oxidases—Congresses. 2. Oxidoreductases—
Congresses. 3. Oxidation, Physiological—Congresses.
I. Singer, Thomas Peter, 1920-   II. Ondarza, Raul
N. III. Title. IV. Series.
QP603.08I57 1977   574.1'9258   78-1706
ISBN 0-444-00265-0

Manufactured in the United States of America

CONTENTS

LIST OF PARTICIPANTS

BRIAN A.C. ACKRELL
University of California
San Francisco

S.P.J. ALBRACHT
University of Amsterdam
Amsterdam

GERARDO AVILA
Metropolitan Autonomous University
Mexico City

HELMUT BEINERT
University of Wisconsin
Madison

ALEJANDRO BLANCO LABRA
National Autonomous University of Mexico
Mexico City

VICTOR MANUEL MAGDALENO
Mexican Institute of Social Security
Mexico City

ARMANDO CINTORA GOMEZ
Metropolitan Autonomous University
Mexico City

FELIX CORDOBA
Center of Biological Studies
La Paz

DALE E. EDMONDSON
University of California
San Francisco

EDGARDO ESCAMILLA
National Autonomous University of Mexico
Mexico City

SERGIO ESTRADA-ORIHUELA
Metropolitan Autonomous University
Mexico City

JAMES A. FEE
University of Michigan
Ann Arbor

HARVEY F. FISHER
Veterans Administration Hospital
Kansas City

ANTONIO GARCIA TREJO
Metropolitan Autonomous University
Mexico City

SANDRO GHISLA
University of Konstanz
Konstanz

CARLOS GOMEZ-LOJERO
Center of Research and Advanced Studies
National Polytechnical Institute
Mexico City

J.W. GORROD
Chelsea College
London

CONCEPCION GUTIERREZ
Metropolitan Autonomous University
Mexico City

JAVIER GUTIERREZ
Autonomous University of Chihuahua
Chihuahua

PETER HEMMERICH
University of Konstanz
Konstanz

BARBARA HOSEIN
University of the West Indies
Kingston

ix

EDNA B. KEARNEY
Veterans Administration Hospital and
University of California,
San Francisco

WILLIAM C. KENNEY
University of California
San Francisco

EVERARDO LOPEZ-ROMERO
Center of Research and Advanced Studies
National Polytechnical Institute
Mexico City

GUSTAVO MARTINEZ ZEDILLO
Mexican Institute of Social Security
Mexico City

JAIME MARTUSCELLI
National Autonomous University of Mexico
Mexico City

VINCENT MASSEY
University of Michigan
Ann Arbor

JOSE BERNARDO MIGUEZ
University of Pamplona
Pamplona (Columbia)

JAIME MORA
National Autonomous University of Mexico
Mexico City

L.E. MORTENSON
Purdue University
West Lafayette,

ENRIQUE G. OESTREICHER
Federal University of Rio de Janeiro
Rio de Janeiro

RAUL N. ONDARZA
National Council of Science and Technology
and
National Autonomous University of Mexico
Mexico City

NORMAN OPPENHEIMER
University of California
San Francisco

W.H. ORME-JOHNSON
University of Wisconsin
Madison

CARLOS M. OROPEZA SALIN
National Autonomous University of Mexico
Mexico City

RAFAEL PALACIOS
National Autonomous University of Mexico
Mexico City

ISRAEL PECHT
Weizmann Institute of Science
Rehovot

LESTER PACKER
University of California
Berkeley

GRAHAM PALMER
Rice University
Houston

JACK PEISACH
Yeshiva University
The Bronx

JOHN POSTGATE
University of Sussex
Sussex

MANUEL ROBERT
National Autonomous University of Mexico
Mexico City

M.A. ROSEMEYER
University College
London

JOSE RUIZ-HERRERA
Center for Research and Advanced Studies
National Polytechnical Institute
Mexico City

ESTELA SANCHEZ DE JIMENEZ
National Autonomous University of Mexico
Mexico City

HEINER SCHIRMER
Max Planck Institute for Medical Research
Heidelberg

LEWIS M. SIEGEL
Duke University
Durham

THOMAS P. SINGER
Veterans Administration Hospital and
University of California
San Francisco

DANIEL STEENKAMP
University of California
San Francisco

MIREYA TORO
Metropolitan Autonomous University
Mexico City

VICTOR MANUEL VALDES
National Autonomous University of Mexico
Mexico City

PARTAB T. VARANDANI
Fels Research Institute
Yellow Springs

CHRISTOPHER WALSH
Massachusetts Institute of Technology
Cambridge

CHARLES H. WILLIAMS, JR.
Veterans Administration Hospital
and University of Michigan
Ann Arbor

TAKASHI YONETANI
University of Pennsylvania
Philadelphia

PREFACE

This book represents the proceedings of the International Symposium on Mechanisms of Oxidizing Enzymes, held in La Paz, Baja California Sur, Mexico on December 5-7, 1977. Some comments may be in order both in regard to the site of the meeting and the subject selected.

Situated in a beautiful bay, at the southern tip of Lower California, near the confluence of the Pacific and the Sea of Cortez, La Paz is a quiet, historic city with a tranquil atmosphere and ideal climate. The Organizing Committee felt that these peaceful surroundings would provide an inspiring setting for a brief but intensive workshop, away from the distractions of big cities. Moreover, the far-sighted administration of the State of Baja California Sur and its Governor, Lic. Angel Cesar Mendoza-Aramburo, recognizing the importance of cultural and scientific factors to the economic well-being and development of the region, have actively and generously supported the Symposium. Holding the meeting in La Paz was also timely, because the National Council of Science and Technology of Mexico have recently established one of their research centers here, adding to their network of existing research centers in Ensenada, Saltillo, and San Cristobal de las Casas.

The subject of the workshop was not one of the traditional themes around which national and international meetings are held at recurring intervals. Thus, while workers in such subspecialties as flavoproteins, copper enzymes, iron enzymes (especially P-450 systems) have opportunities to meet periodically for discussions, analysis of data, and exchange of information, and such meetings have been largely responsible for the impressive advances in these fields in recent years, there is seldom time at such meetings to consider advances in even closely allied fields. As a result, specialists in flavin chemistry and flavoenzymes often find difficulty in keeping up with impressive advances in understanding how DPN- and TPN- linked enzymes work or with the important lessons to be learned from structural and mechanistic studies on cytochrome oxidase, although such information may be of immediate relevance to their current research. The Organizing Committee therefore thought that it would be useful to bring together a small group of experts representing all the subspecialties of the subject of biological oxidation mechanisms in a workshop where some examples of important recent advances, of current thinking, and of unsolved problems in each field would be illustrated, in the hope of crossing the barriers among these fields, of applying knowledge accrued in allied fields, and of focussing the attention of experts in other subspecialties to unresolved questions in one's own field. In order to achieve these goals the organizers obviously had to limit both the number of participants and of presentations, so as to allow considerable time for discussion.

The workshop was preceded by an intensive, advanced course on enzymes, held at the Faculty of Chemistry, Department of Biochemistry, University of Mexico, Mexico City, in order to take advantage of the presence of noted enzymologists from other countries. The course was presented by six lecturers and was attended by 35 postgraduate students from Mexico and other Latin American countries.

The meeting in La Paz appears to have been very successful, as judged by the fact that all participants, students, postdoctorals, and recognized experts in their fields felt that they learned much from both the formal presentations and the unstructured discussions. There was just the right balance between intensive work and relaxation, highlighted by a spectacular luncheon-reception by the Governor of the State and a fascinating after-dinner lecture by Dr. Santiago Genovés on his three Transatlantic crossings aboard rafts. The only disappointment was the inability of a few of the invited participants to come because of illness, including Drs. R. Abeles, T.C. Bruice, N.O. Kaplan, N.R. Orme-Johnson, and J. Shore.

The Organizing Committee is most grateful to all the sponsors of the Symposium, whose list appears at the front of this volume. Special thanks are due to Lic. Angel Cesar Mendoza-Aramburo, Governor of the State, Dr. Edmundo Flores, Director General of the National Council of Science and Technology, and Dr. Guillermo Soberón, Rector of the National University of Mexico, who had the vision to appreciate the importance of basic scientific research to the future of Mexico and, at a time of severe economic retrenchment, lent generous financial aid and considerable personal efforts to assure the success of this meeting, to the Symposium Committee of the I.U.B. and its chairman, Peter Campbell, who provided the seed money and the prestige of their sponsorship which assured support by other organizations, and to the Organization of American States who made it possible for young scientists from various corners of Mexico, the Caribbean, and South America to participate in the meeting. Mrs. Helen Dowe de Mendoza and Miss Aracely Cacho were primarily responsible for the efficient arrangements and the smooth running of the Symposium.

DISULFIDE REDUCTASES

FUNCTIONAL ROLES OF THE TWO ACTIVE SITE CYSTEINE RESIDUES GENERATED ON REDUCTION OF LIPOAMIDE DEHYDROGENASE AND GLUTATHIONE REDUCTASE

CHARLES H. WILLIAMS, JR., COLIN THORPE AND L. DAVID ARSCOTT
Veterans Administration Hospital and the Department of Biological Chemistry,
University of Michigan, Ann Arbor, Michigan 48105 (U.S.A.)

ABSTRACT

Lipoamide dehydrogenase and glutathione reductase are flavoproteins each containing an oxidation-reduction active cystine residue. Two electron reduction generates two cysteine residues. It has been shown with pig heart lipoamide dehydrogenase that the active site cysteine residue nearer the amino-terminus is at least 13-fold more reactive toward iodoacetamide than is its redox active partner [Thorpe, C. and Williams, C.H., Jr. (1976) *J. Biol. Chem. 251*, 3553-3557]. Analogous modification of yeast glutathione reductase also results in the essentially exclusive labeling of the cysteine residue nearer the amino-terminus in this highly homologous active site region. Whereas in lipoamide dehydrogenase the alkylation causes loss of the charge transfer interaction between thiolate and FAD, in glutathione reductase this interaction continues. Both modified enzymes catalyze electron transfer between reduced pyridine nucleotides and artificial acceptors but cannot interact with their respective disulfide substrates.

The spectral changes seen upon the addition of NAD^+ to lipoamide dehydrogenase monoalkylated at the amino-terminal active site cysteine residue have been compared with those of model compounds, and it has been suggested that NAD^+ induced the formation of an adduct at the C(4a) position of FAD [Thorpe, C. and Williams, C.H., Jr. (1976) *J. Biol. Chem. 251*, 7726-7728]. The presumed substituent in the adduct is the active center cysteine residue which is not alkylated. Binding studies show that only one NAD^+ is bound per dimer. In contrast two molecules of aminopyridine adenine dinucleotide bind per enzyme dimer. Aminopyridine adenine dinucleotide does not bleach the monoalkylated derivative, but rather induces charge transfer spectrally similar to that seen in complexes of the two electron reduced unmodified enzyme with NAD^+.

These results suggest that the nascent thiol nearer to the carboxyl-terminus functions in electron transfer to the FAD through covalent catalysis and that the thiol nearer to the amino-terminus functions in interactions with the disulfide substrate probably through a mixed disulfide.

INTRODUCTION

Lipoamide dehydrogenase and glutathione reductase are closely related flavoproteins which catalyze chemically similar reactions[1] as shown*:

$$\text{Lip(SH)}_2 + \text{NAD}^+ \rightleftharpoons \text{Lip(S)}_2 + \text{NADH} + \text{H}^+$$

$$\text{G-S-S-G} + \text{NADPH} + \text{H}^+ \rightleftharpoons 2 \text{ G-S-H} + \text{NADP}^+$$

Lipoamide dehydrogenase is part of the 3 enzyme complex catalyzing the oxidative de-carboxylation of α-ketoacids, pyruvate or α-ketoglutarate. The tripeptide glutathione is present in high concentrations in most cells. It is kept largely in its reduced form by glutathione reductase. Glutathione is responsible for maintaining thiol-disulfide homeostasis in the cell. Both enzymes are quite specific for their respective pyridine nucleotide substrates.

In addition to FAD, each enzyme contains a redox active cystine residue.[1] Therefore both can accept a total of 4 electrons when reduced anaerobically by dithionite.

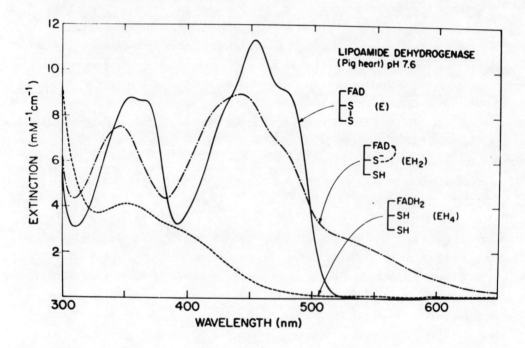

Fig. 1. Spectra of pig heart lipoamide dehydrogenase in the E, EH_2 and EH_4 redox states.

Figure 1 shows spectra of oxidized, 2-electron-reduced and 4-electron-reduced lipo-amide dehydrogenase. The analogous spectra for glutathione reductase are almost identical. The spectra for the oxidized and 4-electron-reduced enzyme are typical for FAD and $FADH_2$ having little absorbance beyond 500 nm. The 2-electron-reduced

---

Abbreviations: E, EH_2 and EH_4, oxidized, 2-electron-reduced and 4-electron-reduced enzyme respectively; G-S-H, reduced glutathione; G-S-S-G, oxidized glutathione; EHR, 2-electron-reduced enzyme alkylated with iodoacetamide; Lip(SH)_2, dihydrolipoamide; Lip(S)_2, oxidized lipoamide; AAD^+, aminopyridine adenine dinucleotide.

4

enzyme, which will be referred to as EH_2, retains the 2-banded spectrum of FAD but also has a third band at longer wavelengths. This spectrum has been interpreted by Kosower[2] and by Massey and Ghisla[3] as a charge transfer complex in which the donor is thiolate and FAD is the acceptor. EH_2 is generated by reduction with either NADH or dihydrolipoamide. It was shown to be a catalytic intermediate in the pioneering studies of Massey, Gibson and Veeger[4,5]. Dithionite titrations of lipoamide dehydrogenase proceed in 2 phases as would be expected from the spectra: E to EH_2 and EH_2 to EH_4. The sequences of amino acid residues around the active site cystine have been determined for *Escherichia coli* and pig heart lipoamide dehydrogenase[6-11], yeast and erythrocyte glutathione reductase[12,13] and for the closely related thioredoxin reductase[6,14,15]. Not only is there homology between the prokaryote and eukaryote lipoamide dehydrogenases but there is extensive homology between these and glutathione reductase. With these sequences we are able to distinguish the 2 thiols produced upon 2-electron-reduction of the enzyme in active site modifications.

Two protons as well as 2 electrons are involved in the formation of EH_2 from E at pH values below 7.9[16,17]. If, as we believe, the donor in the charge transfer is thiolate, then the second proton must be taken up by a base as shown:

The protonated base functions to stabilize the thiolate in the known hydrophobic active site by analogy with the proposed thiolate-imidazolium ion pair in papain and glyceraldehyde-3-phosphate dehydrogenase[18-21]. The model would predict loss of charge transfer upon lowering of the pH and this is observed[16]. A further role has been postulated for the base in lipoamide dehydrogenase[17] as indicated below:

pH dependent reaction mechanism   pH independent reaction mechanism

By analogy with model thiol-disulfide interchange reactions[22], the formation of EH_2 would be expected to proceed via initial attack of a deprotonated substrate molecule on the cystine bridge to form a mixed disulfide intermediate. Since the first ionization of dihydrolipoamide has a pK of 9.4, one would predict a pH *d*ependent reduction of the enzyme by dihydrolipoamide, as shown in the left panel. In fact the reduction is pH *independent* between pH 5 and 8. As shown in the right panel, this observation is accounted for by deprotonation of the substrate by the base.[17]

Figure 2 presents our "working hypothesis" mechanism for lipoamide dehydrogenase and Figure 3 lays out an analogous scheme for glutathione reductase. Oxidized lipoamide dehydrogenase is structure I at the upper left showing the redox active disulfide, the FAD and the base. Moving clockwise (the physiological direction although the reaction is freely reversible at neutral pH) dihydrolipoamide binds, is deprotonated by the base and attacks the disulfide to form a mixed disulfide. This is a presumed intermediate since we have not been able to isolate it. A second thiol-disulfide interchange releases the oxidized lipoamide. The enzyme operates by a ping-pong mechanism. The rate-limiting step is in this half reaction. Intermediate III is EH_2, the charge transfer species. Although the thiolate donor in the charge transfer complex is stabilized by its juxtaposition with the protonated base, we envisage that the alternative ion pair is catalytically significant in the reaction of EH_2 with oxidized lipoamide, i.e., EH_2 is an equilibrium mixture of at least 2 species[17]. All of the subsequent steps are very fast. NAD^+ combines and induces the formation of a covalent adduct of the thiolate and the flavin at the C(4a) position. Preliminary evidence for this species has been given[23] and will be amplified below. The step at the bottom completes the electron transfer to FAD. The elements of a hydride ion are transferred from the reduced flavin to NAD^+ in the next step and finally NADH dissociates regenerating structure I. It can be seen from Figs. 2 and 3 that the two catalytically essential cysteines play different roles in catalysis. Thus, we have represented the "upper" thiol as participating in charge transfer and covalent adduct formation with the FAD, whereas the "lower" thiol interacts with the substrate in mixed disulfide bond formation. Data are presented in the RESULTS section indicating that the electron transfer thiol is nearer the carboxyl-terminus while the interchange thiol is nearer the amino-terminus in the primary sequence.

RESULTS

Pig heart lipoamide dehydrogenase

Differential reactivity of thiols at the active site. We are accumulating evidence for the catalytic roles of both cysteine residues in the active site sequence from chemical modification studies on lipoamide dehydrogenase and glutathione reductase. We have demonstrated that the 2 cysteines generated on reduction of pig heart lipoamide dehydrogenase differ widely in their reactivity toward iodoacetamide. This enabled a derivative of EH_2 to be prepared which is alkylated at the active site cysteine residue closest to the amino-terminus of the protein as shown below. Alkylation of EH_2 is accompanied by a reappearance of a spectrum of oxidized bound flavin[24], and is associated with release of a proton (Thorpe and Williams, unpublished). Participation of the active center disulfide bridge in catalysis is thus

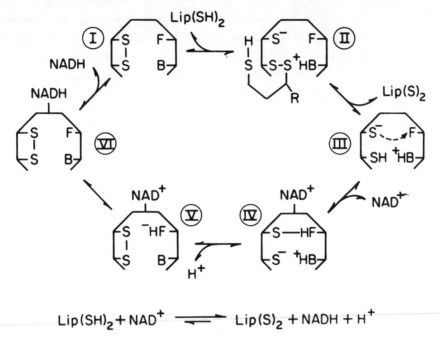

Fig. 2. Working hypothesis mechanism for lipoamide dehydrogenase

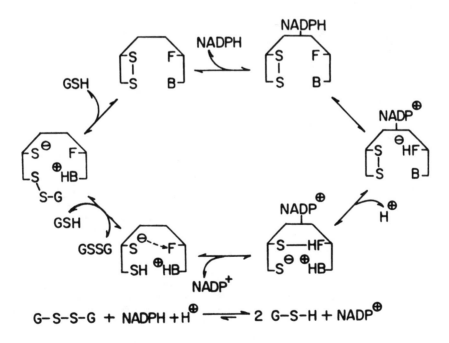

Fig. 3. Working hypothesis mechanism for glutathione reductase.

prevented by alkylation of a single residue; the enzyme is trapped at the 2-electron
reduced level. Reduction of EHR by sodium dithionite results in bleaching of the
flavin upon the uptake of only 2 electrons. EHR is catalytically inactive toward
reduced or oxidized lipoamide. However, the derivative retains a catalytically
competent pyridine nucleotide binding site since it exhibits both transhydrogenase
and diaphorase activities[24]. In addition, EHR exhibits long wavelength absorption
bands on reduction with NADH, similar to those observed between reduced flavin and
NAD^+ in the native enzyme[25].

$$CH_2C^*ONH_2 \qquad\qquad CH_2CONH_2$$
$$|\qquad\qquad\qquad\qquad\qquad |$$

Asn-Glu-Thr-Leu-Gly-Gly-Thr-Cys-Leu-Asn-Val-Gly-Cys-Ile-Pro-Ser-Lys

↑
Chymotrypsin

An adduct of thiolate and FAD at the C(4a) position. Recently we have observed
that NAD^+ binding induces a very rapid reversible partial bleaching of the flavin
chromophore of EHR at 450 nm with a concomitant rise in absorption at 390 nm[23].
Figure 4 shows a titration conducted at pH 7.6 and 15°. The changes are monophasic,

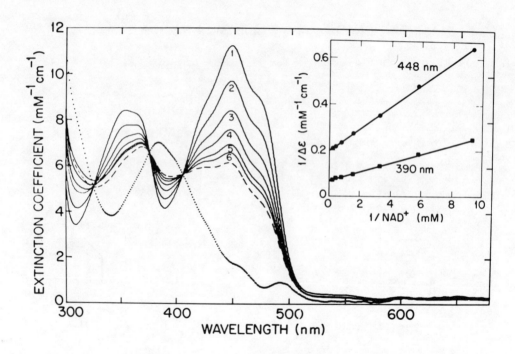

Fig. 4. Spectral changes induced on the addition of NAD^+ to EHR. Curve 1: 20.5 μM
enzyme FAD in 100 mM phosphate buffer, 0.3 mM EDTA, pH 7.6 at 15°. NAD^+ was added
to reference and sample cuvettes to give final concentrations of 0.11 mM, curve 2;
0.32 mM, curve 3; 0.67 mM, curve 4; 1.34 mM, curve 5; and 2.63 mM, curve 6. Intermediate points have been omitted for clarity. The dashed line is a spectrum extrapolated to infinite NAD^+ concentration and the dotted line is an estimate of the
spectrum of the modified flavin species (see text).

with linear double reciprocal plots (see inset) of $\Delta\epsilon$ at 448 nm and at 390 nm versus NAD^+ concentration, yielding a K_D of 240 µM. The dashed line is a spectrum
extrapolated to infinite ligand concentration using the intercept values obtained
from the inset. This extrapolated spectrum exhibits about one-half of the original
448 nm absorbance and retains certain of the spectral features of EHR (i.e., shoul-
ders at 470 nm and 425 nm and a maximum at 448 nm). Thus under these conditions,
NAD^+ only effects a partial conversion of the flavin into a species responsible for
the increase in absorbance at 390 nm. This conclusion is supported by NAD^+ titrations monitored by flavin fluorescence emission. On saturation with NAD^+, one-half
of the flavin fluorescence remains (results not shown). The extent of bleaching at
448 nm at infinite NAD^+ changes little over the pH range 6 to 8.8, although the K_D
shows a complex pH dependence in this range (Thorpe and Williams, unpublished).

Direct measurement of NAD^+ binding by the gel filtration method of Hummel and
Dreyer[26] showed that the partial bleaching observed reflects a stoichiometry of 1.2
molecules of NAD^+ bound per dimer at pH 8.8 with a K_D of 39 µM[23]. We have repeated these experiments at pH 7.6, where the binding of NAD^+ is considerably weaker,
and the results are displayed as a Scatchard plot in Fig. 5. A similar stoichiometry
is obtained of 1.1 molecules NAD^+ per dimeric EHR. The K_D of approximately 170 µM is
in fair agreement with that obtained spectrally (Fig. 4). The discrepancy may reflect the inherent limitations of the gel filtration method in measuring comparatively weak binding, and the sensitivity of the K_D to small pH changes in this region.

The dotted line in Fig. 4 is an estimate of the spectrum of the species induced
on NAD^+ binding. It was constructed assuming that the dashed line represents the
binding of 1.1 molecules NAD^+ per dimer, and that these changes would increase

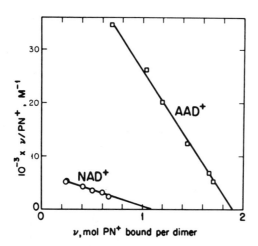

Fig. 5. Scatchard plot of the binding of NAD^+ and AAD^+ to EHR in 0.1 M phosphate
buffer, 0.3 mM EDTA, pH 7.6 at 15°. Binding of NAD^+ to EHR was determined essentially as described earlier[23] using [carbonyl-^{14}C]NAD^+. AAD^+ concentrations were
measured spectrophotometrically.

proportionately until a stoichiometry of 2 per dimer were reached and complete
bleaching effected. The spectrum obtained shows a maximum at 384 nm (ε_{384}; 7000
$M^{-1}cm^{-1}$), and is very similar to that observed during the interaction of several reduced flavoproteins with molecular oxygen, e.g., melilotate[27], bacterial luciferase[28],
and p-hydroxybenzoate hydroxylases [29]. These oxygenated intermediates are believed
to carry an oxygen substituent at the C(4a) position of the isoalloxazine ring[29-32].
We have therefore proposed, by analogy, that the species induced on NAD^+ binding
represents a C(4a) sulfur adduct[23]. Indeed several workers have postulated that a
C(4a) sulfur adduct, formed by attack of a thiolate on oxidized flavin, is an intermediate in lipoamide dehydrogenase catalysis[33-39]. An abbreviated scheme for the internal redox reaction between the reduced active site disulfide residues and FAD at
the EH_2 level of the native enzyme is shown below:

charge transfer   4-a adduct   reduced flavin

Formation of the 4a-adduct provides the mechanism by which the cystine bridge can reform via nucleophilic attack from the second thiolate. The reduced flavin (right
structure) then is reoxidized by bound NAD^+. We do not yet know when proton release
occurs in the half-reaction;

$$EH_2 + NAD^+ \rightleftharpoons E + NADH + H^+$$

it is represented in Fig. 2 as occurring between the left and middle species. In
EHR, alkylation of the lower thiol residue would still allow adduct formation, but
would prevent subsequent steps leading to reduction of NAD^+, since the disulfide
bridge cannot reform. These results support the predicted involvement of a C(4a)
adduct as a catalytic intermediate in lipoamide dehydrogenase, and further suggest
that its formation is promoted by NAD^+ binding during the pyridine nucleotide half-
reaction of the native enzyme.

The analog 3-aminopyridine adenine dinucleotide has proved to be very useful in the study of pyridine nucleotide binding in *Escherichia coli* lipoamide dehydrogenase[40]. Although AAD[+] is formally an oxidized derivative, substitution of the 3-carboamide group by an amino function confers additional electron density to the pyridinium ring, resulting in an analog exhibiting some NADH character[41,42]. AAD[+] is not reducible enzymatically[41,42]. When added to EHR at pH 7.6, AAD[+] effects spectral changes clearly different from those induced by NAD[+] (Fig. 6): notably the appearance of a broad absorption band centered at about 580 nm, a red shift in the principal maximum from 448 nm to 453 nm, and little rise at 390 nm. The final spectrum obtained is reminiscent of that exhibited by complexes between NAD[+] and 2-electron reduced native enzyme[43]. The inset shows the extinction changes at 448 nm plotted by the method of Stockell giving a value of 37 µM for the apparent K_D. The intercept on the ordinate indicates that these spectral changes are associated with the binding of 1 molecule of AAD[+] per FAD (i.e., 2 per dimer) in marked contrast to the NAD[+] results. The observation was confirmed by determining AAD[+] binding directly by gel filtration as shown in Fig. 5. Both the K_D value of 34 µM and the binding stoichiometry of 1.9 per dimer are in good agreement with those determined spectrophotometrically. The linearity of both Stockell and Scatchard plots suggests that

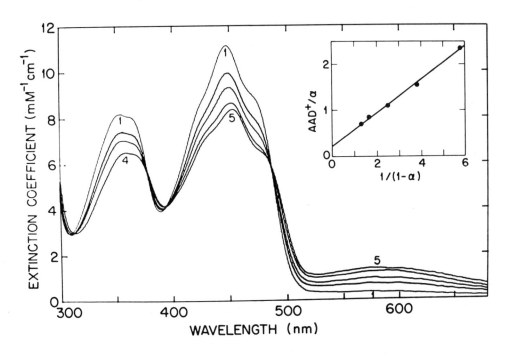

Fig. 6. The effect of AAD[+] on the visible spectrum of EHR. Curve 1: 22 µM enzyme FAD in 70 mM phosphate buffer, 0.3 mM EDTA, pH 7.6, 15°. AAD[+] was added to the reference and sample cuvettes to concentrations of 33 µM, curve 2; 65 µM, curve 3; 194 µM, curve 4; and 1 mM, curve 5. Intermediate points have been omitted for clarity.

EHR binds AAD$^+$ at 2 equivalent or near equivalent sites. Our working hypothesis is that the dimer EHR contains 2 pyridine nucleotide binding sites which are equivalent or near equivalent in the absence of NAD$^+$. NAD$^+$ binding induces adduct formation in one subunit and promotes changes in the other subunit which preclude tight binding of a second molecule of pyridine nucleotide. The binding of NAD$^+$ induces C(4a) adduct formation in that subunit.

Yeast glutathione reductase

Differential reactivity of the thiols at the active site. The 2 nascent thiols generated upon reduction of glutathione reductase to the EH$_2$ level react with iodo-acetamide at markedly different rates as is the case with lipoamide dehydrogenase. These thiols are contained in a sequence which is highly homologous with that of lipoamide dehydrogenase shown above. The very small spectral changes associated with the alkylation are shown in Fig. 7. This is in marked contrast to the complete loss of charge transfer interaction observed with lipoamide dehydrogenase (see above). Radioactive counting and amino acid analysis of the dialyzed mixture showed clearly that one mole of S-carboxamidomethylcysteine per FAD had been formed. The small in-creases in absorbance both at 530 nm and at 450 nm could be due to comproportionation

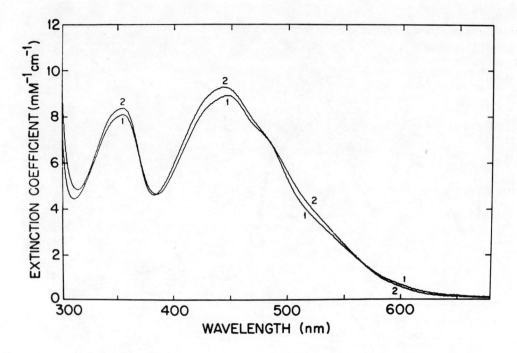

Fig. 7. Spectral changes during the alkylation of yeast glutathione reductase re-duced with 2 electron equivalents of sodium dithionite. Spectrum 1: 110 μM oxidized enzyme in 0.13 M phosphate, 0.3 mM EDTA, pH 7.2. Spectrum 2: 2.5 h after the addi-tion of 27 fold molar excess of iodoacetamide, at 25°; concentrations are based on FAD.

as alkylation of EH_2 progressed ($E + EH_4 \rightleftharpoons 2\ EH_2$). The modified enzyme catalyzes the transfer of electrons from NADPH to ferricyanide but is unreactive with glutathione; in fact large amounts of oxidized glutathione failed to decrease the charge transfer interaction of EHR. Thus, the pyridine nucleotide half-reaction is intact, but the disulfide half-reactions are abolished just as is the case with lipoamide dehydrogenase. The addition of $NADP^+$ or NADPH to EHR results in shifts in the charge transfer band exactly like those seen with EH_2[44].

There are 4 cysteines per FAD in addition to the redox active cystine in yeast glutathione reductase; there are no other cystines[6,12]. The free thiols in oxidized glutathione reductase are quite unreactive toward iodoacetamide. In order to demonstrate the differential reactivity of the nascent thiols produced upon careful titration with dithionite to the EH_2 level and to identify the thiol reacted in the sequence, EH_2 was reacted with radioactive iodoacetamide; the reaction was quenched with cysteine and dialyzed. The remaining thiols were reacted with non-radioactive iodoacetamide in guanidinium chloride. The protein was digested with trypsin. The left panel of Fig. 8 is an analytical electropherogram of the tryptic digest. When long digestion times are used, 2 radioactive peaks appear, a neutral and a cation in addition to the core material at the origin. Shorter digestion times result in about 75% of the radioactivity in the cationic peak. It is known from previous work[12] that the tryptic peptide containing the redox active cystine is cationic. The neutral material arises from a secondary hydrolysis of a quite sensitive bond between the half-cystines. On a preparative scale the intact tryptic peptide was purified by ion-exchange chromatography on DEAE-cellulose and SE-Sephadex. It migrated at the position of the cation in the left panel of Fig. 8 and had the expected amino acid composition (see the Table). It had a specific activity one half that of the iodoacetamide used in the alkylation. When this pure peptide was digested with chymotrypsin, the major radioactive product is a neutral peptide as shown in the right hand panel of Fig. 8. This same peptide as well as its non-radioactive partner were

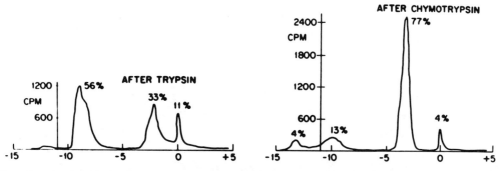

Fig. 8. Analytical electropherograms. Left panel: Tryptic digest of yeast glutathione reductase alkylated after 2-electron reduction; digestion was in 0.5% ammonium bicarbonate for 17 h at 37° with 2% w/w additions of TPCK-trypsin at 0, 2, and 6 h; electrophoresis was at 3000 v for 75 min at pH 6.5. Right panel: Chymotryptic digest of active center tryptic peptide; digestion was in 0.5% ammonium bicarbonate for 2 h at 37° with 2% w/w chymotrypsin; conditions for electrophoresis were the same.

TABLE

AMINO ACID ANALYSIS OF THE TRYPTIC AND CHYMOTRYPTIC PEPTIDES

Amino acid	Tryptic peptide	Neutral chymotryptic peptide	Cationic chymotryptic peptide
S-carboxymethyl cysteine	1.8	0.9	0.9
Aspartic acid	1.0	0.7	0
Threonine	1.0	0.9	0
Proline	1.1	0	1.1
Glycine	3.0	2.0	1.0
Alanine	1.1	1.0	0
Valine	2.8	0.8	1.7
Leucine	1.1	1.1	0
Lysine	1.4	0	0.9

purified from the chymotryptic digest by ion-exchange chromatography on SE-Sephadex and by preparative high voltage electrophoresis at pH 1.9. There is some evidence of heterogeneity in the chymotryptic hydrolysis (fractional values for Asp and Val); however the amino acid analyses of the peptides correspond reasonably to the expected compositions based on the sequence given below (see the Table). Again the thiol with approximately 15-fold higher specific activity is the thiol nearer the amino-terminus. Thus, the results are entirely analogous with those obtained with lipoamide dehydrogenase.

$$S-CH_2C^*ONH_2 \qquad S-CH_2CONH_2$$

$$\mid \qquad\qquad\qquad\qquad\qquad \mid$$

Ala-Leu-Gly-Gly-Thr-Cys-Val-Asn-Val-Gly-Cys-Val-Pro-Lys

$$\uparrow$$

Chymotrypsin

DISCUSSION

There is a large body of structural and mechanistic evidence indicating that glutathione reductase and lipoamide dehydrogenase operate by essentially identical mechanisms. Given the high degree of homology in the sequence of amino acid residues surrounding the redox active cystine, we assume that analogous nascent thiols in the 2 enzymes will perform analogous functions. In the alkylation of glutathione reductase, the charge transfer interaction is maintained and we therefore ascribe this function to the thiol closest to the carboxyl-terminus, i.e., the unmodified residue. By analogy the charge transfer donor of the EH_2 level of lipoamide dehydrogenase has not been alkylated but the modification sufficiently disrupts the active

14

center, by either raising the pK of the charge transfer thiol or perturbing its orientation to the FAD. It should be noted that long wavelength absorption bands are seen on the addition of AAD^+ to EHR (Fig. 6). No long wavelength absorption is induced on binding AAD^+ to the oxidized enzyme (Wilkinson, Matthews and Williams unpublished results). We also suggest that the carboxyl-terminal thiol participates in the covalent adduct formation with FAD induced by the addition of NAD^+ to EHR. Thus the carboxyl-terminal thiol interacts closely with the flavin via charge transfer interaction and covalent bond formation and would be expected to be within bonding distance.

Incubation of glutathione reductase EHR with a large excess of oxidized glutathione does not lead to loss of the charge transfer absorption, and this suggests that the thiol necessary for disulfide interchange has been blocked by alkylation. Further the pH profiles for the reoxidation of lipoamide dehydrogenase EH_2 by lipoamide and the inactivation by iodoacetamide are very similar again suggesting that the amino-terminal thiol residue interacts with the disulfide substrates. Confirmation of this assignment will rest on the isolation of a mixed disulfide intermediate involving the amino terminal cysteine residue.

Acknowledgements: We are grateful for the skilled technical assistance of Mr. W. Horton. Many of the ideas in this work are of vague origin within our group; but we especially wish to thank Drs. R. Matthews, K. Wilkinson, D. Ballou, B. Entsch and of course V. Massey. This work was supported by the Medical Research Service of the Veterans Administration and in part by Grant GM-21444 from The National Institute of General Medical Sciences, National Institutes of Health, Public Health Service.

REFERENCES

1. Williams, C.H., Jr. (1975) in *The Enzymes*, Boyer, P.D. ed., Academic Press, New York, Vol. 13, pp. 89-173.

2. Kosower, E.M. (1966) in *Flavins and Flavoproteins*, Slater, E.C. ed., Elsevier, Amsterdam, pp. 1-14.

3. Massey, V., and Ghisla, S. (1974) *Ann. N.Y. Acad. Sci. 227*, 446-465.

4. Massey, V., Gibson, Q.H., and Veeger, C. (1960) *Biochem. J. 77*, 341-351.

5. Massey, V., and Veeger, C. (1961) *Biochim. Biophys. Acta 48*, 33-47.

6. Williams, C.H., Jr., Burleigh, B.D., Jr., Ronchi, S., Arscott, L.D., and Jones, E.T. (1971) in *Flavins and Flavoproteins,* Kamin, H., ed., University Park Press, Baltimore, pp. 295-311.

7. Burleigh, B.D., Jr., and Williams, C.H., Jr. (1972) *J. Biol. Chem. 247*, 2077-2082.

8. Brown, J.P., and Perham, R.N. (1972) *FEBS (Fed. Eur. Biochem. Soc.) Lett. 26*, 221-224.

9. Williams, C.H., Jr., and Arscott, L.D. (1972) *Z. Naturforsch 27b*, 1078-1080.

10. Brown, J.P., and Perham, R.N. (1974) *Biochem. J. 138*, 505-512.

11. Matthews, R.G., Arscott, L.D., and Williams, C.H., Jr. (1974) *Biochim. Biophys. Acta 370,* 26-38.

12. Jones, E.T., and Williams, C.H., Jr. (1975) *J. Biol. Chem. 250*, 3779-3784.

13. Krohne-Ehrich, G., Schirmer, R.H., and Untucht-Grau, R. (1977) *Eur. J. Biochem.* *80,* 65-71.

14. Ronchi, S., and Williams, C.H., Jr. (1972) *J. Biol. Chem. 247,* 2083-2086.

15. Thelander, L. (1970) *J. Biol. Chem. 245,* 6026-6029.

16. Matthews, R.G., and Williams, C.H., Jr. (1976) *J. Biol. Chem. 251,* 3956-3964.

17. Matthews, R.G., Ballou, D.P., Thorpe, C., and Williams, C.H., Jr., (1977) *J. Biol. Chem. 252,* 3199-3207.

18. Polgar, L. (1974) *FEBS (Fed. Eur. Biochem. Soc.) Lett. 47* , 15-18.

19. Lewis, S.D., Johnson, F.A., and Shafer, J.A. (1976) *Biochemistry 15,* 5009-5017.

20. Polgar, L. (1975) *Eur. J. Biochem. 51,* 63-71.

21. Polgar, L. (1977) *Int. J. Biochem. 8,* 171-176.

22. Foss, O. (1961) in *Organic Sulfur Compounds,* Karasch, N. ed., Pergamon Press, New York, Vol. 1, pp. 83-96.

23. Thorpe, C., and Williams, C.H., Jr. (1976) *J.Biol. Chem. 251,* 7726-7728.

24. Thorpe, C., and Williams, C.H., Jr. (1976) *J. Biol. Chem. 252,* 3553-3557.

25. Massey, V., and Palmer, G. (1962) *J. Biol. Chem. 237,* 2347-2358.

26. Hummel, J.P., and Dryer, W.J. (1962) *Biochim. Biophys. Acta 63,* 530-532.

27. Strickland, S., and Massey, V. (1973) *J. Biol. Chem. 248,* 2953-2962.

28. Tu, S.-C., and Hastings, J.W. (1975) *Biochemistry 14,* 1975-1979.

29. Entsch, B., Ballou, D.P., and Massey, V. (1976) *J. Biol. Chem. 251,* 2550-2563.

30. Ghisla, S., Hartmann, U., Herrerich, P., and Müller, F. (1973) *Liebigs. Ann. Chem.,* 1388-1415.

31. Ghisla, S., Massey, V., Lhoste, J.M., and Mayhew, S.G. (1974) *Biochemistry 13,* 589-597.

32. Ghisla, S., Entsch, B., Massey, V., and Husein, M. (1977) *Eur. J. Biochem. 76,* 139-148.

33. Walker, W.H., Hemmerich, P., and Massey, V. (1970) *Eur. J. Biochem. 13,* 258-266.

34. Hamilton, G.A. (1971) in *Progress in Bioorganic Chemistry,* Kaiser, E.T., and Kézdy, J.J., eds. Wiley-Interscience, New York, Vol. 1, pp. 83-157.

35. Bruice, T.C. (1975) in *Progress in Bioorganic Chemistry,* Kaiser, E.T., and Kézdy, F.J., eds., Wiley-Interscience, New York, Vol. 4., in press.

36. Hemmerich, P. (1976) in *Progress in Natural Product Chemistry,* Grisebach, H. ed., Springer-Verlag, New York, Vol. 33, pp. 451-526.

37. Loechler, E.L., and Hollocher, T.C. (1975) *J. Am. Chem. Soc. 97,* 3235-3237.

38. Yokoe, I., and Bruice, T.C. (1975) *J. Am. Chem. Soc. 97,* 450-451.

39. Fischer, J., Spencer, R., and Walsh, C. (1976) *Biochemistry 15,* 1054-1064.

40. Wilkinson, K.D., and Williams, C.H., Jr., manuscript in preparation.

41. Anderson, B.M., Ciotti, C.J., and Kaplan, N.O. (1959) *J. Biol. Chem. 234,* 1219-1225.

42. Fisher, T.L., Vercelloti, S.V., and Anderson, B.M. (1973) *J. Biol. Chem. 248,* 4293-4299.

43. Matthews, R.G., Wilkinson, K.D., Ballou, D.P. and Williams, C.H., Jr. (1976) in *Flavins and Flavoproteins,* Singer, T.P. ed., Elsevier, Amsterdam, pp. 464-472.

44. Williams, C.H., Jr., Arscott, L.D., and Jones, E.T. (1976) in *Flavins and Flavoproteins,* Singer, T.P. ed., Elsevier, Amsterdam, pp. 455-463.

STRUCTURAL STUDIES ON CRYSTALLINE GLUTATHIONE REDUCTASE FROM HUMAN ERYTHROCYTES

EMIL F. PAI, R. HEINER SCHIRMER, and GEORG E. SCHULZ

Max-Planck-Institut für Medizinische Forschung, Jahnstr. 29, 6900 Heidelberg (Germany) - Abteilung Biophysik -

ABSTRACT

The crystal structure of the dimeric enzyme has been analysed at 0.3 nm (=3Å) using X-ray diffraction techniques. The electron density map reveals the course of the polypeptide chain (465 C_α-atoms per subunit) and the conformations of prosthetic group (FAD) and bound substrates. The structure of the catalytic site is consistent with the established concept that reduction equivalents move along the following pathway: Reduced nicotinamide moiety of NADPH $\longrightarrow$ isoalloxazine ring of FAD $\longrightarrow$ redox-active cysteine residues $\longrightarrow$ disulfide bridge of GSSG. - Structural details of the enzyme mechanism will be discussed.

INTRODUCTION

Glutathione reductase[1] maintains a high value for the ratio 2 [GSH] / [GSSG] in the cytosol. This is of particular importance for the integrity of those cells which do not synthesize macromolecules any longer, such as fibres of the eye lens and mature erythrocytes[2]. After Worthington and Rosemeyer[3] published an effective isolation procedure for crystalline glutathione reductase from human red blood cells, physico-chemical[4], immunological[5] and catalytic properties[1,6] of this enzyme have been studied in great detail (Table 1). Recently a large-scale isolation procedure has been described which yields 200 mg enzyme within less than a week[7]. The availability of such quantities should facilitate studies of the enzyme by methods of nuclear magnetic resonance and fast kinetics as well as the analysis of the amino acid sequence[7]. The analysis of the crystal structure of the enzyme was started in collaboration with Drs. Rosemeyer and Worthington[8] and has recently yielded an electron density map at 0.3 nm resolution[9]. This map shows structural features of the catalytic centre in atomic detail. Furthermore, it allows tracing of the polypeptide chain (465 C_α-atoms per subunit). The known sections of the primary structure (Untucht-Grau, Schirmer, and Wittmann-Liebold, unpublished results) could be fitted to the electron density map.

TABLE 1

PROPERTIES OF HUMAN GSSG-REDUCTASE

---

a) The Enzyme

1.1. Catalysed reaction (EC 1.6.4.2)   GSSG + NADPH + H$^+$ $\rightleftharpoons$ 2 GSH + NADP

1.2. K_M of substrates, K_i of products[6]   65μM   8.5μM   6mM   70μM

1.3. Concentrations in erythrocytes[6]   150μM   50 μM   3mM   3-30μM

1.4. Equilibrium constant[23] $$K = \frac{[\text{NADP}] \cdot [\text{GSH}]^2}{[\text{NADPH}] \cdot [\text{GSSG}] \cdot [\text{H}]^+} = 5 \cdot 10^{12}$$

1.5. Enzyme activity in[7] lysed erythrocytes   3.7 U/g protein ≈ 3.7 U/g haemo-globin

1.6. Specific activity of the isolated enzyme[3,5,7]   240 U/mg protein

1.7. Turnover number[3,7]   200 s^{-1}

b) The Protein (a dimer composed of two identical subunits)

Data for one subunit:

1.8. Molecular weight 50,000[4]; size 5.5 nm x 5.0 nm x 4.5 nm[12]

1.9. ≈460 amino acid residues[4,7]: Asx_{36}, Thr_{28}, Ser_{27}, Glx_{42}, Pro_{21}, Gly_{43}, Ala_{40}, Val_{43}, Met_{15}, Ile_{28}, Leu_{34}, Tyr_{13}, Phe_{14}, His_{16}, Lys_{32}, Arg_{16}, Cys_{10}, Trp_4

1.10. One catalytic site containing the isoalloxazine ring of FAD and one disulfide[1,4,7]; sequence of the redox-active segment (= glupo-segment):

Leu-Gly-Gly-Thr-Cys-Val-Asn-Val-Gly-Cys-Val-Pro-Lys[1,7]

1.11. Symbols: E = oxidized enzyme

FAD
├─S "proximal" sulfur ⎫
│   ⎬ of the redox segment
├─S "distal" sulfur   ⎭

EH$_2$ = reduced enzyme

FAD
├─S$^-$
├─SH

18

One should also try to fit the primary structure of proteins which are believed to be homologous with human glutathione reductase to this map. Possible candidates are GSSG-reductase from yeast and E.coli and lipoamide dehydrogenases from various sources[1]. A common starting point could be the sequence around the redox-active cystine. This glupo sequence (1.10 in Table 1) is identical in at least 9 out of 13 positions when glutathione reductases and lipoamide dehydrogenases are pairwise compared[1]. Although thioredoxin reductase contains a redox-active segment with unrelated sequence[1], it might belong to the same protein family as GSSG-reductase. The function of thioredoxin reductase can be carried out *in vivo* by a system based on glutaredoxin glutathione and glutathione reductase[10]. Thus a duplicate of a primordial gene could have developed the disulfide-loop arrangement of thio-redoxin reductase while a GSSG-reductase-like enzyme would have pre-served ancestral functions. This hypothesis can be checked by com-paring the three-dimensional structures of the two enzymes because chain-folds of homologous proteins were particularly well conserved during evolution[11].

PROPERTIES OF HUMAN GLUTATHIONE REDUCTASE IN THE CRYSTALLINE STATE

Enzyme crystals are studied at 4° either in deionized water or in 2M $(NH_4)_2SO_4$ - 0.1M phosphate buffer, pH 7.0. The structure of the enzyme in the two crystal forms[12] is almost identical which indicates that this is also the enzyme's shape at physiological ionic strength.

The crystallized enzyme - as well as the enzyme in solution - is a dimer (Fig. 1). The finding that there is one subunit per crystal-lographic asymmetric unit proves that the two subunits are exactly identical.

In deionized water the enzyme crystals are rather unstable when exposed to ions or substrates. Consequently most binding studies[12,13] were carried out at 2M $(NH_4)_2SO_4$. The enzyme activity in solution is strongly inhibited (90%) at high ionic strength $(I > 1\mu)$ when GSSG and NADPH are used as substrates[6]. At 2M $(NH_4)_2SO_4$ the crystallized enzyme can be reduced by a variety of agents including NADPH, NADH, dithionite and $NaBH_4$[13]. The reduction was followed by monitoring crystal colour. As these studies were done under aerobic conditions reoxid-ation occurred spontaneously but its rate was significantly enhanced by addition of GSSG.

These results as well as spectroscopic comparisons of the crystal-lized enzyme with the enzyme in solution[13] suggest that the active centre region of the enzyme is not altered by crystallization.

19

DESCRIPTION OF THE CRYSTALLIZED ENZYME

The dimensions of a subunit are approximately 5.0 nm x 4.5 nm x 5.5 nm and the dimensions of the dimer (2 x 5.0) nm x 4.5 nm x 5.5 nm. Each subunit can be subdivided into three structural domains[11], i.e. into three geometrically separable regions: the globular FAD-binding domain which also contains the redox-active disulfide, the globular NADPH-binding domain, and the GSSG-binding domain, which is also called interface domain as it forms the interface between the subunits[12]. The dinucleotides FAD (the prosthetic group) and NADP (a product) are bound to their respective domains in extended conformations (Fig.1). The catalytic site is formed by the binding site for the nicotinamide-containing moiety of NADPH, the iso-alloxazine ring of FAD, the cysteines of the glupo-segment (1.10. in Table 1), residues from other chain segments, and the binding site for GSSG.

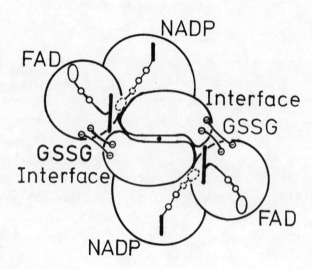

Fig. 1. Sketch of the dimeric enzyme glutathione reductase. The central point denotes an exactly twofold axis. The FAD-binding-, the NADP-binding-, and the interface domain are indicated. GSSG binds between subunits. FAD is in an extended conformation with the iso-alloxazine ring pointing to GSSG. In the crystal only the adenosine and pyrophosphate moieties of NADP are tightly attached to the enzyme and therefore visible. But there is space for the nicotinamide moiety to reach the isoalloxazine ring as inserted with dashed lines.

COMMENTS ON THE CATALYTIC MECHANISM OF GLUTATHIONE REDUCTASE AND LIPOAMIDE DEHYDROGENASE

The structure of the active site is consistent with the concept that FAD receives the reduction equivalents from reduced nicotinamide and passes them to the redox-active cysteine residues of the protein which in turn hand them over to the disulfide bridge of glutathione[1,14-17]. Thus the structural studies confirm structure predictions which were based on indirect methods. Non-crystallographic techniques have also provided facts and working hypotheses concerning structural <u>details</u> which may be relevant for the catalytic mechanism of disulfide reductases. Some examples are listed below - the oxidized form of the enzyme (either GSSG-reductase or lipoamide dehydrogenase) is designated E and the reduced enzyme EH_2:

a) Correlation between structure and function of the redox-active glupo-segment (1.10. and 1.11. in Table 1). One sulfur atom of the disulfide bridge in E is postulated[1] to be closer to the iso-alloxazine ring ("proximal" sulfur) and the other one to be closer to the disulfide substrate ("distal" sulfur).

b) Conformational changes at the catalytic site accompanying the transition $E \rightleftharpoons EH_2$.

c) The existence of a transitory adduct between the proximal thiol(ate) and the C(4α) of FAD during the $E \rightleftharpoons EH_2$ transition[18].

d) The nature of the charge transfer complex between the iso-alloxazine ring and the proximal thiol(ate) in EH_2[19,20].

e) The nature of the postulated base which may be unprotonated in E but protonated in EH_2 and its possible role in stabilizing a thiolate at pH values as low as 5.5[21].

f) Aspects referring to the accessibility of the active site thiol(ate)s in EH_2, e.g. the different chemical reactivities of the proximal and distal thiol(ate)s[22] and the possible roles of these groups for enzyme inactivation and aggregation[6].

We look forward to the opportunity of discussing these and other points concerning the mechanism of disulfide reductases in La Paz.

ACKNOWLEDGEMENT

Our work is supported by the Deutsche Forschungsgemeinschaft.

The help of Frau Anke Hennemann in preparing the camera-ready manuscript is gratefully acknowledged.

REFERENCES

1. Williams,C.H.,jr. (1976) in The Enzymes, vol. 13 (Boyer,P.D., ed.) 3rd edn. pp. 89-173, Academic Press, New York.

2. Krohne-Ehrich,G. and Untucht-Grau,R. (1976) Biol. in unserer Zeit, 6, 175-182.

3. Worthington,D.J. and Rosemeyer,M.A. (1974) Eur. J. Biochem. 48, 167-177.

4. Worthington,D.J. and Rosemeyer,M.A. (1975) Eur.J.Biochem. 60, 459-466.

5. Nakashima,K., Miwa,S., and Yamauchi,K. (1976) Biochim.Biophys. Acta, 445, 309-323.

6. Worthington,D.J. and Rosemeyer,M.A. (1976) Eur.J.Biochem. 67, 231-238.

7. Krohne-Ehrich,G., Schirmer,R.H., and Untucht-Grau,R. (1977) Eur. J.Biochem. 80, 65-71.

8. Schulz,G.E., Zappe,H., Worthington,D.J., and Rosemeyer,M.A.(1975), FEBS-Lett. 54, 86-88.

9. Pai,E.F. (1978) Ph.D. thesis, University of Heidelberg.

10. Holmgren,A. (1976) Proc. Natl. Acad. Sci. U.S.A. 73, 2275-2279.

11. Schulz,G.E. and Schirmer,R.H. (1978) Principles of Protein Structure, Springer Verlag, Berlin-Heidelberg-New York.

12. Zappe, H., Krohne-Ehrich,G., and Schulz, G.E. (1977) J. Mol. Biol. 113, 141-152.

13. Pai,E.F., Sachsenheimer,W., Schirmer,R.H., and Schulz,G.E. (1977) Hoppe-Seyler's Z. Physiol. Chem. 358, 1255 (Meeting Abstract)

14. Searls,R.L. (1960) Federation Proc. 19, 36.

15. Massey,V. and Veeger,C. (1961) Biochim. Biophys. Acta 48, 33-47.

16. Sanadi,D.R. (1963) in The Enzymes (Boyer,P.D., Lardy,H., and Myrback,K. eds.) Vol. 7, pp. 275-306, Academic Press, New York.

17. Mannervik,B. (1976) Flavins and Flavoproteins, V, 485-491.

18. Thorpe,C. and Williams,C.H.,jr. (1976) J.Biol.Chem. 251, 7726-7728.

19. Searls,R.L., Peters,J.M., and Sanadi,D.R.(1961) J.Biol.Chem.236, 2317-2322.

20. Kosower,E.M. (1966) in Flavins and Flavoproteins (Slater,E.C.,ed.) pp. 1-14, Elsevier, Amsterdam.

21. Matthews,R.G. and Williams,C.H.,jr. (1976) J.Biol.Chem. 251, 3956-3964.

22. Thorpe,C. and Williams,C.H.,jr. (1976) J.Biol.Chem. 251, 3553-3557.

23. Massey,V. and Ghisla,S. (1974) Ann.N.Y.Acad.Sci. 227, 446-465.

COMPARISON BETWEEN PHYSICAL PROPERTIES OF GLUTATHIONE REDUCTASE AND NADP-LINKED DEHYDROGENASES

M.A. ROSEMEYER, P. COHEN, B.M.F. PEARSE and D.J. WORTHINGTON

Department of Biochemistry, University College London, Gower Street, London WC1E 6BT, England

ABSTRACT

Isolation (from human erythrocytes) and characterization of glucose-6-phosphate dehydrogenase, 6-phosphogluconate dehydrogenase and glutathione reductase show extensive similarities in the general physical properties of these enzymes. The physical and catalytic properties suggest a common structural pattern which corresponds to the common metabolic function of the enzymes.

INTRODUCTION

The maintenance of cellular thiols depends on the presence of glutathione, which occurs in mM concentrations[1] and is kept reduced by glutathione reductase (GR):

$$GSSG + NADPH + H^+ \longrightarrow 2GSH + NADP^+$$

The NADPH for this reaction _in vivo_ is provided by the enzymes of the hexosemonophosphate pathway, namely glucose-6-phosphate dehydrogenase (G6PD) and 6-phosphogluconate dehydrogenase (6PGD). Therefore, the three enzymes may be considered as a metabolic system using glucose-6-phosphate, cycling through the reduced and oxidized forms of the coenzyme, and generating reduced glutathione. It would be an economy of design if this common metabolic role was achieved by common structural properties of the enzymes.

The study of each enzyme took the following sequence.

Purification was carried out using general procedures to separate proteins according to differences in charge and size.

Homogeneity was assessed on the basis of charge and size, and also the specific activity obtained.

Catalytic properties were investigated for optimal assay conditions, and to measure the kinetic parameters.

Subunit Composition was deduced by dissociation.

Molecular Associations were inferred from sedimentation-equilibrium experiments in various solvent conditions.

METHODS

G6PD. The methods of isolation and investigation are given by Cohen and Rosemeyer[2-4].

6PGD. The methods of isolation and investigation are given by Pearse and Rosemeyer[5-7].

GR. The methods of isolation and investigation are given by Worthington and Rosemeyer[8-10].

All assays were at 25°C.

RESULTS AND DISCUSSION

Purification. There were four basic operations in the isolation of the enzymes[2,5,8]. These are separation using (i) DEAE-Sephadex (ii) CM-Sephadex (iii) gel-filtration on Sephadex G-200, and (iv) ammonium sulphate fractionation. These procedures can be arranged to give a concurrent yield of the three enzymes.

The purification started with 30-40 litres of human blood and gave:

G6PD   25 mg, 25% yield, 50,000-fold purification[2];

6PGD   50 mg, 6% yield, 5,000-fold purification[5];

GR   40 mg, 40% yield, 50,000-fold purification[8].

The separation of 6PGD from the others relies on its precipitation at higher concentrations of ammonium sulphate - 70% saturation rather than 55% saturation. G6PD is retained on DEAE-Sephadex at pH 6.5 I 0.06, while GR is not bound. Otherwise the operational conditions are very similar and indicate that the three enzymes are isoelectric near pH 6 to 6.5

6PGD showed some instability and its yield can be increased to 200 mg (25%) by including 10% (v/v) glycerol in the buffers[7].

GR shows a close similarity in behaviour to catalase from which it was separated by gel-filtration. It also forms aggregates in the absence of 2-mercaptoethanol[8].

Affinity chromatography on 2'5'-ADP-Sepharose 4B has been used to isolate GR[11]. Following extraction with organic solvents, GR from human erythrocytes has been purified by affinity chromatography (Krohne-Ehrich et al[12]). The affinity method[13] may be adapted to the retention of all three enzymes on the basis of a common coenzyme-binding site.

Homogeneity.

(i) Charge. Electrophoresis of G6PD and 6PGD on starch gel indicates that essentially a single species was present and there was a correspondence between protein and activity stains. A single band was observed if sufficient NADP was incorporated in the gel. At pH 8.6 the mobility of G6PD was slightly more than, and of 6PGD slightly less than, a haemoglobin marker[5].

Electrophoresis of GR on starch gel or polyacrylamide gave diffuse bands, probably due to aggregation. There was a correspondence between protein and activity stains[8].

(ii) <u>Size</u>. In each case sedimentation showed a homogeneous entity. 6PGD had an $s_{20,w}$ of 5.8S and molecular weight of 104,000[6] and GR had an $s_{20,w}$ of 5.6S and molecular weight of 100,000[9]. However, G6PD showed a discrete species of 9.0S, molecular weight 210,000, which dissociated to an entity half this size ($s_{20,w}$ of 5.6S, molecular weight 105,000). The degree of dissociation depended on the solvent conditions[3].

Therefore, the common molecular entity is a species of molecular weight in the range 100,000-105,000.

<u>Catalytic Properties</u>. G6PD was assayed at pH 9 I 0.1. The optimum for 6PGD was at pH 8.0 I 0.1, and for GR pH 7.0 I 0.3.

The ionic strength optimum[2] for G6PD may be explained as follows. Increase in ionic strength promotes release of the product and increases V_{max}. When substrate-binding is the limiting factor increase in ionic strength leads to an increase in K_m.

The attachment of substrate and coenzyme can be random especially as the site contains FAD and must be extensive[10]. This is confirmed by structural information for this enzyme (Zappe et al.[14]).

The inactivation of GR in the presence of NADPH is of some interest. There is a corresponding aggregation which also requires the presence of air. The inactivation and aggregation are prevented and reversed by GSH[10].

The activity of GR in the erythrocyte depends on the cellular level of FAD[8,15].

<u>Subunit Composition</u>. On treatment with dissociating agents G6PD, 6PGD and GR gave molecular entities of 53,000, 52,000 and 50,000 respectively. These species were seen on polyacrylamide electrophoresis in the presence of sodium dodecylsulphate[6,8,9]. G6PD and 6PGD were also dissociated by treatment with maleic anhydride[3,6].

The results indicate that the stable active species for each enzyme is a dimer. A basic dimeric structure for G6PD was also indicated by hybridisation of human variants[16] of the enzyme and of various mammalian G6PDs[17]. Dissociation occurred in ammonium sulphate solution at low pH with removal of NADP. The enzyme was re-associated at neutral pH in the presence of NADP.

A dimeric structure for GR was also suggested by its FAD content[8], and the structural symmetry indicates identity of the subunits[18].

<u>Molecular Associations</u>. Although the enzymes were highly purified there were indications of size-heterogeneity due to molecular associations. This was most evident for G6PD which showed an equilibrium between tetramer and dimer in the pH range 6 to 8 at moderate ionic strength[3]. This enzyme associated to larger aggregates (hexamers or octamers) at pH 6 and low ionic strength.

6PGD is predominantly dimeric but at pH 8, or on storage, shows traces of a larger aggregate that corresponds to a dodecamer[6]. Also maleylation to dissociate the dimer results in trace amounts of a hexameric species.

GR is also predominantly dimeric but shows traces of a tetramer[9]. Although the species can be reduced by 2-mercaptoethanol its persistence is suggested by the spread of bands on electrophoresis. Further aggregates (hexamers or octamers) can also occur in the presence of substrate or coenzyme[10].

CORRELATION OF PROPERTIES   The physical and catalytic properties of the enzymes are summarized in Table 1.

TABLE 1

SUMMARY OF PHYSICAL PROPERTIES

PROPERTY	G6PD	6PGD	GR
$A_{280}^{1\%}$	12.2	12.5	13.5
Specific activity (U/mg)	220	15.0	240
Turnover number (s^{-1})	190	13	200
K_m (substrate) (μM)	100	20	65
K_m (coenzyme) (μM)[a]	10	30	8.5
Partial specific volume (ml/g)	0.734	0.737	0.74
Subunit molecular weight	53,000	52,000	50,000
Molecular weight[b]	210,000	104,000	100,000
$s_{20,w}$ (S)	9.0	5.8	5.6
$D_{20,w}$ ($\mu m^2/s$)	41	51	52

[a] The coenzyme is NADP for the two dehydrogenases and NADPH for the reductase.
[b] Of discrete entity in sedimentation-equilibrium experiments.

Under conditions in the erythrocyte the potential activity of each enzyme is less than the maximal activity owing to divergence from the pH and ionic strength optima of the enzymes. Also GR may be less than fully saturated with FAD[8,15]. From the yields of enzymes an assessment of the quantities and potential activities under physiological conditions is given in Table 2.

TABLE 2

OCCURRENCE OF G6PD, 6PGD AND GR IN HUMAN ERYTHROCYTES

ENZYME	QUANTITY (mg/l)	SUBUNIT CONCN. (μM)	ACTIVITY (U/l)[a]
G6PD	5	0.1	550
6PGD	50	1.0	500
GR	6	0.1	1,000

[a] Estimated potential activity at pH 7.2, I = 0.15.
Units are defined as μmoles of NADPH produced/minute.

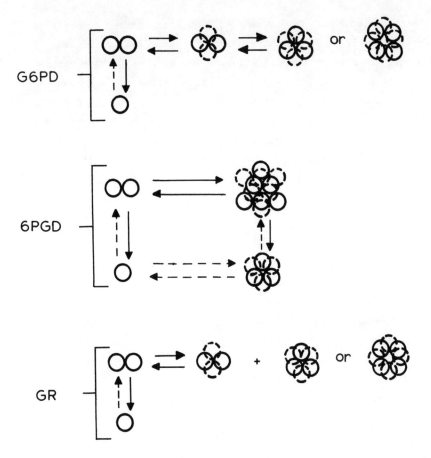

Fig. 1. The subunit structures and molecular associations of glucose-6-phosphate dehydrogenase, 6-phosphogluconate dehydrogenase and glutathione reductase. Observed steps are given in full lines and assumed steps are in dashed lines.

In human erythrocytes G6PD, 6PGD and GR are present in the approximate molar ratio of 1:10:1, while the potential activities are in the ratio of 1:1:2 respectively. The latter is the ratio expected for efficient cycling of the coenzyme, whereby the NADPH produced by the two dehydrogenases is used by the reductase.

Yoshida[19] has pointed out that the deleterious effects in abnormal G6PDs arise from changes in K_m for NADP rather than in V_{max}. The K_ms of the three enzymes towards their coenzymes are of the order of 10^{-5}M. At low levels of coenzyme the activity depends on a pseudo-first-order rate constant, namely V_{max}/K_m. The activities under physiological conditions and the K_ms suggest a correspondence in the rate constants for the three enzymes. The balance in kinetic properties of the system would be disrupted by changes in the K_m of one of the enzymes.

The subunit compositions and molecular associations are shown in Fig. 1. The enzymes are based on a similar pattern with subunits of approximately the same size forming an active dimeric species. All the enzymes show a tendency to form further molecular associations. In the case of GPGD this further association leads to a dodecamer. A dodecameric pattern is often the basis of larger molecular aggregates[20]. From the similarities in properties of these enzymes and their tendency to polymerize there remains a possibility of inter-enzyme associations. Their occurrence in a ratio of 1:10:1 corresponds to the quantities required for a multiple dodecameric structure, with a balance of activities under physiological conditions. Thus, the physical properties would allow the enzymes to function as a multienzyme complex, although there is no direct evidence as yet for such a complex.

REFERENCES

1. Jocelyn, P.C. (1972) Biochemistry of the SH Group, Academic Press, London, pp. 240-278.

2. Cohen, P. and Rosemeyer, M.A. (1969) Eur. J. Biochem., 8, 1-7.

3. Cohen, P. and Rosemeyer, M.A. (1969) Eur. J. Biochem., 8, 8-15.

4. Cohen, P. and Rosemeyer, M.A. (1975) Methods in Enzymology, 41, 208-214.

5. Pearse, B.M.F. and Rosemeyer, M.A. (1974) Eur. J. Biochem., 42, 213-223.

6. Pearse, B.M.F. and Rosemeyer, M.A. (1974) Eur. J. Biochem., 42, 225-235.

7. Pearse, B.M.F. and Rosemeyer, M.A. (1975) Methods in Enzymology, 41, 220-226.

8. Worthington, D.J. and Rosemeyer, M.A. (1974) Eur. J. Biochem., 48, 167-177.

9. Worthington, D.J. and Rosemeyer, M.A. (1975) Eur. J. Biochem., 60, 459-466.

10. Worthington, D.J. and Rosemeyer, M.A. (1976) Eur. J. Biochem., 67, 231-238.

11. Mannervik, B., Jacobsson, K. and Boggaram, V. (1976) FEBS Lett., 66, 221-224.

12. Krohne-Ehrich, G., Schirmer, R.H. and Untucht-Grau, R. (1977) Eur. J. Biochem., 80, 65-71.

13. Brodelius, P., Larsson, P.O. and Mosbach, K. (1974) Eur. J. Biochem., 47, 81-89.

14. Zappe, H.A., Krohne-Ehrich, G. and Schulz, G.E. (1977) J. Mol. Biol., 113, 141-152.

15. Beutler, E. (1969) J. Clin. Invest., 48, 1957-1966.

16. Yoshida, A., Steinmann, L. and Harbert, P. (1967) Nature, 216, 275-276.

17. Cohen, P. (1969) Ph.D. Thesis, Univ. of London.

18. Schulz, G.E., Zappe, H., Worthington, D.J. and Rosemeyer, M.A. (1975) FEBS Lett., 54, 86-88.

19. Yoshida, A. (1973) Science, 179, 532-537.

20. Pearse, B.M.F. (1976) J. Mol. Biol., 103, 785-798.

MECHANISTIC AND STRUCTURAL ASPECTS OF GLUTATHIONE-INSULIN TRANSHYDROGENASE
(PROTEIN-DISULFIDE INTERCHANGE ENZYME)*

PARTAB T. VARANDANI

Fels Research Institute and Biological Chemistry Department, Wright State University,
School of Medicine and College of Science and Engineering, Yellow Springs, Ohio 45387

ABSTRACT

   Work from several laboratories on the structural and mechanistic features of
glutathione-insulin transhydrogenase (thiol:protein-disulfide oxidoreductase,
E.C. 1.8.4.2) is reviewed. The enzyme in vitro catalyzes sulfhydryl-disulfide
interchange in a wide range of proteins of unique and diverse properties. Thus,
its action, on the one hand, includes the inactivation of some hormones (insulin,
oxytocin, vasopressin), and, on the other hand, the activation of inactive scrambled forms (i.e., incorrectly-paired disulfide bonds) of proinsulin and ribonuclease. Kinetic analyses of the mechanism of insulin degradation show that the
enzyme functions in a complex manner by either a random or an allosteric mechanism.
The enzyme has multiple distinct sites of reaction, both catalytic and allosteric,
and its activity is modulated by different types of thiols, certain phospholipids,
certain chelators and the ionic environment. The purified enzyme is strikingly
similar to purified ribonuclease-reactivating enzyme (thiol:protein-disulfide
isomerase, E.C. 5.3.4.1), in physical, chemical and enzymic properties. Attempts
to resolve the two catalytic activities into separate enzyme species have been
unsuccessful. It is concluded that the oxidoreductase and isomerase activities
are both catalyzed by a single enzyme species containing multiple sites of reaction. It is suggested that the few discrepancies observed between the two
catalytic activities in the rat microsomal fraction are explainable on the basis
of differences in permeability and affinity attributable to the nature, conformation and macromolecular size of the substrates used, insulin and scrambled
ribonuclease, and to the masking of the sites of reaction (in particular, allosteric site or sites) by membrane structures.

INTRODUCTION

   The purpose of this paper is to discuss briefly the structural and mechanistic
features of glutathione-insulin transhydrogenase (GIT**, thiol:protein-disulfide
oxidoreductase, E.C. 1.8.4.2). The physiological aspects of this enzyme and its
regulatory mechanisms have been recently reviewed elsewhere[1]. The presentation

---

* This is paper XXIV in the series entitled, "Insulin Degradation."
** Abbreviation used: GIT, glutathione-insulin transhydrogenase.

is divided into two major parts. In the first part, mechanistic features of the enzyme are presented. In the second part, the structural aspects of GIT are discussed and the enzyme is compared with another enzyme, ribonuclease-reactivating enzyme (thiol:protein-disulfide isomerase, E.C. 5.3.4.1). These two enzymes are strikingly similar and a review of recent evidence strongly suggests that a single enzyme containing multiple distinct sites may be responsible for the two catalytic activities.

MECHANISTIC FEATURES OF GIT

GIT _in vitro_ in the presence of a thiol compound promotes sulfhydryl-disulfide interchange[2,3] in a wide range of proteins with diverse activities (Table 1). The process is not an oxidation, but rather a rearrangement of inter- and intramolecular disulfide bonds in a given protein. In the presence of a thiol compound the enzyme causes the cleavage of disulfide bonds in certain native polypeptides, insulin[4,5], proinsulin[6], oxytocin[3,7,8], and vasopressin[3], and results in their inactivation.

TABLE 1

SUBSTRATES OF THIOL:PROTEIN-DISULFIDE OXIDOREDUCTASE/ISOMERASE

The conformation of the substrate is a major factor in determining the degree of specificity of the enzyme[10].

Inactivation/Disulfide cleavage	Reactivation/Disulfide formation
Insulin	Scrambled Ribonuclease
des-Asn-des-Ala-Insulin	Scrambled Lysozyme[47]
des-Ala-Insulin	Scrambled Trypsin Inhibitor[47]
desoctapeptide-Insulin	Immunoglobulin[48]
Proinsulin	Scrambled Proinsulin
Oxytocin	
Vasopressin	

Numerals given as superscripts are references for these substrates; references for other substrates are presented in the text.

Reactivation/Folding of Scrambled Proteins

GIT _in vitro_ can also accelerate the transformation of the inactive scrambled forms (i.e., proteins containing incorrectly paired disulfide bonds) of proinsulin[6] and ribonuclease[3,9,11] to their biologically active and apparently native forms. The conformation of the substrate seems to be a major factor that determines the degree of specificity for the enzyme[10]. For example, the enzyme will cause disulfide interchange in scrambled ribonuclease, which is a thermodynamically unstable form, but it does not promote interchange in native ribonuclease which is a thermodynamically stable form[10]. An interesting feature is that the enzyme _in vitro_ can serve

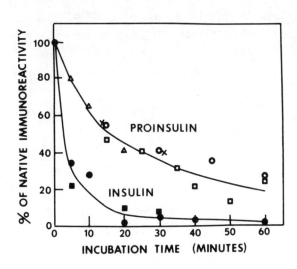

Fig. 1. Inactivation of native insulin and proinsulin by GIT as a function of time at 37°. From ref. 6.

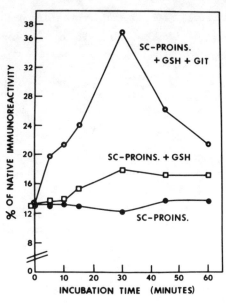

Fig. 2. Activation of scrambled proinsulin by GIT as a function of time at 37°. Sc = scrambled. From ref. 6.

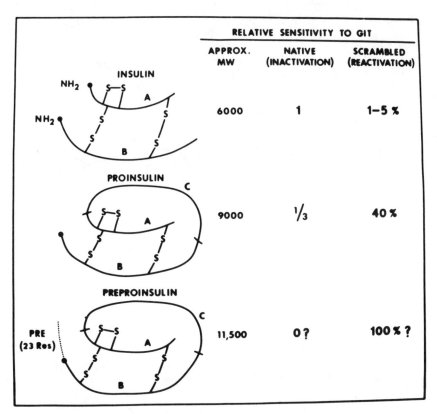

Fig. 3. Schematic illustration for the use of GIT as a probe to monitor the degree of stability of disulfide bonds in a given protein. Cf. ref. 6.

as a useful probe to monitor the degree of thermodynamic stability of the disulfide bonds in a given protein. This point can best be illustrated by comparing the insulin-proinsulin system with ribonuclease. The enzyme causes inactivation of proinsulin at a much slower rate than insulin[6] (Fig. 1). This indicates that the proinsulin disulfide bonds (which are in the insulin moiety) are relatively more stable than those in free insulin, but they are still less stable than the disulfide bonds in the native ribonuclease. We may therefore anticipate that in preproinsulin[12], which is proinsulin with an aminoterminal extension of 23 amino acid residue, the disulfide bonds in the same insulin moeity would be either more stable than in proinsulin or, like native ribonuclease, might be insusceptible to the action of GIT. When refolding of scrambled forms is studied, while the enzyme produces little or no activation of insulin[6,13,14], it first produces about 40% activation in the case of proinsulin followed by inactivation[6] (Fig. 2). Again, whether, with preproinsulin, the enzyme will produce >40% of activation of the scrambled form and will not cause subsequent inactivation of the native form remains to be seen. These events are illustrated in Fig. 3.

Inactivation/Unfolding of Native Proteins

The enzyme in the presence of·GSH rapidly cleaves the disulfide bonds of insulin and forms A and B chains (Scheme I); actually, B chain appears as an aggregate randomly cross-linked with A chain[2].

SCHEME I

$$\text{insulin} + \text{GSH} \rightleftharpoons \text{GSSG} + \text{A chain} + \text{B chain}$$

Random/Allosteric Mechanism of Action. Kinetic analysis of the mechanism of insulin degradation by GIT using alternate thiol substrates and products as inhibitors showed that the enzyme functions in a complex manner by either a random or an allosteric mechanism[15]. This conclusion was arrived at by the fitting of data to various equations using a nonlinear least-squares method[15]; graphically, this evidence is as follows. Lineweaver-Burk plots for insulin as a variable substrate at several fixed levels of GSH were linear (Fig. 4), while these reciprocal plots with GSH as a variable substrate were parabolic (Fig. 5). The non-parallel nature of the reciprocal plots shows that the enzyme does not follow a ping-pong mechanism. All of the products tested (GSSG, $ASSO_3$, or $BSSO_3$) showed competitive inhibition with both substrates; plots of inhibition by GSSG of the reaction at variable levels of insulin are shown in Fig. 6 and at variable levels of GSH are shown in Fig. 7. Since none of the inhibitor-substrate pairs were uncompetitive, an "ordered" mechanism does not seem operative. Clearly, the reaction must have alternate possible sequences for the addition of substrates and the order of release of products must

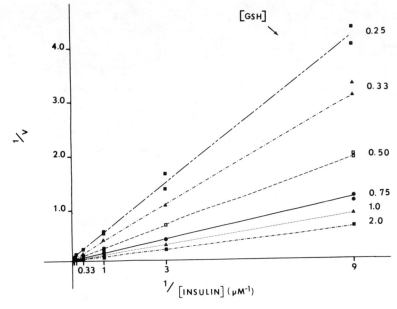

Fig. 4. The initial velocity of insulin degradation catalyzed by GIT with insulin as variable substrate at the indicated fixed levels of GSH (mM). From ref. 15.

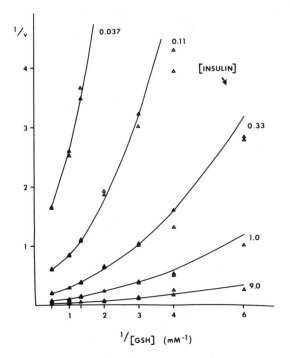

Fig. 5. Insulin degradation by GIT with GSH as variable substrate at the fixed concentrations of insulin (µM). From ref. 15.

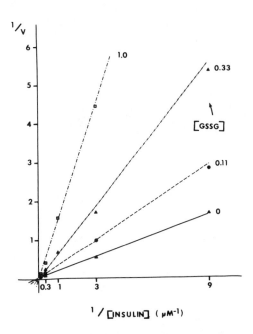

Fig. 6. Inhibition by GSSG (mM) of insulin degradation by GIT with insulin as a variable substrate in the presence of 0.33 mM GSH. From ref. 15.

also have alternate possible pathways. Thus, these studies indicate that the enzyme probably follows a random mechanism which is depicted in Scheme II.

SCHEME II

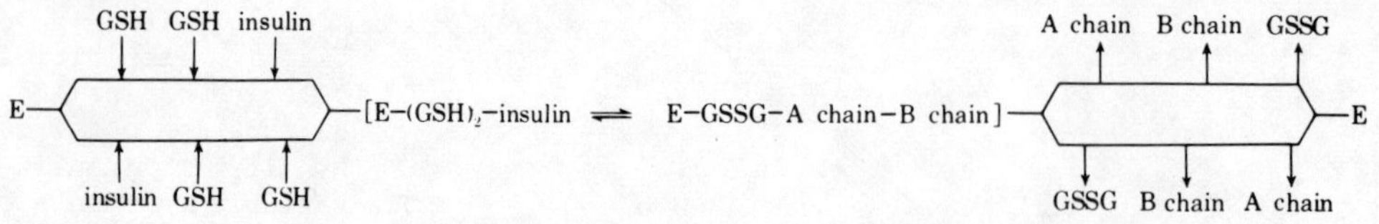

Multiple Distinct Sites of Reaction. Several lines of evidence indicate that GIT has multiple distinct sites of reaction.

(1) Besides GSH, the enzyme utilizes other thiol compounds and certain thiol proteins[16]. Lineweaver-Burk plots for each of the thiols studied as varied substrates have been found to be nonlinear[15]: the plots for monothiols (GSH and mercaptoethanol) were parabolic with a Hill coefficient of about 1.5 for GSH (Fig. 5); for low molecular weight dithiols (dithiothreitol, dihydrolipoic acid, and 2,3 dimercaptopropanol) the plots were apparently linear (as judged from Hill coefficient values which were about 1) but were modified by substrate inhibition (for illustration Fig. 8 shows data with dihydrolipoic acid); and plots for protein polythiols were parabolic with a Hill coefficient of about 2 and superimposed by substrate inhibition.

(2) Of several phospholipids tested, lysolecithin and phosphatidic acid have been found to markedly inhibit the activity of the purified enzyme[17] (Fig. 9). With rat liver microsomal fractions, low levels of these phospholipids increased the GIT activity (Fig. 10). At higher concentrations, lysolecithin caused inhibition as it did with the purified enzyme, but, phosphatidic acid, which inhibited the purified enzyme with a potency far greater than lysolecithin, caused only slight inhibition of the microsomal fraction activity. The inhibitions by lysolecithin and phosphatidic acid were noncompetitive versus both substrates. Considering the nonsimilarity in structures between these phospholipids and substrates or products, and the noncompetitive nature of inhibition, these agents probably interact at an allosteric site(s) on the enzyme.

(3) Likewise, Triton X-100 and deoxycholate which greatly increase the enzyme activity in the microsomal fraction[9,18] are strong inhibitors of the purified enzyme (Fig. 11). Clearly, these data show that the enzyme contains at least two distinct sites, one catalytic and one allosteric, which in microsomes are apparently compartmentalized or spatially separated by membrane components such as phospholipids[17,19]. Detergents or detergent-like agents seem to enhance the enzyme activity

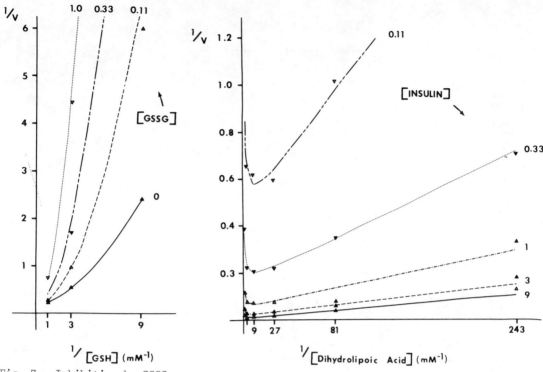

Fig. 7. Inhibition by GSSG (mM) of insulin degradation by GIT with GSH as a variable substrate in the presence of 0.33 μM insulin. From ref. 15.

Fig. 8. Insulin degradation by GIT with dihydrolipoic acid as thiol substrate at fixed levels of insulin (in μM). From ref. 15.

by solubilization of certain membrane structures resulting in the unmasking of the catalytic site(s), whereas drastic conditions such as multiple steps of purification are required to unmask the allosteric site(s), where phosphatidic acid, Triton and deoxycholate seem to interact in the purified enzyme[17].

STRUCTURAL FEATURES OF GIT

The structural features of GSH-insulin transhydrogenase (protein disulfide oxido-reductase, E.C. 1.8.4.2) are presented together with those of protein disulfide isomerase (E.C. 5.3.4.1) since the two enzymes are very similar and probably identical. From the viewpoint of history, the reductive cleavage of the disulfide bonds of insulin by an extract of rat liver was first reported in 1959 by Narahara and Williams[20]. In the same year Tomizawa and Halsey[21] reported the isolation of a homogenous preparation of insulin-degrading enzyme from beef liver. Three years later, this enzyme was shown to degrade insulin by promoting the cleavage of its disulfide bonds[4,5]. Very low rates of reactivation of reduced proteins led to the isolation in 1963 by Venetianer and Straub[22] and by Anfinsen's group of an enzyme capable of accelerating reactivation of reduced ribonuclease[23,24]. However, it became soon apparent that the two enzymes, i.e., insulin-degrading enzyme and ribonuclease-reactivating enzyme, are very similar if not identical[3,9,11,25,26].

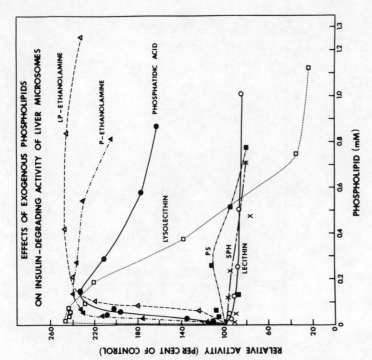

Fig. 10. Effect of phospholipids on insulin degradation by rat liver microsomes. Abbreviations are explained in Fig. 9. Phosphatidic acid, which inhibited markedly the purified enzyme (Fig. 9), caused only slight inhibition of the microsomal fraction activity. From ref. 17.

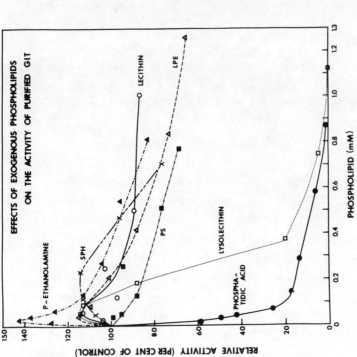

Fig. 9. Effect of phospholipids on insulin degradation by purified GIT. Only two phospholipids, phosphatidic acid and lysolecithin, caused inhibition; all other phospholipids were without effect. Abbreviations are: PS, phosphatidyl serine; LPE, lyso-phosphatidyl ethanolamine; SPH, sphingomyelin. From ref. 17.

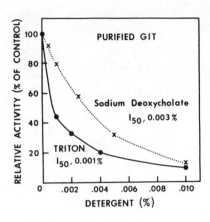

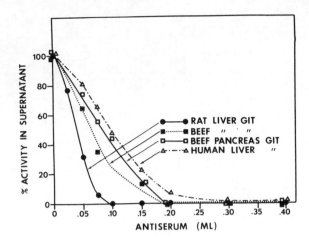

Fig. 11. Inhibition by detergents, Triton X-100 and deoxycholate of insulin degradation by purified GIT. I_{50} = the concentration of detergent causing 50% inhibition of the enzyme activity.

Fig. 12. Antigen-antibody titrations of rat liver, beef liver, human liver and beef pancreas GITs against antiserum to rat liver GIT. From ref. 32.

As will become apparent, the distinction between the two enzymes is strictly operational. Investigators interested in the degradation of polypeptide hormones use insulin as a substrate and refer to this enzyme as transhydrogenase or oxidoreductase (E.C. 1.8.4.2), while investigators interested in the folding or synthesis of proteins use reduced or scrambled forms of ribonuclease as a substrate and refer to this enzyme as reactivating enzyme or isomerase (E.C. 5.3.4.1). The properties of the two enzymes are summarized in Table 2. Both enzymes are widely distributed and are primarily located in the microsomal fraction of the liver. Both catalyze thiol-disulfide interchange, both act upon the same substrates and produce the same products, and both contain a highly reactive thiol group that becomes available only after a reaction with a thiol. Both enzymes have the same sedimentation coefficient; however, there were differences in the dependence of sedimentation coefficient upon protein concentration, although these differences might be because of the different media used.

Hawkins and Freedman[27] report that upon isoelectric focusing the purified beef liver enzyme shows multiple protein bands with a pI value of 4.65 for both transhydrogenase and isomerase activities; as shown in Table 2, similar pI values have been found in our laboratory for 4 GITs isolated from beef, human and rat livers and beef pancreas[28]. Until recently, it was felt that there were differences in the molecular weights of the two enzymes since these were approximately 60,000 daltons for GITs isolated from several different sources and 42,000 daltons for

TABLE 2

COMPARISON OF PROPERTIES OF GLUTATHIONE-INSULIN TRANSHYDROGENASE AND RIBONUCLEASE-REACTIVATING ENZYME

PROPERTY	TRANSHYDROGENASE	REACTIVATING ENZYME
Tissue Distribution	Ubiquitous	Ubiquitous
Subcellular Location	Microsomal	Microsomal
Reaction	SH/SS interchange	SH/SS interchange
Substrate Specificity	wide range of proteins	wide range of proteins
Active Site	Reactive -SH	Reactive -SH
$S_{20,w}$	3.27, independent of protein concentration	3.26, dependent on protein concentration
Mol. Wt.	~60,000	~42,000 (Anfinsen)[24] ~61,250 (Freedman)[27]
pI	4.65 ⎤ Beef liver (4.5- ⎬ Beef pancreas ⎪ Rat liver 4.85) ⎦ Human liver	4.65 Beef Liver
Absorption Spectra	Tryptophan, tyrosine chromophores only	Tryptophan, tyrosine chromophores only
Amino Acid Composition	Beef liver ⎤ very similar Beef pancreas ⎬ with minor Rat liver ⎦ differences	⎡ Beef liver
Immunological Properties	Beef liver ⎤ Beef pancreas ⎪ very similar Rat liver ⎬ with minor Human liver ⎪ differences Human kidney ⎦	

beef ribonuclease-reactivating enzyme[24]. However, recent studies by Hawkins and Freedman report average molecular weights of about 61,000 daltons for this enzyme[27].

The absorption spectra between 600 and 250 nm show that both enzymes have no other chromophores than tryptophan and tyrosine.

The amino acid compositions of three preparations of GIT from rat liver[11], beef pancreas[26] and beef liver[29], and one preparation of ribonuclease-reactivating enzyme[30] from beef liver are strikingly similar (Table 3). There are some minor quantitative differences which might occur because of organ and species differences, since immunological studies with various GITs have shown such differences[31,32] (Fig. 12). Of course, immunological differences could also occur because of differences in carbohydrate composition since the two GITs have been shown to be glycoproteins[26,29].

Perhaps the strongest evidence in favor of a single enzyme thesis comes from the work of Hawkins and Freedman[27] since all attempts to separate the two activities into two distinct protein species were unsuccessful. Both activities coeluted at

TABLE 3

COMPARISON OF AMINO ACID COMPOSITIONS OF GSH-INSULIN TRANSHYDROGENASE AND RIBONUCLEASE REACTIVATING ENZYME

| | Beef Pancreas GIT | | Rat Liver GIT | | Beef Liver GIT | Beef Liver Ribonuclease Reactivating Enzyme | |
	Conc.	(Res.)	Conc.	(Res.)	(Res.)	Conc.	(Res.)
Glu	13.22	(64)	14.50	(70)	(66)	15.84	(76)
Asp	12.21	(59)	12.38	(59)	(55)	11.46	(55)
Lys	9.70	(47)	10.70	(51)	(48)	10.37	(50)
Ala	8.55	(41)	9.36	(45)	(44)	9.40	(45)
Leu	8.56	(41)	9.22	(44)	(42)	9.05	(43)
Phe	6.65	(32)	6.51	(31)	(31)	7.00	(34)
Gly	6.86	(33)	6.18	(30)	(29)	6.46	(31)
Val	5.24	(25)	5.44	(26)	(28)	5.16	(25)
Ile	4.60	(22)	4.37	(21)	(19)	4.47	(22)
Ser	5.62	(27)	4.38	(21)	(24)	4.47	(22)
Thr	4.60	(22)	4.81	(23)	(21)	4.23	(20)
Pro	5.02	(24)	3.97	(19)	(21)	3.85	(19)
His	2.34	(11)	2.68	(13)	(12)	2.16	(10)
Tyr	2.28	(11)	2.30	(11)	(11)	2.26	(11)
Arg	1.86	(9)	2.10	(10)	(11)	1.98	(10)
1/2 Cys	1.65	(8)	1.39	(7)	(7)	0.93	(5)
Met	1.03	(5)	0.93	(5)	(4)	0.83	(4)
Total	99.99	(481)	101.22	(486)	(473)	99.92	(477)
Try		(9)		---	---		---

Concentration is in μmol/100 μmoles of total amino acids. Values given in parentheses are residues based on mol. wt. of 54,000 for all enzyme preparations (see reference 26) except for beef liver GIT preparation which is based on mol. wt. 60,000 (ref. 29).

each step of the purification. However, there were differences for the two activities in their degrees of purification at different steps and in response of the impure enzyme preparations (microsomal fractions) to heat denaturation and to deoxycholate[27,33]. Nevertheless, these differences in the two catalytic activities in the microsomal fraction are explainable on the basis of differences in the affinity and permeability of the substrates (insulin and scrambled ribonuclease) which must occur because of differences in their structure, molecular size and conformation and because of the occurrence of multiple reaction sites on the enzyme and their masking (in particular of allosteric site or sites) by membrane structures[9,17,19]. This possible explanation is also supported by the fact that the GIT activity which was apparently purified to a much lower degree was never found in any of the discarded fractions[27] and the purified enzyme is markedly inhibited by deoxycholate (this paper). In view of the foregoing, it is our opinion that the thiol:protein disulfide oxidoreductase and isomerase activities are probably catalyzed by a single enzyme with multiple sites of reaction.

CONCLUDING REMARKS

Two different physiological roles can be assigned to thiol:protein disulfide oxidoreductase/isomerase enzyme: the cleavage of disulfide bonds in the biodegradation of disulfide-containing hormones or the folding of nascent polypeptide chains in the biosynthesis of disulfide-linked proteins. For the reactivation of scrambled proteins in vitro, generally it has been necessary to use quite high enzyme concentrations to substrate (scrambled ribonuclease) concentration, approaching stoichiometric. "Some purely physical action of this microsomal protein is suggested by the finding that it accelerates the rate of renaturation of urea-denatured B. subtilis α-amylase, which has no disulfide bonds"[34] (Wetlaufer and Ristow[35]). Epstein[36] states that "it is still impossible to decide whether this system has any related physiological function in vivo" in protein folding. On the contrary, the enzyme rapidly inactivates insulin (and other peptide hormones), and this process is the obligatory first step in the catabolism of the hormone[37-39]. The ratio of the enzyme to substrate (insulin) is quite low and is in keeping with the characteristics of enzyme-catalyzed reactions. The insulin-degrading rates of the enzyme in vitro far exceed the rates of insulin synthesis in vivo[25], so, rigid control over the enzyme must exist. In fact several regulatory controls over GIT activity have been identified. For example, the concentration of GIT in liver is under feedback control by the level of circulating insulin[40-45]. The activity of GIT in vitro is modulated by glucagon and growth hormone[10], by certain phospholipids[17], and by certain chelators[46]. The mechanism of action of the enzyme itself is complex; the presence of multiple sites of reaction modified by the membrane organization further adds to the complexity of the system. It is possible that different membrane components provide different compartments and microenvironments thereby imparting the necessary selectivity and orientation to the enzyme (thiol:protein disulfide oxidoreductase/isomerase) for varied functions of the enzyme which of course merits further study.

ACKNOWLEDGEMENTS

This work was supported, in part, by U.S. Public Health Service Grant AM-03854 from the National Institute of Arthritis, Metabolic and Digestive Diseases.

REFERENCES

1. Varandani, P.T. (1977) Proc. IX Congress of International Diabetes Federation (New Delhi, India, 1976), Excerpta Medica, in press.
2. Varandani, P.T. (1966) Biochim. Biophys. Acta 118, 198-201.

3. Katzen, H.M. and Tietze, F. (1966) J. Biol. Chem. 241, 3561-3570.

4. Tomizawa, H.H. (1962) J. Biol. Chem. 237, 428-431.

5. Katzen, H.M. and Stetten, D. Jr. (1962) Diabetes 11, 271-280.

6. Varandani, P.T. and Nafz, M.A. (1970) Arch. Biochem. Biophys. 141, 533-537.

7. Ferrier, B.M., Johns, T., Say, A., and Branda, L.A. (1973) Can. J. Biochem. 51, 1555-1558.

8. Small, C.W. and Watkins, W.B. (1974) Nature 251, 237-239.

9. Varandani, P.T. (1973) Biochim. Biophys. Acta 304, 642-659.

10. Varandani, P.T., Nafz, M.A. and Chandler, M.L. (1975) Biochemistry 14, 2115-2120.

11. Ansorge, S., Bohley, P., Kirschke, H., Langner, J., Marquardt, I., Wiederanders, B., and Hanson, H. (1973) FEBS Lett. 37, 238-240.

12. Chan, S.J., Keim, P. and Steiner, D.F. (1976) Proc. Natl. Acad. Sci. USA 73, 1964-1968.

13. Katzen, H.M., Tietze, F. and Stetten, D. (1963) J. Biol. Chem. 238, 1006-1011.

14. Varandani, P.T. (1967) Biochim. Biophys. Acta 132, 10-14.

15. Chandler, M.L. and Varandani, P.T. (1975) Biochemistry 14, 2107-2115.

16. Chandler, M.L. and Varandani, P.T. (1973) Biochim. Biophys. Acta 320, 258-266.

17. Varandani, P.T. and Nafz, M.A. (1976) Biochim. Biophys. Acta 438, 358-369.

18. Ansorge, S., Bohley, P., Kirschke, H., Langner, J., Wiederanders, B. and Hanson, H. (1973) Eur. J. Biochem. 32, 27-35.

19. Varandani, P.T., Darrow, R.M. and Nafz, M.A. (1977) Proc. Soc. Exp. Biol. Med. 156, 123-126.

20. Narahara, H.T. and Williams, R.H. (1959) J. Biol. Chem. 234, 71-77.

21. Tomizawa, H.H. and Halsey, Y.D. (1959) J. Biol. Chem. 234, 307-310.

22. Venetianer, P. and Straub, F.B. (1963) Biochim. Biophys. Acta 67, 166-168.

23. Goldberger, R.T., Epstein, C.J. and Anfinsen, C.B. (1963) J. Biol. Chem. 238, 628-635.

24. Fuchs, S., de Lorenzo, F. and Anfinsen, C.B. (1967) J. Biol. Chem. 242, 298-302.

25. Tomizawa, H.H. and Varandani, P.T. (1965) J. Biol. Chem. 240, 3191-3194.

26. Varandani, P.T. (1974) Biochim. Biophys. Acta 371, 577-581.

27. Hawkins, H.C. and Freedman, R.B. (1976) Biochem. J. 159, 385-393.

28. Hern, E.P. and Varandani, P.T. unpublished results.

29. Carmichael, D.F., Morin, J.E. and Dixon, J.E. (1977) J. Biol. Chem. 252, 7163-7167.

30. de Lorenzo, F., Goldberger, R.F. and Anfinsen, C.B. (1966) J. Biol. Chem. 241, 1562-1567.

31. Varandani, P.T. and Nafz, M.A. (1970) Int. J. Biochem. 1, 313-321.

32. Varandani, P.T. (1972) Biochim. Biophys. Acta 286, 126-135.

33. Ibbetson, A.L. and Freedman, R.B. (1976) Biochem. J. 159, 377-384.

34. Yutani, K., Yutani, A., Isemura, T. (1967) J. Biochem. Tokyo, 62, 576-583.

35. Wetlaufer, D.B. and Ristow, S. (1973) Ann. Rev. Biochem. 42, 135-158.

36. Epstein, C.J. (1972) Aspects of Protein Synthesis, Part A, Anfinsen, C.B., ed., Academic Press, New York, p. 463.

37. Varandani, P.T., Shroyer, L.A. and Nafz, M.A. (1972) Proc. Natl. Acad. Sci. USA 69, 1681-1684.

38. Varandani, P.T. (1973) Biochim. Biophys. Acta 320, 249-257.

39. Varandani, P.T. and Nafz, M.A. (1976) Diabetes 25, 173-179.

40. Varandani, P.T., Nafz, M.A. and Shroyer, L.A. (1971) Diabetes 20, 342.

41. Varandani, P.T. (1974) Diabetes 23, 117-125.

42. Thomas, J.H. and Wakefield, S.M. (1973) Biochem. Soc. Trans. 1, 1179-1182.

43. Cudworth, A.G. and Barber, H.E. (1975) Eur. J. Pharmacology 31, 23-28.

44. Kohnert, K.D., Hahn, H.J., Zühlke, H., Schmidt, S. and Fiedler, H. (1974) Biochim. Biophys. Acta 338, 68-77.

45. Varandani, P.T. and Nafz, M.A. (1976) Biochim. Biophys. Acta 451, 382-392.

46. Varandani, P.T. and Nafz, M.A. (1976) Endocrinology (suppl.) 98, 300.

47. Steiner, R.F., de Lorenzo, F. and Anfinsen, C.B. (1965) J. Biol. Chem. 240, 4648-4651.

48. Corte, E.D. and Parkhouse, R.M.E. (1973) Biochem. J. 136, 597-606.

FLAVOENZYMES

MECHANISTIC STUDIES ON THE BACTERIAL FLAVOPROTEIN: THIAMINE DEHYDROGENASE

DALE E. EDMONDSON AND CARLOS GOMEZ-MORENO

Department of Biochemistry and Biophysics, University of California and Molecular Biology Division, Veterans Administration Hospital, San Francisco, California, U.S.A.

ABSTRACT

Thiamine dehydrogenase catalyzes the 4-electron oxidation of thiamine to thiamine acetic acid. The enzyme contains one mole of 8α-[N(1)-histidyl]-FAD in a covalent linkage to one mole (45,000 daltons) of enzyme. The studies presented here show thiamine dehydrogenase might be more correctly termed thiamine oxidase since the enzyme readily forms a sulfite adduct, forms an anionic flavin semiquinone, and reacts with O_2 to form H_2O_2 rather than O_2^-. This latter property is due to the fact that the redox potentials for the PFI/PFI· and PFI·/PFIH$_2$ couples are too high (+80 and +30 mV respectively) to reduce O_2 by a 1-electron mechanism. The enzyme can accomodate 2 electrons either from the substrate or as measured by spectrocoulometry in the potential range of +300 to -320 mV. Reductive titrations with dithionite, however, show stoichiometries of ~ 3 moles per mole of active enzyme and 1 mole per mole of inactivated enzyme. These data suggest the presence of an unknown group(s), capable of reacting with dithionite, on the functional enzyme in addition to the flavin. Steady state and pre-steady state kinetic data show reduction of the enzyme by thiamine to proceed at a rate faster than catalytic turnover.

INTRODUCTION

Thiamine dehydrogenase (TD) was first isolated and partially characterized by Neal[1] in his studies on the bacterial catabolism of thiamine. The enzyme is isolated from an unidentified soil bacterium grown on thiamine as its sole source of carbon and nitrogen. The reaction catalyzed is the 4-electron oxidation of the hydroxy ethyl side chain of thiamine to thiamine acetic acid as shown in Figure 1. Presumably the reaction occurs via two sequential 2-electron oxidation steps in which

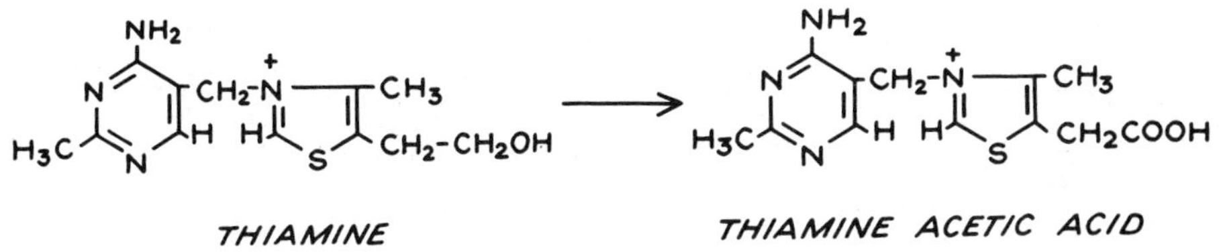

THIAMINE   THIAMINE ACETIC ACID

Fig. 1. The reaction catalyzed by thiamine dehydrogenase.

the hydroxyl group is oxidized to an aldehyde and the resulting aldehyde is further oxidized to a carboxyl group. No evidence has been found for the release of the intermediate aldehyde during catalysis[1]. Our interest in the catalytic mechanism of this enzyme is due to the unusual nature of the reaction catalyzed, which is unique among the simple flavoenzymes, and to the previous finding in this laboratory that the flavin (FAD) moiety is covalently bound to the polypeptide chain via the 8α-position of the flavin to the N(1) position of the imidazole ring of a histidine[2,3]. The structure of the flavin coenzyme is shown in Figure 2. Since no other redox-active groups (such as

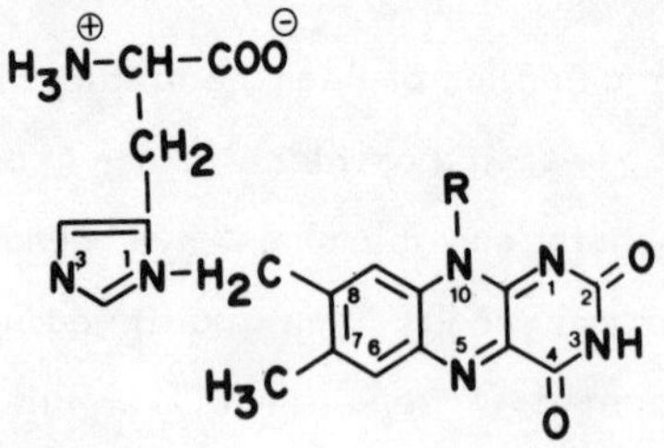

Fig. 2. Structure of 8α-[N(1)-histidyl]-FAD. R is the rest of the FAD moiety.

hemes, iron-sulfur centers, etc.) with absorbance in the visible spectral region exist in the enzyme in addition to the flavin, this enzyme is thus a good system to study, without interference, the properties and catalytic roles of the 8α-substituted flavin by spectroscopic techniques.

RESULTS AND DISCUSSION

The enzyme is purified using a procedure modified from that of Neal's[1] which results in a 40% yield of electrophoretically homogeneous material. The purified enzyme has an A_{280}/A_{447} spectral ratio of 12.0 which is quite high as compared to the ratios of other flavoenzymes and suggests a large proportion of aromatic amino acids in its amino acid composition.

Enzymic activity can be measured either spectrophotometrically by measuring the rate of reduction of 2,6-dichlorophenolindophenol (DCIP) in the coupled dye assay with phenazine methosulfate (PMS) as the immediate electron acceptor or polarographically by monitoring the rate of disappearance of O_2. DCIP does not function as an electron acceptor catalytically in the absence of PMS. Experiments designed to determine the stoichiometry of the catalyzed reactions show 2 moles of DCIP reduced per mole of thiamine under anaerobic conditions whereas in the oxidase assay 1.99 moles of O_2 are reduced to H_2O_2 per mole of thiamine oxidized to thiamine acetic acid. No evidence for the intermittent formation of O_2^- in the oxidase reaction was found as there is no O_2 mediated reduction of cytochrome c and the presence of superoxide dismutase has no effect on the rate of O_2 consumption whereas catalase decreases the rate by a factor of two.

The above results suggest that TD may be more accurately termed a flavoprotein oxidase in accord with the general properties elucidated by Massey and co-workers[4,5]. If so, the covalently bound flavin of TD should react with sulfite to form an adduct at the N(5) position as do most other flavoprotein oxidases[4]. Figure 3 shows that the addition of sulfite to TD results in substantial bleaching of the flavin absorbance. An extrapolated value ($\Delta\varepsilon_{447}$) of 9,430 M^{-1}-cm^{-1} was determined from a Benasi-Hildebrand plot[6] of the data in Figure 3. A K_D value of 0.117 ± 0.01 mM is calculated which shows the enzyme to have a high affinity for sulfite. The bottom spectrum in

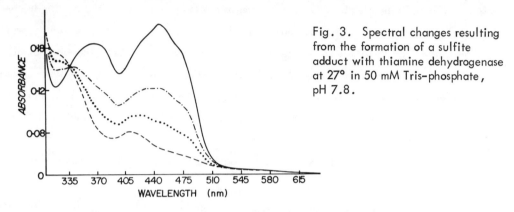

Fig. 3. Spectral changes resulting from the formation of a sulfite adduct with thiamine dehydrogenase at 27° in 50 mM Tris-phosphate, pH 7.8.

Figure 3 is that of the pure enzyme-sulfite adduct obtained by computer manipulation of the data to eliminate any spectral contributions by the uncomplexed enzyme. The observed peak at 410 nm (Figure 3) is not seen with the sulfite adducts of other flavoprotein oxidases or of 8α-[N(1)-histidyl]-riboflavin[6]. A similar spectrum is seen with the sulfite adduct of glycolate oxidase which, however, has an absorbance maximum at 425 nm[7]. Schuman and Massey[7] have suggested this absorption band to be due to the presence of another chromophore in the enzyme preparation which was later found

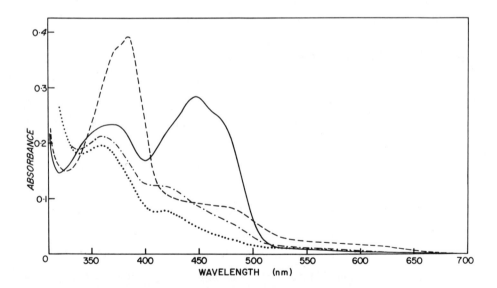

Fig. 4. Absorption spectra of thiamine dehydrogenase in its oxidized form (———); semiquinone form (‑ ‑ ‑); substrate-reduced form (‑•‑•‑); and after reduction by 3 molar equivalents of $S_2O_4^=$ (••••). All solutions contained 50 mM Tris-phosphate, pH 7.8.

to be non-stoichiometric quantities of 6-hydroxy-FMN[8]. Whether this proves to be the case with TD awaits further investigation.

Figure 4 shows the absorption spectra of the various redox forms of TD. As found with other flavoprotein oxidases[4], TD forms an anionic flavin semiquinone on irradiation in the presence of EDTA or on partial reduction by $S_2O_4=$. Electron spin resonance spectral data confirm this spectral species to be due to the flavin semiquinone. Reduction of the enzyme with substrate decreases the absorbance at 447 nm by 70% whereas the reduced form produced by $S_2O_4=$ reduction decreases the A_{447} by 82%. The more extensive bleaching by $S_2O_4=$ reduction is not due to the presence of non-functional enzyme as the difference spectrum between the two reduced species shows a broad, featureless shape with a maximum at 408 nm rather than 447 nm if additional flavin were reduced on $S_2O_4=$ reduction.

To determine how many moles of substrate are required to fully reduce the enzyme-bound flavin,

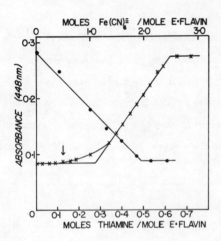

Fig. 5. Anaerobic titration of thiamine dehydrogenase with thiamine (●) and subsequent reoxidation with ferricyanide (x). The arrow denotes the point at which reducing equivalents due to excess thiamine are consumed by ferricyanide.

anaerobic titrations were performed with thiamine as the reductant. The results of such an experiment are shown in Figure 5. A stoichiometry of 0.5 moles of thiamine are required to reduce 1 mole of enzyme-bound flavin. Reoxidation of the resulting reduced enzyme with $Fe(CN)_6^{-3}$ requires 2 molar equivalents (after correction for the slight excess of thiamine present in the solution). No evidence was found for any flavin semiquinone formation during the thiamine titration, however, the non-linear increase in absorbance at 447 nm on reoxidation by $Fe(CN)_6^{-3}$ suggests the formation of substantial quantities of semiquinone (Figure 5). This experiment shows the enzyme can accommodate 2 electrons from the substrate and that the intermediate thiamine aldehyde is capable of reducing the enzyme. It is not known whether thiamine aldehyde dissociates from the reduced enzyme and reduces an oxidized enzyme molecule or if the reduced enzyme-thiamine aldehyde complex is capable of reducing an oxidized enzyme molecule. The anaerobic addition of 0.5 moles of thiamine to TD results in a rapid reduction of the enzyme in the time required to measure the spectrophotometric changes in the flavin absorbance. The distribution of 4 electrons from 0.5 molar equivalents of thiamine among the enzyme molecules in solution is thus a rapid process.

Since the absorption spectrum of the $S_2O_4^=$ reduced enzyme is different than that of the substrate reduced enzyme, it was of interest to determine if TD is reduced by $S_2O_4^=$ with the same electron stoichiometry as with substrate. Figure 6 shows that approximately 3 moles of $S_2O_4^=$ per mole of

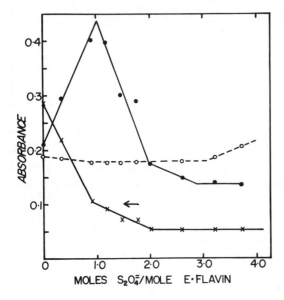

Fig. 6. Absorption spectral changes occurring during the anaerobic titration of thiamine dehydrogenase with $S_2O_4^=$. (●) absorption changes at 384 nm, (x) absorption changes at 447 nm, and (o) absorption changes at 339 nm. The arrow denotes the absorption value at 447 nm on substrate reduction of the enzyme.

enzyme are required before excess titrant is observed in the solution. The endpoint in the titration is taken at 339 nm (an isosbestic point in the absorption spectrum of the enzyme) where an increase in absorption due to excess $S_2O_4^=$ appears. The absorption changes at 384 nm show that substantial amounts of anionic flavin semiquinone are formed.

To provide an independent measurement of the number of electrons that can be accomodated by TD, spectrocoulometric experiments were performed using an apparatus described by Wilson[9]. Reducing or oxidizing equivalents are generated by applying a potential between a reference electrode and an optically transparent SnO_2 electrode. Methyl viologen and $Fe(CN)_6^{-3}$ are used as mediator-titrants to facilitate electron transfer from the electrode to the enzyme. By determination of the charge required (Q in coulombs) to reduce the enzyme, the number of electrons (n) can be calculated by the following equation:

$$Q = n \; F \, (\text{moles Enzyme})$$

where F is the Faraday constant. The data in Figure 7 show that 2 electrons are required to reduce the enzyme in the potential range of +300 mV to –320 mV. The redox changes observed in Figure 7

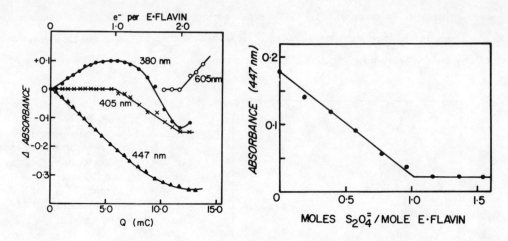

Fig. 7. (Left). Spectrocoulometric titration of thiamine dehydrogenase (28.3μM in 0.1 M sodium phosphate- 1mM EDTA, pH 7.2. The increase in absorption at 605 nm is due to the appearance of the methyl viologen radical cation.

Fig. 8. (Right). Anaerobic titration of inactivated thiamine dehydrogenase with $S_2O_4^=$. The enzyme was inactivated by freezing a solution of the enzyme in 0.1 M sodium phosphate, pH 6.5 for 3 days which also results in precipitation. After centrifugation, the protein precipitate was redissolved in 0.1 M Tris-Cl⁻, pH 8.5 and the titration experiment performed.

were fully reversible in that the same number of coulombs were required to reoxidize the enzyme. The reason for the anomalous consumption of electrons using $S_2O_4^=$ as reductant is not known at present. One possibility is that $S_2O_4^=$ reacts with an unknown low-potential group or groups on the enzyme which would not be observed in the potential range investigated in the spectrocoulometric titration experiments. Inactivation of the enzyme by freezing at pH 6.5 results in a form that is reduced by 1 mole of $S_2O_4^=$ or 2 electrons with no formation of an intermediate semiquinone during reduction (Figure 8). This observation suggests that the group(s) responsible for consumption of the additional reducing equivalents from $S_2O_4^=$ are able to do so only when the enzyme is in a catalytically active form.* Further work is underway to determine the structure and catalytic role of this unknown group(s). Since TD contains no metals, the unknown group must be organic in nature.

The oxidation-reduction potential of the flavin of TD could readily be measured during the spectrocoulometric studies using a Pt electrode and a Ag/AgCl reference electrode. The PFl/PFl· couple and the PFl·/PFlH₂ couple were found to be +80 mV and +30 mV respectively (Figure 9) with

*This would imply that the unknown group(s) is either unable to accept electrons directly from $S_2O_4^=$ but only via the flavin in its conformationally active form or that the unknown group(s) is dissociated from the enzyme and lost on replacement of the buffer. Further work is required to distinguish between these two possibilities.

n values corresponding to 1-electron couples. The 2-electron potential for 8α-[N(1)-histidyl]-riboflavin has been found to be −165 mV[2] which shows that the environment of the flavin on the enzyme results in a shift in potential of +200 to +250 mV. The increase in potential does not reflect

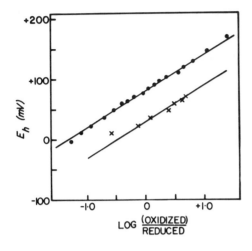

Fig. 9. Oxidation-reduction potential titration of thiamine dehydrogenase. (●)ΔA_{447} and (x)ΔA_{380} using an extrapolated value for the A_{380} of 100% semiquinone. The same buffer as in Figure 7 was used.

an increase in binding affinity of the protein for either the flavin semiquinone or hydroquinone inasmuch as the flavin is covalently bound to the protein. The above redox potential data provide an explanation for the failure to observe O_2^- in the catalytic oxidase assay. The O_2/O_2^- redox couple is known to be −330 mV[10]. Thus, in accord with thermodynamic principles, the 1-electron redox couples of the flavin in TD are far too positive to be able to reduce O_2 by a 1-electron mechanism. TD has, however, the necessary reducing potential to reduce O_2 to H_2O_2 since the O_2/H_2O_2 redox couple is +270 mV[11].

From the observed stoichiometry in which one mole of substrate is oxidized by two moles of the two-electron acceptors (O_2 or PMS/DCIP), one might expect complex steady state kinetic behavior. The Lineweaver-Burk plots shown in Figure 10 show this, in fact, to be the case. The thiamine oxidase data in Figure 10A show catalytic activity to be inhibited by high concentrations of O_2 with a concurrent increase in the K_m for thiamine. Upon increasing the [O_2] from 0.237 mM to 0.653 mM, the maximal velocity decreases from 1000 moles O_2 consumed/min/mole enzyme to 840, while the K_m value for thiamine increases from 2.3×10^{-5} M to 1.6×10^{-3} M. Due to the observed O_2 inhibition at concentrations higher than air-saturated, the maximal velocity at infinite concentrations of thiamine and O_2 as well as the K_m value for O_2 has not been determined with any degree of accuracy. An extrapolated maximal velocity of 1500 to 2000 moles O_2/min/mole enzyme and a K_m value for O_2 of $\sim 10^{-4}$ M have been estimated from secondary plots of the data in Figure 10A. These values should, however, be regarded as upper limits based on the data presently available.

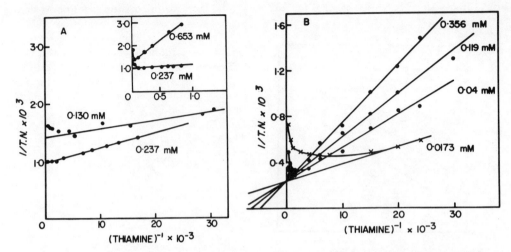

_Fig. 10.   A)   Lineweaver-Burk plots of thiamine oxidase activity at various O_2 concentrations at 30° in 50 mM sodium phosphate- 1 mM EDTA, pH 7.2.   T.N. is expressed as moles O_2 consumed/ min/mole enzyme.   The ordinate and abscissa of the inset have the same units respectively as those of the figure.   B)   Lineweaver-Burk plots of thiamine-PMS/DCIP reductase activity at various indi- cated concentrations of PMS under the same conditions as A.   T.N. is expressed as moles DCIP re- duced/min/mole enzyme.   The assay buffers were purged with N_2 to minimize complications due to oxidase activity.

Steady state kinetic studies using PMS as electron acceptor (and coupled with DCIP as terminal electron acceptor) shows converging reciprocal plots with a maximal velocity of 4,167 moles DCIP reduced/min/mole enzyme (Figure 10B).   The plots are complex in that substrate inhibition by thia- mine becomes more predominant at low concentrations of PMS.   Similarly, the K_m value for thia- mine increases as the concentration of PMS is increased.   The Lineweaver-Burk plots obtained re- semble those exhibited by enzymes in which there is a double-competitive substrate inhibition in a ping pong mechanism as described in Cleland's review[11].   More detailed kinetic studies are re- quired before a definitive statement regarding the sequence of addition of substrate and release of products can be made.

Since the maximal velocity found in the PMS/DCIP assay is ∿ 3 times that found in the oxidase assay, the rate of substrate reduction of the enzyme could not be the rate-limiting step in catalysis. Furthermore, the temperature dependence of the PMS/DCIP assay is found to have an E_a = 12 Kcal/ mole while that of the oxidase assay is 16.8 Kcal/mole (Figure 11).   These data suggest a different rate-limiting step in each catalytic reaction since identical energies of activation would be ob- served if the same rate-limiting step were operative in each case.

Anaerobic stopped-flow kinetic studies (Figure 12) show TD to be reduced by thiamine at times much faster than turnover in either catalytic assay.   A half-time of 5.5 msec is found from the

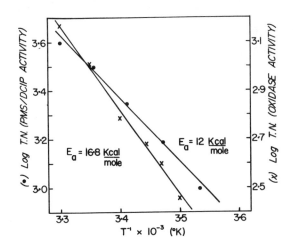

Fig. 11. Arrhenius plot of the temperature dependence of catalytic activity in the oxidase assay (x) and in the PMS/DCIP assay(•). The thiamine concentration in the oxidase assay was 7.4 mM and 2.5 mM in the PMS/DCIP assay. Air-saturated solutions were used in the oxidase assays while the PMS concentration was 0.356 mM in the dye assays.

linear first-order plot of the rate of flavin reduction (Figure 12). The calculated rate constant of 7,500 min^{-1} is faster than the observed maximal velocities in either catalytic assay which shows that the rate of reduction of TD is not the rate-limiting step in catalysis. No spectral intermediates absorbing at wavelengths longer than the spectral range of flavin absorption were observed during reduction.

The results presented above show TD to exhibit all of the properties expected for a flavoprotein oxidase. The relatively high 1-electron redox couples of the covalently-bound flavin provide a

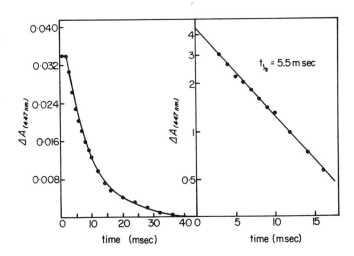

Fig. 12. (Right) Time course for the reduction of thiamine dehydrogenase (8.4 μM) by thiamine (2.5 mM) under anaerobic conditions in 50 mM sodium phosphate- 1 mM EDTA, pH 7.2 at 30°. The concentrations stated are those after mixing in the stopped-flow apparatus. (Left) First order plot of the data.

rationale for the inability of the enzyme to reduce O_2 by a 1-electron mechanism. Elucidation of the structure and catalytic role of an unknown redox-active group on the enzyme as well as the detailed mechanism for the oxidation of thiamine to thiamine acetic acid remain for future work. Studies are currently underway to investigate these fascinating problems.

ACKNOWLEDGEMENTS

This work was supported by NIH Program Project Grant HL-16251 and by NSF Grant PCM 74-03264. The authors wish to acknowledge the excellent technical assistance of Ms. Margaret Choy and to thank Dr. G.S. Wilson, University of Arizona, for his help and hospitality in performing the spectrocoulometric and redox potential experiments.

REFERENCES

1. Neal, R.A. (1970) J. Biol. Chem. 245, 2599-2604.

2. Edmondson, D.E., Kenney, W.C., and Singer, T.P. (1976) Biochemistry 15, 2937-2945.

3. Kenney, W.C., Edmondson, D.E., and Seng, R. (1976) J. Biol. Chem. 251, 5386-5390.

4. Massey, V., Muller, F., Feldberg, R., Schuman, M., Sullivan, P.A., Howell, L.G., Mayhew, S.G., Mathews, R.G., Foust, G.P. (1969) J. Biol. Chem. 244, 3999-4006.

5. Massey, V., Strickland, S., Mayhew, S.G., Howell, L.G., Engel, P.C., Mathews, R.G., Schuman, M., and Sullivan, P.A. (1969) Biochem. Biophys. Res. Commun. 36, 891-897.

6. Benasi, H.A., and Hildebrand, J.H. (1949) J. Am. Chem. Soc. 71, 2703-2707.

7. Schuman, M. and Massey, V. (1971) Biochim. Biophys. Acta 227, 500-520.

8. Mayhew, S.G., Whitefield, C., Ghisla, S., and Schuman-Jorns, M. (1974) Eur. J. Biochem. 44, 579-591.

9. Wilson, G.S., in Methods in Enzymology. (S. Fleischer and L. Packer, eds.) Academic Press, New York, in press.

10. Wood, P.M. (1974) FEBS Lett. 44, 22-24.

11. George, P. (1965) in Oxidases and Related Redox Systems (T.E. King, H.S. Mason, and M. Morrison, eds.) Vol. I, pp. 3-36; J. Wiley and Sons, New York.

12. Cleland, W.W. (1970) in The Enzymes (P.D. Boyer, ed.) Vol. II, 3rd Edition, pp. 1-65, Academic Press, New York.

ON THE REACTION MECHANISM OF THE FLAVOENZYME LACTATE OXIDASE: EVIDENCE
FOR THE OCCURRENCE OF COVALENT INTERMEDIATES DURING CATALYSIS

SANDRO GHISLA and VINCENT MASSEY

Fachbereich Biologie der Universität, D-775 Konstanz, Germany, and Department of
Biological Chemistry, University of Michigan, Ann Arbor, 48109, MI USA

ABSTRACT

The flavoenzyme lactate oxidase has been found to distinguish between the two
hydrogen atoms at the α-carbon atom of glycollate. Tritium and deuterium labels were
introduced at this carbon atom in the Re- and Si- configurations by reaction of glyoxylate with NAD^2H and NAD^3H and the stereospecific enzymes, pig heart L-lactate dehydrogenase and *L. leichmanii* D-lactate dehydrogenase. The catalytic activity of
lactate oxidase is shown to be associated with the abstraction of the Re-H to form a
labile R-glycollyl adduct with the enzyme flavin at position N(5). The enzyme is
gradually inhibited during turnover due to the diastereomer, formed by abstraction of
the Si-H of glycollate, leading to a stable S-glycollyl-N(5)-flavin adduct. The proton abstraction step is at least partially rate determining, as evidenced by substantial deuterium isotope effects.

INTRODUCTION

Flavoprotein oxidases and dehydrogenases catalyze, as the primary event, the oxidation of "C-H substrates". During this process, reducing equivalents are transferred from substrate to the flavin coenzyme, which in turn can transfer them to suitable acceptors:

$$E\text{-}Fl_{ox} \ + \ >C<^H_X \ \longrightarrow \ E\text{-}Fl_{red} \ + \ >C=X$$

In recent years, the molecular mechanism of this process has been the subject of
extensive investigation dealing with enzyme as well as with chemical model reactions.
From these studies different mechanistic proposals have emerged, which are in part
still contradictory and controversial. They focus on the mode of breakage of the
substrate C-H bond, and on the successive or concomitant transfer of redox equivalents, and have variously been named "hydride", "radical", "carbanion" and "group
transfer" mechanisms. For detailed and comprehensive discussions of these mechanisms
we refer to the original literature[1-6].

We have focused our studies of the past years on the mechanism of the flavoenzyme
lactate oxidase (lactate 2-monooxygenase; L-lactate:oxygen 2-oxidoreductase (de-
carboxylating), EC 1.13.12.4). This enzyme catalyzes the oxidation of L-lactate
(R = CH_3) to acetate, CO_2 and water: $R\text{-}CHOH\text{-}COO^- + O_2 \longrightarrow R\text{-}COO^- + CO_2 + H_2O$

Glycollate (R = H) was found[6] to be a similar substrate; we therefore proposed that its oxidation leads to formation of formate (R = H), CO_2 and water[6]. A detailed kinetic study of the course of enzyme reduction however, demonstrated substantial differences between L-lactate and glycollate as substrate. While in the first case the anaerobic reduction proceeds without observable intermediates to form the reduced enzyme-pyruvate complex[7], with glycollate two metastable species could be demonstrated[6]. These two species are clearly differentiated from free reduced enzyme, or from a reduced enzyme glyoxylate complex by their particular spectral properties. Both species have similar absorption spectra (λ_{max} ~ 360 nm) and strong fluorescence emission (λ_{max} = 465 nm). In contrast reduced enzyme-keto acid complexes appear to have no fluorescence while free reduced enzyme has about 20 percent the fluorescence intensity, and a different emission (λ_{max} = 510 nm)[6].

The outstanding difference between both intermediates consists in their chemical stability. One intermediate appears to be in a fast equilibrium with the reduced enzyme glyoxylate complex, as demonstrated by the reactivity of the system with oxygen[6]. The second species, on the other hand, can be isolated by Sephadex G-25 chromatography at 0-4°; it is unreactive towards oxygen, but decays slowly at 25° mainly to oxidized enzyme and presumably glycollate. Clues to the structure of these adducts came from two sets of observations: a) The absorption and fluorescence spectra of the intermediates correspond closely to those of covalent flavocoenzyme N(5)-adducts, which were obtained from lactate oxidase either by photo-alkylation[8] or photoreduction[9]. b) Oxidized enzyme forms with tartronate a tight complex which undergoes a rapid photochemical reaction to form an adduct[6]. This adduct in turn has properties indistinguishable from those of the more stable adduct obtained either from oxidized enzyme and glycollate, or from reduced enzyme and glyoxylate (cf. also Scheme 2 below)[6]. The photoreaction of oxidized lactate oxidase with activated dicarboxylic acids is known to yield covalent flavin-N(5)-adducts[9,10], the process involving decarboxylation of the dicarboxylic acid and addition of the residue to the coenzyme. On the basis of these experimental observations we proposed that both intermediates in question are flavocoenzyme N(5)-glycollyl adducts[6]. Their different chemical properties were suggested to originate from their having diastereomeric structures (cf. structures below). The labile, catalytically competent adduct would thus be derived from reaction of an enzyme glycollate "Re-complex", i.e. a complex in which the glycollate Re-hydrogen is undergoing oxidation; this adduct will be called "R-adduct". (Note that the Re-hydrogen in glycollate has the same chiral position as the α-hydrogen in L-lactate). The more stable adduct on the other hand would thus have the enantiomeric S-structure (S-adduct) (Scheme 1). It should be noted that the denominations "R,S-adducts" are merely formal, denoting the adducts formed by removal respectively of the Re- or the Si-hydrogen, and might not represent the absolute configuration of the adducts, as the reaction(s) leading to their formation might involve inversion at the glycollate α-carbon. For sake of clarity the denominations

(Re,Si), which denote two enantiotopic faces of a molecule, will be used also for differentiating the position of ^3H and ^2H in glycollate, as label in the Re face will correspond to (R)*H-CHOH-COO$^-$ and in the Si face to (S)*H-CHOH-COO$^-$.

Scheme 1

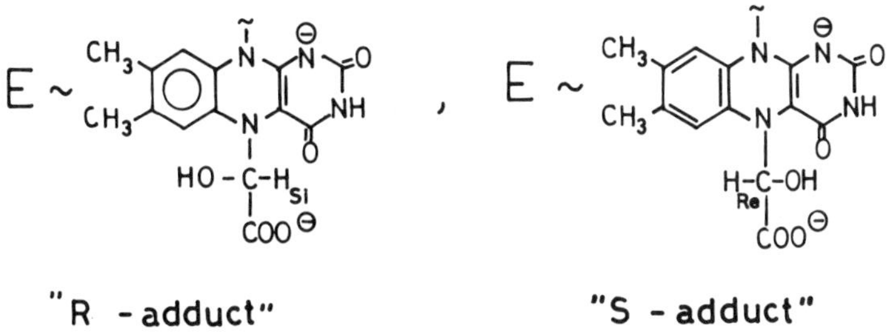

"R - adduct"   "S - adduct"

The experiments discussed in this work indicate that this hypothesis is indeed correct, and that covalent adducts can be formed directly from enzyme and substrate, thus supporting the concept of "covalent catalysis".

RESULTS AND DISCUSSION

Modes of formation, decay, and structure of the stable glycollyl flavoenzyme adduct

As already mentioned in the introduction, the relatively stable adduct can be formed by three different ways as shown in Scheme 2.

Scheme 2

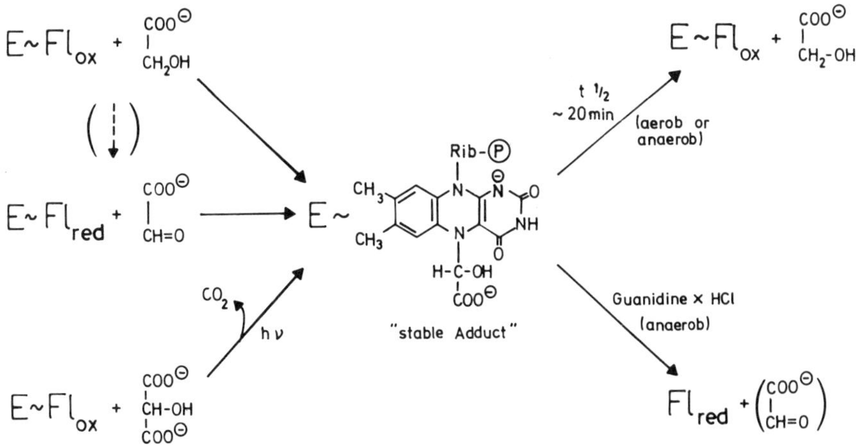

Both the reaction of oxidized enzyme with 1-^{14}C-glycollate, and of reduced enzyme with U-^{14}C-glyoxylate lead to incorporation of label in the stable adduct. The results of such experiments will be detailed later (S. Ghisla and V. Massey, in preparation). Similarly, when the photoreaction of the oxidized enzyme tartronate complex was carried out as outlined earlier with α-^{3}H tartronate (using different samples having specific activities in the range 360 to 1000 dpm/nMole) the isolated stable adduct showed the expected incorporation of approximately 1 equivalent of label. (The average of five experiments indicated a specific activity for the isolated adduct of approximately 110% that of tartronate; error $\pm$ 40%.)

For an analysis of the role of the adducts in catalysis it is of obvious importance to demonstrate that they are indeed adducts of the glycollate α-carbon to position N(5) of the flavin. When the photochemically formed stable adduct is **denatured** by anaerobic addition of guanidine hydrochloride, its fluorescence disappears immediately and a species is formed which has the typical spectrum of reduced FMN. Subsequent admission of oxygen leads to rapid and complete formation of oxidized FMN, thus indicating that reduced coenzyme is indeed released upon denaturation of the complex. The second requirement for structural proof, i.e. the demonstration that glycollate is formed during the slow non-denaturing decay process will be discussed below.

Rate of formation of oxidized lactate oxidase from reduced enzyme and glyoxylate

In order to locate the intermediates in the catalytic cycle and to assess their function, it is of importance to estimate the rate of conversion of reduced to oxidized enzyme by glyoxylate. In the case of pyruvate, the product of anaerobic oxidation of lactate, the rate of this reverse reaction is very slow[7], and can be estimated as 0.2 min^{-1} (S. Ghisla and V. Massey, unpublished). In the case of glycollate, however, steady state kinetic measurements indicate that this rate is not negligible[6]; calculations show that a rate of the order of 20 min^{-1} is required to account for the steady state kinetics. A direct estimation of this value is possible by measurement of the rate of appearance of inactivated enzyme, when the experiment is carried out in the presence of α-hydroxybutynoic acid (Fig. 1). This inhibitor rapidly and irreversibly traps oxidized enzyme[11]. It should be pointed out, however, that the value found of k ~ 1.8 min^{-1} will be a lower limit one. Glyoxylate in fact exists in solution mainly as the hydrated form instead of the keto form, which would be preferentially the one leading to addition to the reduced flavin. As a result of reduced enzyme binding both species of glyoxylate, but reacting with only one of them, the observed rate would therefore be lower than the true reverse rate.

Demonstration that the decay of the stable adduct leads to formation of glycollate

From the slow decay of the stable (S)-adduct of lactate oxidase the formation of one equivalent of glycollate is expected. This glycollate is a substrate of the enzyme and therefore it should be oxidized readily to formate and CO_2. That the oxidation of glycollate by lactate oxidase indeed leads to formation of formate and CO_2

was demonstrated in two sets of experiments: Lactate oxidase ($3 \cdot 10^{-5}$ M) was incubated at 25° and at oxygen concentrations of 1.3×10^{-3} M (100% O_2 atmosphere) or $2.5 \cdot 10^{-4}$ M (air) with $2 \cdot 10^{-4}$ M aliquots of glycollate. Five such aliquots were added over a period of 3 hrs and then the mixture was analyzed. Such a procedure was chosen in order to avoid an oxygen deficiency in the system, which would lead to production of glyoxylate[6]. A high concentration of enzyme is needed as lactate oxidase is inactivated by aerobic turnover with glycollate[6], (due to the accumulation of the stable S-adduct). Formation of 90 $\pm$ 20% of the theoretically possible amount of formate was demonstrated with an enzymatic test similar to that described by Höpner and Knappe[12], but using a yeast formate dehydrogenase (Böhringer). In similar experiments, but using U-^{14}C-glycollate as substrate, approximately 50% of the radio-activity was released as ^{14}C-CO_2 and adsorbed by quaternary organic base (Soluene 350, Packard).

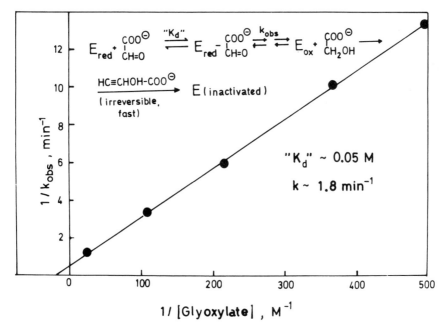

Fig. 1. Formation of oxidized lactate oxidase from reduced enzyme and glyoxylate. Lactate oxidase, 2.5×10^{-5} M in pH 7.0 universal buffer was made anaerobic and then reduced with 1.3 equivalents of LiLactate. The concentrations shown of glyoxylate, and α-hydroxybutynoate to a final concentration of 0.016 M, were added anaerobically at 25°. The formation of the inactivated enzyme was monitored either by detection of the fluorescence increase at 502 nm or of the absorbance increase at 320 nm, which both accompany inactivation[11].

The inset shows the sequence of the occurring reactions.

Similar experiments were carried out starting with isolated, stable (S) adduct, after its complete decay had occurred. With the formate dehydrogenase test formation of formate was clearly demonstrated; the amounts of formate produced in an average experiment (<8 nMoles), however, proved to be too low for reliable quantitation by this method (cf. 12). In 6 experiments, from 40% to 240% of the expected amounts of formate, with an average of 105% were found. During the decay of stable adduct, which was prepared from reduced lactate oxidase and U-^{14}C-glyoxylate (cf. Scheme 2), formation of ^{14}C-CO_2 was also found; the determination of the stoichiometry of this reaction is being presently carried out.

Determination of the stereochemistry of the glycollate formed by decay of the stable lactate oxidase glycollyl adduct

When the stable adduct was prepared starting from ^{3}H-labelled tartronate, the expected amount of radioactivity was incorporated (cf. Legend to Fig. 2A). In the decay process leading to formation of glycollate one proton will be incorporated from the solvent (Scheme 3). If the stable adduct indeed has the postulated "S" structure, then the solvent hydrogen will be incorporated in the glycollate Si position, while the label will be found in the Re position. Thus the glycollate formed will be chirally labelled, the position of the label reflecting the stereochemistry of the stable adduct (cf. Scheme 4, below).

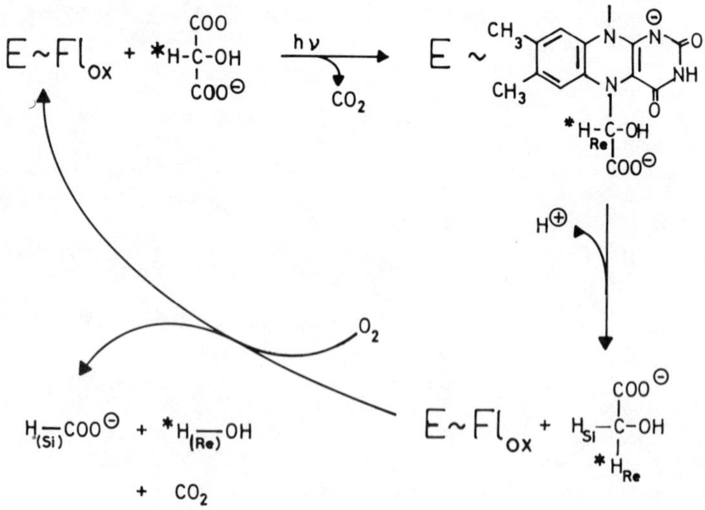

Scheme 3

The determination of the stereochemistry of this glycollate is straightforward and relies on comparison of its reactivity with those of stereochemically pure Re or Si ^{3}H-labelled glycollates. The latter were synthesized enzymatically from NAD[3]H and glyoxylate with L- and D-lactate dehydrogenase as outlined in the Legend of Fig. 2A. The stereochemistry of the label has been determined by neutron diffraction[11]; it is verified by the experiments shown in Fig. 2, and also by oxidation with glycollate oxidase from *pisum sativum* (G. Fendrich and S. Ghisla, to be published), which has been reported as being L-specific[12]. In Fig. 2A it is shown that the Re-hydrogen of glycollate and the ^{3}H-label of the glycollate produced by the decay of the stable adduct are abstracted by lactate oxidase (see Scheme 3 for the reaction sequence). In these experiments the total concentrations of authentic Re or Si labelled glycollate were chosen as to be similar to that expected from the decay of the stable adduct, i.e. around 3 x 10[-5] M. The rate of oxidation of [3]H$_{Re}$-glycollate depends

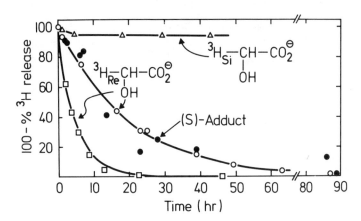

Fig. 2A. Determination of the stereochemistry of the stable (S)-adduct. a) Formation of the adduct: In a typical experiment lactate oxidase (78 nMoles) was incubated with 1060 nMoles α-³H tartronate (~360 dpm/nMole) in 1.0 ml pH 6.0 of the universal buffer described elsewhere[9]. The photochemical formation of the stable adduct was carried out at 0° as described earlier[6]. The reaction mixture was then separated over a Sephadex column (G-25 medium, void vol. ~6 ml, separation time 10-15 min, 0-4°) equilibrated with pH 7.0 universal buffer. The protein fraction (1.5 ml) contained 39 nMoles stable adduct as determined from the amount of reoxidation occurring within 12 hrs (specific activity of (S)-adduct ~300 dpm/nMole). This adduct solution was allowed to decay at 22°. At the intervals shown aliquots containing 10-15 nMoles of adduct were applied to a Dowex 1x2 AG column (OH⁻ form, bead vol. 0.5 ml). The column was washed with distilled water until no ³H was present in the eluate (6 ml), then with 6 ml 1 N HCl. 1.5 ml samples of the eluate were counted in 15 ml liquid scintillator (ELS 294, Koch and Light). The vials containing appreciable amounts of ³H were standardized internally, the counting efficiency being 15-20%. Curve (●——●) shows the percent of radioactivity eluted with water (as ³HOH) (eluted ³HOH + eluted ³H-COOH = 100, whereby > 95% of the counts applied to the column were eluted). One preparation of (S)-adduct allowed the determination of 2-3 points.

b) ³H-(Re or Si) glycollate (2 x 10⁵ dpm/nMol) was prepared from glyoxylate and NAD³H respectively with pig heart L-lactate dehydrogenase, or with D-lactate dehydrogenase from *Lactobacillus leichmanii* (both from Böhringer), NAD³H was prepared from α³H-ethanol (NEN) and NAD⁺ with yeast alcohol dehydrogenase (Böhringer).

Curve (Δ——Δ) shows the percentage of ³H release (as ³H₂O) when 3.3 nMoles lactate oxidase were incubated with 9 nMoles (Si)-³H-glycollate in 100 μl pH 7.0 universal buffer, at the times shown. Curve (□——□): incubation of 14.3 nMoles (Re)-³H glycollate with 15.5 nMoles lactate oxidase in 500 μl buffer pH 7.0. Curve (O——O); incubation of 16.7 nMoles (Re)-³H-glycollate with 3.1 nMoles lactate oxidase in 500 μl universal buffer pH 7.0.

on the concentration of enzyme present in the incubation (Fig. 2). The difference in oxidation rates found for the stable adduct decay solution (enzyme conc. was an average ~ 3 x 10⁻⁵ M) as compared to the decay rate of ³H$_{Re}$-glycollate incubated with 3 x 10⁻⁵ M lactate oxidase (Fig. 2B) originates in the fact that the specific activity of the enzyme used for preparation of the adduct had decreased during illumination from 2000 to 360-400 turnover units[7]. As the curve (●——●) of Fig. 2 was constructed from different preparations of stable adduct, the relatively large scatter probably originates in different degrees of inactivation of the enzyme. On the other hand, employing a concentration of intact enzyme corresponding to the average activity present in the adduct decay solutions (~6 x 10⁻⁶ M) the curve for ³H removal in the latter could be reproduced satisfactorily (Fig. 2).

61

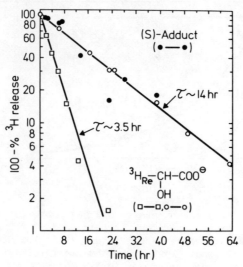

Fig. 2B. Rates of oxidation of (Re)-^{3}H-glycollate and of (S)-adduct decay product by lactate oxidase. The data of Fig. 2A were plotted semilogaritmically.

As required by our hypothesis, the label in ^{3}H$_{Si}$-glycollate is not released into water under similar conditions: ^{3}H$_{Si}$-glycollate is expected to be oxidized to ^{3}H-COOH, which is not a substrate of lactate oxidase (Fig. 2A, Scheme 3). This experiment also yields the necessary control, that the formate C-hydrogen is not released into solvent under the present conditions. Formate has also been reported not to exchange its hydrogen in 1% NaOH, at 185° over a period of 4 hrs[13]. The initial, relatively rapid release of approximately 5% of the total label of ^{3}H$_{Si}$-glycollate, might be attributed to oxidation of glycollate to ^{3}H-glyoxylate, which is then further oxidized to ^{3}HOH and oxalate[6], or, alternatively, to contamination of the ^{3}H$_{Si}$-glycollate with ~5% of the ^{3}H$_{Re}$-isomer.

Half times of 3-15 hrs appear, at first sight, to be quite large for enzymatic reactions. It should be kept in mind, however, that the rates will reflect an ^{3}H/^{1}H isotope effect of ~6, as the ^{2}H/^{1}H isotope effect is 4.2 (see below). Furthermore, the K$_m$ for glycollate (5 mM) is comparatively high. This is reflected in the anaerobic reduction rate of the enzyme with equimolar concentrations of glycollate, which at concentrations similar to those of Fig. 2 (~2 x 10^{-5} M) was found to be ~3 hrs.

Kinetics of adduct formation and deuterium isotope effects

The kinetic traces for adduct formation under anaerobic conditions also support the concept that glycollate forms two distinctive complexes with lactate oxidase. As detailed earlier[6], under anaerobic conditions, the rate of adduct formation (measured spectrophotometrically or fluorimetrically) is closely similar to the maximum catalytic turnover number with glycollate. It was therefore proposed that under these conditions the major product must be the catalytically active adduct, the stable adduct accumulating only slowly through the series of equilibria shown in Scheme 4. In accordance with this scheme, the kinetic trace of adduct formation is distinctly biphasic. Typical records are shown in Fig. 3.

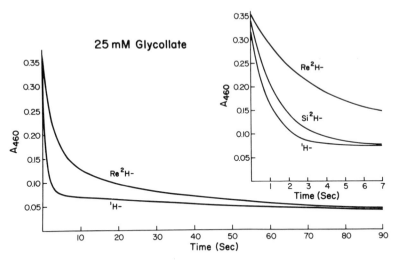

Fig. 3. Effect of stereospecifically monodeuterated glycollates on the rate of adduct formation. Lactate oxidase (1.9×10^{-5} M) was reacted at 25° under anaerobic conditions in 0.01 M imidazole acetate buffer, pH 7.0, with 25 mM glycollate (concentrations after mixing). The formation of adduct (predominantly R-adduct[6]) was followed by the decrease in absorbancy at 460 nm in a stopped-flow spectrophotometer; similar results are obtained if the characteristic fluorescence of the adduct is monitored[6]. The A_{460} values shown are for the 2 cm light path of the stopped-flow optical cell.

With [1]H-glycollate, approximately 88% of the reaction occurs in a fast phase. Similar results are obtained with [2]H-Si-glycollate; however with [2]H-Re-glycollate the biphasic nature of the reaction is much more marked, with only about 60% occurring in the fast phase. Qualitatively, similar results have been obtained in the kinetics of reduction of lactate oxidase with DL-lactate, and shown to be due to the slower establishment of an equilibrium between oxidized enzyme and D-lactate, which removes temporarily some of the enzyme in a form unreactive with L-lactate, hence slowing the overall reduction process (V. Massey, and S. Ghisla, unpublished results). Abstracting pseudo first order rate constants from such complex reactions is therefore somewhat difficult. In the case of DL-lactate the process was facilitated by the ability to study separately the reaction of the enzyme with D- and L-lactates. In this case it was found that a good approximation to the rate of reduction by L-lactate in the presence of an equimolar concentration of D-lactate could be obtained from the initial slope of semilogarithmic plots. A similar method has been used to analyze the data with glycollate; the results are shown in Fig. 4.

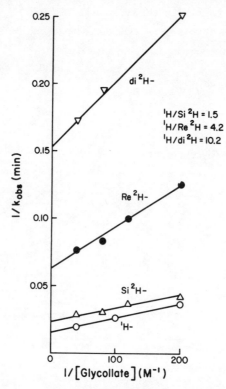

Fig. 4. Dependence of the rate of adduct formation on glycollate concentration and deuterium substitution. Conditions, as in Fig. 3. The k_{obs} values were obtained from the initial slopes of semilogarithmic plots of the absorbancy changes. (o), ^{1}H-glycollate. (Δ), Si ^{2}H-glycollate. (o), Re ^{2}H-glycollate. (∇), dideutero-glycollate.

In accordance with the body of evidence favoring a partially rate-determining proton abstraction from the α-carbon atom of the substrate by an enzyme base[9,10,11,16], a large deuterium isotope effect ($k^1H/k^2H \sim 10$) was found when the enzyme was reacted with dideutero glycollate instead of the normal ^{1}H-form (Fig. 4). In gratifying confirmation of our hypothesis, as outlined in Scheme 4, it was found that very different effects were obtained with monodeuterated glycollates, depending on whether the ^{2}H-label was in the Si- or Re-position. In confirmation of the stereochemistry outlined in Scheme 4, it is found that when the ^{2}H label is in the Si-position there is only a small deuterium isotope effect ($k^1H/k^2H \sim 1.5$). However, when the ^{2}H label is in the Re-position, the isotope effect is quite substantial ($k^1H/k^2H = 4.2$).

CONCLUSIONS

The sequences of reactions leading to formation of the "stable" (S)-adduct and the "labile", catalytically competent (R)-adduct, as well as the estimated reaction rates, are summarized in Scheme 4. The left hand side of the scheme should represent the sequence of reactions which are important in the turnover of the enzyme with (Re)-glycollate, while the right hand side should show the sequence of reactions

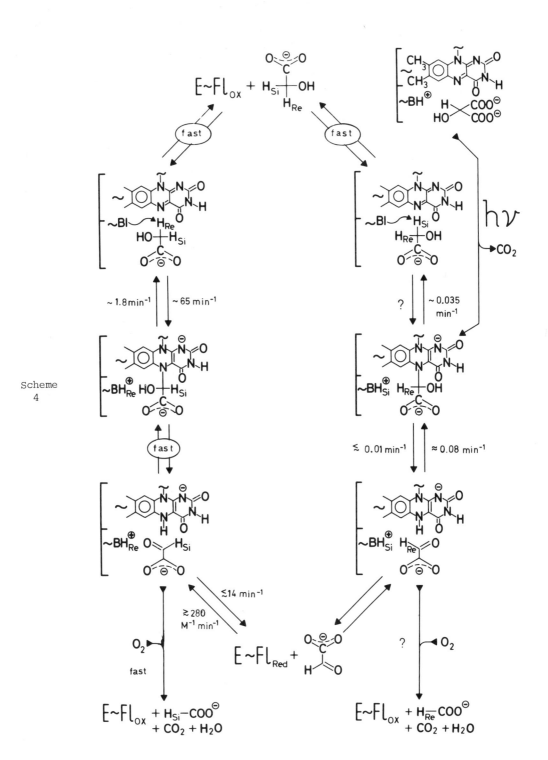

Scheme 4

65

leading to accumulation, or photochemical formation of the stable S-adduct. The studies with selectively labelled deutero glycollate support the assumption that glycollate can bind to the oxidized enzyme to form two distinct diasteromeric complexes in which the prochiral glycollate will react at either its Re or Si face, and that these do not interconvert <u>directly</u>; i.e. decay of one complex must precede formation of its enantiotopic counterpart. Analogously it is reasonable to assume also that reduced enzyme will bind glyoxylate to form two diastereomeric complexes, which will be in equilibrium with free reduced enzyme and glyoxylate. In accord with this interpretation is the formation of R and S adducts starting from reduced enzyme and glyoxylate[6]. In this case, however, we have not direct experimental evidence for answering the question of direct interconvertibility of these complexes.

From the fact that in the normal turnover reaction, in which the R-adduct is on the reaction path or at least in a rapid equilibrium with it, the Re-hydrogen is being removed, while the S-adduct is connected with the abstraction-addition of the Si-proton, we conclude that these two adducts are indeed diastereomers.

We interpret the above discussed results and the spectroscopic results reported earlier[6] as evidence that the S-adduct decays directly to oxidized enzyme and glycollate. Alternative reaction sequences (Scheme 4) cannot account for the experimental results for the following reasons: a) Direct interconversion of the two adducts (S ——> R) and subsequent formation of oxidized enzyme via the "Re-pathway" would lead to incorporation of ^{3}H into the Si-position of the formed glycollate because the enzyme is Re-specific (Fig. 2). b) Interconversion at the level of the reduced enzyme-glyoxylate complexes is very improbable, as the slow reoxidation step over the "Re-pathway" would have to compete with fast oxygen oxidation and also with product release as glyoxylate. c) Reduction of free glyoxylate (formed by S-adduct decay) over the "Re-pathway" is similarly very improbable because of competing fast oxygen reoxidation of free reduced enzyme. d) Sequences c) and b) would also lead to incorporation of label into the Si-position. e) Oxygenation of glyoxylate at the level of the glyoxylate-reduced enzyme complexes to form either labelled or unlabelled formate clearly requires molecular oxygen and is thus not compatible with the oxygen-independent (anaerobic) formation of oxidized enzyme during decay of the S-adduct[6].

The observation that the S-adduct decays <u>directly</u> to oxidized enzyme and glycollate is of uttermost importance. According to the law of microscopic reversibility, this implies that the reverse reaction must have a finite value. In other words, the reaction of oxidized enzyme glycollate Si-complex leads to formation of the S-adduct, which in turn is on the Si-reduction pathway. This interpretation is in accordance with the accumulation of S-adduct during turnover[6]. Furthermore, we propose that the "normal", i.e. the "Re-catalytic pathway" occurs analogously as shown on the left hand side of Scheme 4. This mechanism is thus equivalent with covalent catalysis, a mechanism long advocated by Hemmerich and coworkers[5].

Similar mechanisms have already been suggested by Blankenhorn et al.[17] based on model studies, and by Porter et al.[18,3] from experiments with D-amino acid oxidase and the artificial substrate nitromethane. Williams and Bruice[19] have questioned the occurrence of covalent intermediates for the reversible oxidation of lactate by oxidized flavin in their chemical model system. While their arguments might be valid for the chemical model, they might as well not apply for enzymatic reactions. In fact the experimental evidence in favor of the occurrence of N(5)-covalent adducts is growing rapidly. Starting from the trapping of a covalent adduct with D-amino acid oxidase by Porter et al.[18], a series of L-α-hydroxy acid oxidizing flavoenzymes have been found to form covalent suicide products involving the flavin position N(5)[11,20].

On the other hand it should be pointed out that the results presented here do not allow strict conclusions about the molecular mechanism of formation of the N(5)-covalent adducts. These might be formed via an α-carbanion of glycollate and its direct addition to the flavocoenzyme, a mechanism we favor[6,9]. Radical type mechanisms, such as those proposed by Williams and Bruice[19], cannot be excluded, however.

The difference in reactivity between the R and S-adducts, which allowed the characterization of the latter, most probably results from the different chemical environment of the diastereomeric complexes. Similarly, the fact that with L-lactate no fluorescent intermediates can be observed, can be attributed to their very short life time, which in turn might originate in the different steric requirements of a methyl versus a hydrogen function. In fact the photoreaction of methyltartronate with lactate oxidase leads also to formation of a fluorescent adduct, but which decays much faster than the S-glycollyl adduct to reduced enzyme (S. Ghisla and V. Massey, unpublished).

ACKNOWLEDGEMENTS

We wish to thank Drs. P. Hemmerich and G. Blankenhorn for many helpful discussions and Ms. M. Sappelt for skillful technical assistance. This work was supported by a grant from the Deutsche Forschungsgemeinschaft to SF and from the U.S. Public Health Service (GM 11106) to VM.

REFERENCES

1. Hemmerich, P., Nagelschneider, G. and Veeger, C. (1970) FEBS Letters 8, 69-83.

2. Massey, V. and Hemmerich, P. (1975) in The Enzymes (Boyer, P.D., ed) pp. 191-252, Academic Press, New York.

3. Bright, H.J. and Porter, D.J.T. (1975) in The Enzymes (Boyer, P.D., ed.) pp. 421-505, Academic Press, New York.

4. Bruice, T.C. (1976) in Progress in Bioorganic Chemistry (Kaiser, E.T. and Kezdy, E.J., eds.) Vol. 4, pp. 1-87, J. Wiley and Sons, New York.

5.   Hemmerich, P. (1976) in Progress in the Chemistry of Organic and Natural Products (Grisebach, H., ed.) Vol. 33, pp. 451-526, Springer Verlag, Berlin.

6.   Massey, V. and Ghisla, S. (1975) in Proceedings of the Tenth FEBS Meeting, 145-158.

7.   Lockridge, O., Massey, V. and Sullivan, P.A. (1972) J. Biol. Chem. 247, 8097-8106.

8.   Ghisla, S., Massey, V., Lhoste, J.M. and Mayhew, S.G. (1974) Biochemistry 13, 589-597.

9.   Ghisla, S. and Massey, V. (1977) J. Biol. Chem. 252, 6729-6735.

10.   Ghisla, S. and Massey, V. (1975) J. Biol. Chem. 250, 577-584.

11.   Ghisla, S., Ogata, H., Massey, V., Schonbrunn, A., Abeles, R.H. and Walsh, C.T. (1976) Biochemistry 15, 1791-1797.

12.   Höpner, Th. and Knappe, J. in Methoden der enzymatischen Analyse (Bergmeyer, H.U., ed.) pp. 1596-1600, Verlag Chemie, Weinheim.

13.   Johnson, C.K., Gabe, E.J., Taylor, M.R., and Rose, I.A. (1965) J. Am. Chem. Soc. 87, 1802-1804.

14.   Kerr, M.W. and Groves, D. (1975) Photochemistry 14, 359-362.

15.   Kawazoe, Y. and Ohnishi, M. (1966) Chem. Pharm. Bull. 14, 1413-1418.

16.   Walsh, C.T., Lockridge, O., Massey, V. and Abeles, R.H. (1973) J. Biol. Chem. 248, 7049-7054.

17.   Blankenhorn, G., Ghisla, S. and Hemmerich, P. (1972) Z. Naturforsch. Teil B. 27, 1038-1040.

18.   Porter, D.J.T., Voet, J.G. and Bright, H.J. (1973) J. Biol. Chem. 248, 4400-4416.

19.   Williams, R.F. and Bruice, T.C. (1976) J. Amer. Chem. Soc. 98, 7752-7768.

20.   Walsh, C.T. (1978) Annual Review of Biochemistry, in press.

A STUDY OF DEUTERIUM ISOTOPE EFFECTS ON THE BACTERIAL BIOLUMINESCENCE REACTION

PATRICK SHANNON,[a] ROBERT B. PRESSWOOD,[b] ROBIN SPENCER,[a] JAMES E. BECVAR,[b] J. WOODLAND HASTINGS,[b] and CHRISTOPHER WALSH[a]

[a] Departments of Chemistry and Biology, Massachusetts Institute of Technology, Cambridge, Massachusetts 02139

[b] The Biological Laboratories, Harvard University, Cambridge, Massachusetts 02138

ABSTRACT

Bacterial luciferase catalyzes light production from $FMNH_2$, O_2, and a long chain aldehyde (1-^1H or 1-^2H) with a deuterium isotope effect of 1.7 on the kinetics of light decay, independent of temperature over the range 2-25°C. A larger deuterium isotope effect of 5.4 is seen in stopped-flow kinetic experiments starting from $[1\text{-}^1H]$ or $[1\text{-}^2H]$-aldehyde and 4a-hydrogenperoxyflavin in a step which clearly preceeds formation of the light emitting species. This intrinsic isotope effect is then only partially expressed in light emission. Thus, aldehyde oxidation is necessary for luminescence and the C_1-H bond scission cannot be concomitant with emitter formation.

INTRODUCTION

Bacterial luciferase catalyzes the bioluminescent oxidation of reduced flavin mononucleotide ($FMNH_2$) by molecular oxygen in the presence of a long chain aldehyde to produce oxidized FMN, the long chain acid, and light (490 nm) with a quantum yield of 0.1.[1,2] Reduced FMN is supplied to the luciferase by the NAD(P)H:flavin oxidoreduc-

$$NADH + FMN \xrightarrow{\text{oxidoreductase}} NAD^+ + FMNH_2$$

$$FMNH_2 + O_2 + RCHO \xrightarrow{\text{luciferase}} FMN + RCOO^- + H_2O + 0.1\ h\nu$$

tase,[3] which we have studied with a number of flavin analogs.[4-6]

The first step in the luciferase reaction is believed to be the formation of an enzyme bound 4a-hydroperoxyflavin.[7-10] Reaction of this enzyme-bound intermediate with aldehyde produces a light with a quantum yield equal to that in the overall reaction.[8] In model studies, a low quantum yield chemiluminescence of N_5-ethyl-4a-hydroperoxyflavin and aldehyde has been reported.[11,12] Although considerable effort over the past fifteen years has been devoted to elucidation of the steps in the reaction pathway following formation of the oxygenated flavin intermediate, no other intermediates have been identified. Some seven different mechanisms for the reaction have been proposed[13-19] but the mechanism of formation and identity of the ever elusive emitter is still a matter of debate.[2]

Our approach in the study of the luciferase mechanism has centered on the reaction of the aldehyde with the luciferase intermediate to form the emitter. Replacement of the C-1 proton of the aldehyde with a deuteron results in a kinetic isotope effect on the decay of light produced by luciferase.[20] A detailed study was directed toward construction of a minimal kinetic mechanism which would relate carbon-hydrogen bond scission to production of the emitter, and thereby place contraints on any proposed chemical mechanism.

MATERIALS AND METHODS

Luciferase enzyme from <u>Beneckea</u> <u>harveyi</u> was purified to homogeneity.[21,22] Single turnover assays were carried out essentially as described elsewhere.[23] $[1-^{1}H]$- and $[1-^{2}H]$-decanal, dodecanal, and tetradecanal were synthesized from the corresponding ethyl esters via lithium aluminum hydride (deuteride) reduction to the alcohol and oxidation to the aldehyde using the method of Kornblum.[24]

Rapid reaction studies were performed on a Durrum-Gibson stopped-flow spectro-photometer with a dead time of two milliseconds. Luciferase for these experiments were purified from <u>B</u>. <u>harveyi</u> wild type or mutant M-17[21,25]; the results were the same from either. Absorbance changes during the course of the reaction of the inter-mediates were monitored at 380 nm with a Corning color filter #7-54 forward of the photomultiplier and at 600 nm with a Corning color filter #2-73, effectively eliminating bioluminescence. Aldehydes were prepared as either a 0.01% suspension by sonication and dilution or as a saturated solution in buffer. All solutions were 0.35 $\underline{M}$ in potassium phosphate, pH 7.0. Computer calculations were carried out on a Hewlett-Packard 9825A.

RESULTS

<u>Isotope effects vs. aldehyde chainlength</u>. Figure 1 shows a typical result of light intensity vs. time. Such data may be obtained either with the stopped-flow spectrophotometer or with the standard photometer assay. Plots of log (intensity) vs. time yield k_{decay}, the first order rate constant for light decrease; such plots are linear for more than two orders of magnitude of light intensity.* With purified luciferase from <u>B</u>. <u>harveyi</u>, isotope effects on k_{decay} were measured with $[1-^{1}H]$- and $[1-^{2}H]$-decanal, dodecanal, and tetradecanal, and are summarized in Figure 2. The isotope effects on k_{decay} for C_{10}, C_{12}, and C_{14} aldehydes are 1.7 $\pm$ .1, 1.2 $\pm$ .1, and 1.0 $\pm$ .2,** respectively, under the conditions of Figure 2. The k_{decay} and isotope effects with decanal and dodecanal remain constant over a 100-fold concentration range in these aldehydes. For each of the three aldehyde chain lengths, deuteration

---

*This simple first-order decay is seen only with decanal and dodecanal.

**Tetradecanal does show an isotope effect of 1.2-1.7 in the absence of serum albumin.

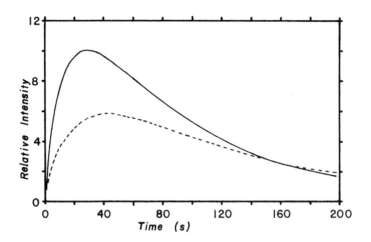

Figure 1. Light production with $[1-{}^{1}H]$-decanal (——) and $[1-{}^{2}H]$-decanal (--) in single turnover assay.

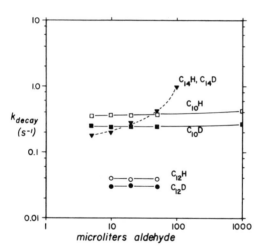

Figure 2. Light decay rate constants vs. aldehyde concentration for C_{10}, C_{12}, and C_{14} protio and deutero aldehydes. Stock solutions are a 0.1% suspension.

at C_1 <u>did</u> <u>not</u> <u>affect</u> <u>their</u> <u>respective</u> <u>quantum</u> <u>yields</u>. This is in contrast to the model work of Kemal and Bruice,[11,15] in which aldehyde C_1 deuterium substitution produces an isotope effect on the amount of light produced (a "product" effect) and no effect on the rate of light decay. The isotope effect of 1.7 observed with decanal is too large to represent anything but a primary kinetic effect on carbon-hydrogen bond scission,[26] and provides unambiguous evidence that the oxidation of the aldehyde is essential to the production of the light emitting species.* Due to its larger isotope effect and higher solubility, decanal was used exclusively in all further experiments.

---

*Studies of the acid production[28] suggest this, but cannot prove an obligatory and coupled reaction with a quantum yield of 0.1.

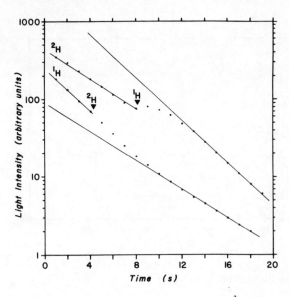

Figure 3. Typical aldehyde exchange results with $[1-^1H]$- and $[1-^2H]$-decanal.
Addition of second aldehyde is at ∇ (see text).

Aldehyde exchange experiments. Since it is the addition of aldehyde to the enzyme-hydroperoxyflavin complex that initiates the light-producing reaction sequence,[9] the isotope effect on k_{decay} may be used to probe the rate and extent of the reversibility of the aldehyde-enzyme-hydroperoxyflavin association. As the scheme suggests, the k_{decay} of a luciferase assay in progress might be altered by the rapid addition of a

$$E-Fl-OOH \cdots RCHO \rightarrow \rightarrow h\nu; \; k_{decay} = .31 \; s^{-1}$$

$$E-Fl-OOH \xleftarrow[RCDO]{RCHO}$$

$$E-Fl-OOH \cdots RCDO \rightarrow \rightarrow h\nu; \; k_{decay} = .23 \; s^{-1}$$

large excess of the oppositely substituted aldehyde. In standard photometer assays containing 10 µl of the aldehyde, at various times, Δt, after reaction initiation, 1 ml of the oppositely substituted aldehyde was injected. Regardless of the order of aldehyde addition (H, then D; or D, then H), k_{decay} became indistinguishable from k_{decay} in assays containing the second (excess) aldehyde alone. Figure 3 shows typical assays. The extent of aldehyde exchange is independent of Δt since k_{decay} will still change to that of the second aldehyde when the reaction is greater than 95% complete (based on light intensity). The reversibility of aldehyde association was suggested in previous exchange experiments employing aldehydes of different chain lengths[27]; the current results show that this reversibility is independent of chain length effects and extends at least up to carbon-hydrogen bond breakage.

Figure 3 shows that the change of k_{decay} in these exchange experiments is not instantaneous. Plots of log (observed light intensity - extrapolated light intensity) vs. time are linear, and give a rate constant $k_{reverse}$ of $0.8 \pm .2 \; s^{-1}$ that is independent of the order of aldehyde addition. Since k_{decay} is much less than the

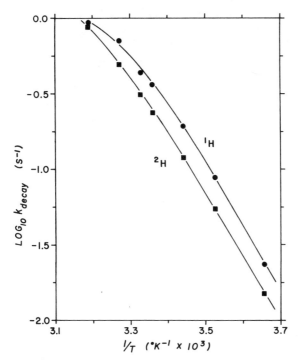

Figure 4. Arrhenius plot of k_{decay} as a function of temperature for decanal.

pseudo-first order rate constant for aldehyde addition (vide infra), $k_{reverse}$ represents the rate of aldehyde dissociation from the E-Fl-OOH complex.

   <u>Temperature effects on k_{decay}</u>. The temperature dependence of k_{decay} with [1-^{1}H]-
and [1-^{2}H]-decanal is shown in the Arrhenius plot of Figure 4. The activation
enthalpy of k_{decay} is 19 kcal regardless of C_1 deuteration, hence the isotope effect
is independent of temperature. In conjunction with our stopped-flow results (<u>vide
infra</u>), this suggests that C-H bond scission and emitter formation (which are
kinetically distinguishable) have near identical activation enthalpies of 19 kcals.
   <u>Stopped-flow absorbance studies</u>. Luciferase intermediate II (the enzymehydro-
peroxyflavin complex) and aldehyde were mixed in the stopped-flow spectrophotometer
at $2 \pm 1°C$. The reaction was monitored at 380 nm, an isosbestic point for E-Fl-OOH
and E + FMN, and at 600 nm, at which neither E-Fl-OOH nor E + FMN have significant
absorbances.[29] In addition, bioluminescence was monitored with the spectrophotometer
beam off. Figures 1 and 5 show typical results, and rate constants for all phases are
listed in Table 1.
   Absorbance changes at 380 nm are characterized by a rapid increase followed by a
biphasic decrease. The initial absorbance increase and the first phase of the
absorbance decrease are rapid (>10 and 3.5 s^{-1}, respectively) and are identical for
[1-^{1}H]- and [1-^{2}H]-decanal (Figure 5A). The second phase of 380 nm absorbance
decrease (Figure 5B) exhibits an isotope effect of 5.4 (0.083 s^{-1} for [1-^{1}H]-decanal

TABLE 1

STOPPED-FLOW RESULTS

	$1\text{-}^1\text{H-decanal}$	$1\text{-}^2\text{H-decanal}$	Assignment in Scheme 1
absorbance at 380 nm	$>10\ \text{s}^{-1}$	$>10\ \text{s}^{-1}$	k_1
	3.5	3.5	k_2
	0.083	0.015	k_3
absorbance at 600 nm	$>10\ \text{s}^{-1}$	$>10\ \text{s}^{-1}$	-
	0.090	-	k_3
	0.013	-	k_4
	-	0.0072	$k_3k_4/(k_3 + k_4)$
bioluminescence	$0.012\ \text{s}^{-1}$	$0.0071\ \text{s}^{-1}$	$k_3k_4/(k_3 + k_4)$

and $0.015\ \text{s}^{-1}$ for $[1\text{-}^2\text{H}]$-decanal) and is much more rapid than the decay of bio-luminescence ($0.012\ \text{s}^{-1}$ and $0.0071\ \text{s}^{-1}$ for $[1\text{-}^1\text{H}]$- and $[1\text{-}^2\text{H}]$-decanal).

A rapid absorbance increase is also observed at 600 nm which is independent of aldehyde deuteration ($>10\ \text{s}^{-1}$). With $[1\text{-}^1\text{H}]$-decanal the decay of 600 nm abosrbance (Figure 5C) is biphasic, with rate constants of $0.090\ \text{s}^{-1}$ and $0.013\ \text{s}^{-1}$; with $[1\text{-}^1\text{H}]$-decanal the decay is a single phase with a rate constant of $0.0072\ \text{s}^{-1}$. The first phase of 600 nm absorbance decay with $[1\text{-}^1\text{H}]$-decanal ($0.090\ \text{s}^{-1}$) corresponds to the decay of 380 nm abosrbance ($0.083\ \text{s}^{-1}$), and the second phase ($0.013\ \text{s}^{-1}$) is coinsident with the decay of bioluminescence ($0.012\ \text{s}^{-1}$). The decay of 600 nm absorbance with $[1\text{-}^2\text{H}]$-decanal ($0.0072\ \text{s}^{-1}$) has no direct counterpart at 380 nm but is also coincident with the decay of emitted light ($0.0071\ \text{s}^{-1}$).

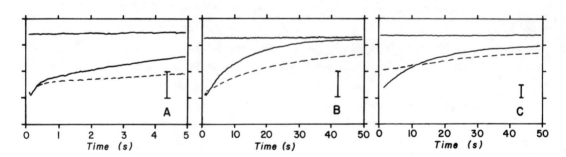

Figure 5.   % Transmittance vs. time for C-1 protio (—) and deutero (--) decanal at 380 nm (A and B) and 600 nm (C) observed in stopped-flow experiments. At time zero, pre-formed enzyme-4a-hydroperoxyflavin is mixed with aldehyde. Insert represents 2% transmittance. Final transmittance attained is also shown.

DISCUSSION

Kinetic mechanism of luciferase. The data presented here place two important restrictions on any proposed mechanism for the bacterial luciferase. First, aldehyde carbon-hydrogen bond scission preceeds formation of the light emitting species. The reactions cannot be concerted. Second, the emitting species must be either an excited singlet or a short-lived triplet (vide infra), so that the emitter does not accumulate in the steady state. Thus Scheme 1 is the minimal kinetic scheme consistent with all of these results. Intermediates I and II follow from earlier studies[9];

SCHEME 1

$$I \xrightarrow{O_2} II \underset{k_{-1}}{\overset{k_1[RCHO]}{\rightleftharpoons}} III \underset{k_{-2}}{\overset{k_2}{\rightleftharpoons}} IV \xrightarrow{k_3} V \underset{slow}{\xrightarrow{k_4}} VI^* \underset{fast}{\xrightarrow{k_5}} E + FMN + h\nu$$

E-FlH$_2$ E-Fl-OOH   C--H   emitter

intermediates III through VI extend this nomenclature where, previously, all species following aldehyde addition were categorized as intermediate IIa. The rate constants are those assigned in Table 1. In this scheme k_3 represents C-H bond breakage, which shows an isotope effect of 5.4 (apparent at 380 nm), and k_4 represents the slow (rate determining) formation of the emitter which has no intrinsic isotope effect. In the approximation that intermediates II, III, and IV are in rapid pre-equilibrium, the observed decay of bioluminescence for [1-^{2}H]-decanal will be $k_{decay} \simeq k_{3[^2H]}k_4/ (k_{3[^2H]} + k_4)$. With the values of Table 1, the calculated k_{decay} is 0.0070 s^{-1}, in good agreement with that observed (0.0070 s^{-1} on hν; 0.0072 s^{-1} at 600 nm). Thus Scheme 1 is fully consistent with the observed isotope effect on light decay in the overall reaction of 1.7. In addition, the analytical solution to the simplified Scheme 2 was determined, and with this solution the production of light was simulated on a digital computer. The simulation is in good agreement with the observed isotope effects on the buildup and decay of light intensity, the maximum intensity achieved, and the time required to reach that maximum intensity; further refinement of these rate constants using the simplified scheme is currently under way.

SCHEME 2

$$II \underset{k'_{-1}}{\overset{k'_1}{\rightleftharpoons}} IV \xrightarrow{k_3} V \xrightarrow{k_4} VI^* \underset{fast}{\xrightarrow{k_5}} h\nu$$

Our enzymatic kinetic isotope effect data are not in conflict with the model system product isotope effect data of Kemal and Bruice.[11] In the model system, most of the 4a-hydroperoxy-5-alkylflavin is solvolized in a dark reaction sequence,[30] while the light-producing reactions are a minor parallel pathway. As shown in Scheme 3, the light intensity in the model system is proportional to the concentration of

SCHEME 3

$$\text{F1-OOH} \quad \overset{\text{RCHO}}{\nearrow} \longrightarrow \longrightarrow \longrightarrow h\nu_{530-570 \text{ nm}}$$
$$\searrow \longrightarrow \longrightarrow \longrightarrow \text{solvolysis (no light)}$$

F1-OOH, so that the rate of disappearance of FL-OOH, and thus k_{decay}, is controlled primarily by $k_{solvolysis}$, which is isotope independent. Therefore, the kinetic isotope effect within the light-producing sequence will be manifest in the total amount of light produced and not on k_{decay}.

Chemical mechanism of luciferase. Since the triplet state of oxidized flavin has an extinction coefficient of ∿2000 $\underline{M}^{-1} cm^{-1}$ at 600 nm,[31] it is intriguing to suggest that the luciferase intermediate V (Scheme 1) that accumulates ahead of the rate determining step is itself the emitter, a long-lived triplet. We can, however, exclude this possibility for two reasons. First, the intrinsic lifetime of the triplet of FMN_{ox} (∿170 msec[32]) is much too short to allow its accumulation; and second, Figure 4 and reference 33 show that the luciferase quantum yield is temperature independent. If intermediate V were a long-lived triplet, increasing temperature should increase its probability of radiationless decay and thereby lower its quantum yield. The emitter must then either be an excited singlet or a short-lived triplet that cannot be observed in the stopped-flow spectrophotometer.

The nature of the 600 nm absorbing intermediates IV and V also remains uncertain. Neither species is likely to be a flavin semiquinone, since (i) the long wavelength absorbance is featureless between 600 and 800 nm (data not shown), and (ii) a solution of luciferase intermediate II, frozen within seconds of decanal addition (while it was emitting enough light to see in daylight) showed no unpaired spin by electron paramagnetic resonance (R. Presswood and J.W. Hastings, unpublished observations). We suggest that intermediates IV and V are flavin charge-transfer species of unknown structure.

In mechanistic terms, our data exclude the recent postulate of Kemal and Bruice[19] that a variant of the Russell reaction[34,35] produces a ring-opened emitter (Scheme 4). In this reaction, production of the excited state is concerted with C-H bond scission,

SCHEME 4

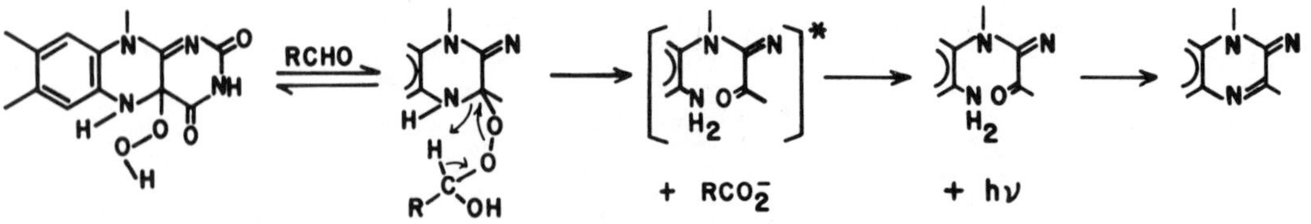

and so is not compatible with our kinetic findings.

Of the structural mechanisms proposed, only that of Lowe et al.[18] has C-H bond scission in a step prior to the formation of the emitter, and creates the emitter in a spin-forbidden 2,2-cycloreversion to produce an excited state (shown in Scheme 5).

SCHEME 5

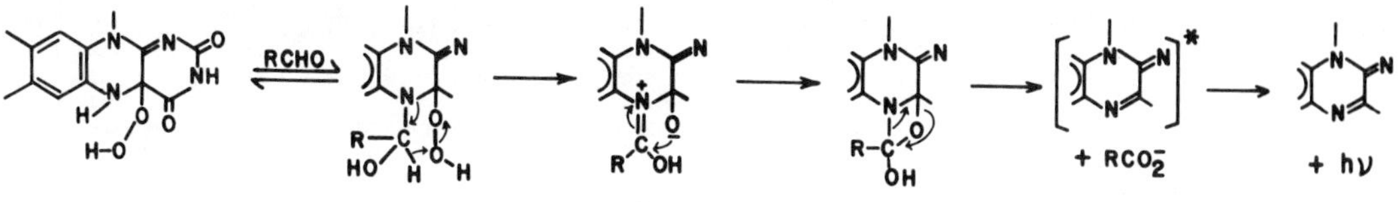

This mechanism proposes an excited singlet of FMN as the emitter which is not compatible with the fact that luciferase-bound FMN is non-fluorescent[22] and free FMN has a fluorescence emission maximum at 530 nm. The emitter could be a short-lived triplet of FMN which may be protonated at N-1 ($pK_a = 5.2$[31]). Flavins protonated at N-1 ($pK_a = 0$ in the ground state) have shown emission maximum near that of bioluminescence.[36]

The mechanism of Lowe et al. does have two other serious discrepancies with experimental results. First, it is not directly compatible with the chemiluminescence of N-5 alkylated flavins[11,15]; it may be possible, however, to form the 1,3-oxazetidine (intermediate II of Scheme 5) without forming the N-5 imine (intermediate I of Scheme 5). Second, the mechanism does not suggest species which would have significant extinction coefficients at 600 nm. We hope to obtain complete spectra of intermediates III through V in the near future, and so relate them to known flavin chromophores.

ACKNOWLEDGEMENTS

We thank Dr. Shiao-Chun Tu for helpful discussions, Professor George Whitesides for providing access to the Hewlett-Packard 9825A, N.I.H. grants #GM 20011 and 21643, and N.S.F. grants #PCM 74-23651 and PCM 77-19917 for partial support of this work.

REFERENCES

1. Becvar, J.E., et al. (1976) in Flavins and Flavoproteins, Singer, T.P. ed., Elsevier Scientific, Amsterdam, pp. 94-100.

2. Hastings, J.W., and Nealson, K.H. (1977) Ann. Rev. Microbiol., 31, 549-595.

3. Duane, W. and Hastings, J.W. (1975) Molecular and Cell Biochemistry, 6, 53-64.

4. Fisher, J., et al. (1976) Biochemistry, 15, 1054-1064.

5. Spencer, R., et al. (1977) Biochemistry, 16, 3594-3602.

6. Walsh, C.T., et al. (1977) Biochemistry, submitted.

7. Hastings, J.W., et al. (1973) Proc. Natl. Acad. Sci. USA, 70, 3468-3472.

8. Hastings, J.W. and Balny, C. (1975) J. Biol. Chem., 250, 7288-7293.

9. Hastings, J.W., et al. (1976) in Flavins and Flavoproteins, Singer, T.P. ed., Elsevier Scientific, Amsterdam, pp. 53-61.

10. Massey, V. and Hemmerich, P. (1976) in The Enzymes, Vol. XII, Boyer, P.D. ed., Academic Press, New York, pp. 191-252.

11. Kemal, C. and Bruice, T.C. (1976) Proc. Natl. Acad. Sci. USA, 73, 995-999.

12. Kemal, C., et al. (1977) Proc. Natl. Acad. Sci. USA, 74, 405-409.

77

13. Eberhard, A. and Hastings, J.W. (1972) Biochem. Biophys. Res. Commun., 47, 348-353.

14. McCapra, F. and Hysert, D.W. (1973) Biochem. Biophys. Res. Commun., 52, 298-304.

15. Keay, R.E. and Hamilton, G.A. (1975) J. Am. Chem. Soc., 97, 6876-6878.

16. Hemmerich, P. (1976) in Fort. Chemie Naturstoffe, Grisebach, H. ed., Springer-Verlag, pp. 451-526.

17. Bruice, T.C. (1976) in Prog. Bioorganic Chem., 4, Kaiser, E.T. and Kezdy, F.J. eds, Wiley, New York, pp. 1-87.

18. Lowe, J.N., et al. (1976) Biochem. Biophys. Res. Commun., 73, 465-469.

19. Kemal, C. and Bruice, T.C. (1977) J. Am. Chem. Soc., 99, 7064-7067.

20. Bentley, D, et al. (1974) Biochem. Biophys. Res. Commun., 56, 865-868.

21. Gunsalus-Miguel, A., et al. (1972) J. Biochem., 247, 398-404.

22. Baldwin, T.O., et al. (1975) J. Biol. Chem., 250, 2763-2768.

23. Tu, S.-C., et al. (1977) Arch. Bioch. Biophys., 179, 342-348.

24. Kornblum, N., et al. (1959) J. Am. Chem. Soc., 81, 4113-4114.

25. Hastings, J.W., et al. (1969) Biochemistry, 9, 4681-4689.

26. Kirsch, J. (1977) in Isotope Effects on Enzyme-Catlayzed Reactions, Cleland, W., O'Leary, M., and Northrop, D. eds., University Park, Baltimore, pp. 100-121.

27. Hastings, J.W., et al. (1966) in Bioluminescence in Progress, Johnson, F.H. and Haneda, Y. eds., Princeton University Press, Princeton, pp. 151-193.

28. Dunn, D.K., et al. (1973) Biochemistry, 12, 4911-4917.

29. Hastings, J.W. (1975) in Energy Transformations in Biological Systems, Ciba Foundation Symposium 31, Wolstenholme, G.E.W. and Fitzsimons, D.W. eds., Associated Scientific, Amsterdam, pp. 125-146.

30. Kemal, C. and Bruice, T.C. (1976) J. Am. Chem. Soc., 98, 3955-3964.

31. Schreiner, S. and Kramer, H.E.A. (1976) in Flavins and Flavoproteins, Singer, T.P. ed., Elsevier Scientific, Amsterdam, pp. 793-799.

32. Sun, M. and Song, P.-S. (1973) Biochemistry, 12, 4662-4669.

33. Lee, J. and Murphy, C.L. (1975) Biochemistry, 14, 2259-2267.

34. Russell, G.A. (1957) J. Am. Chem. Soc., 79, 3871-3877.

35. Hiatt, R.R., et al. (1975) J. Am. Chem. Soc., 97, 1556-1562.

36. Eley, M., et al. (1970) Biochemistry, 9, 2902-2908.

PYRIDINE NUCLEOTIDE DEPENDENT ENZYMES

Pyridine Coenzyme Probes of the Active Site of Dehydrogenases

Norman J. Oppenheimer, Robert M. Davidson, Terry O. Matsunaga and Bernard L. Kam

Department of Pharmaceutical Chemistry
University of California, San Francisco
San Francisco, California 94143

ABSTRACT

The use of N-15 and proton nuclear magnetic resonance is discussed as a probe for the study of pyridine coenzyme binding to dehydrogenases. A one step synthesis of [15]N-1 nicotinamide is described and N-15 NMR chemical shift data are presented for labeled NAD and model compounds. The spectra indicate N-15 to be highly sensitive to the redox state of the coenzyme. Proton nuclear magnetic resonance studies are reported for both intramolecular association in NADH and the interaction of NAD with pig heart lactate dehydrogenase. The use of deuterium labeling to suppress short range dipolar interactions within the coenzyme is discussed as it relates to conformational studies of the coenzyme and to coenzyme-dehydrogenase binding.

INTRODUCTION

Nuclear magnetic resonance, NMR, has proven to be a powerful tool in the study of molecular structure and interaction of small molecules with macromolecules[1]. For the pyridine coenzymes the focus of attention is the nicotinamide ring and its interaction with the active site of dehydrogenases. The ultimate goal is to understand the mechanism by which these enzymes activate both the coenzyme and the substrate in the transition complex in order to facilitate hydride transfer.

The nicotinamide ring contains three different spin 1/2 nucleons with a potential for NMR studies: protons, carbon-13, and nitrogen-15. These nucleons have an ideal property for a probe. They do not alter the fundamental biochemistry of the reaction or markedly effect the nature of the binding interaction. I will first discuss some general considerations for NMR studies and the relative magnetic properties of these three nucleons.

There are two primary considerations in order to assess the feasibility of any NMR study: resolution and sensitivity. Resolution is the ability to observe interpretable signals. It is a function of linewidth and the number of resonances in any given spectral region; hence it depends strongly on the chemical shift range of the nucleons. Sensitivity is the ability to acquire data in a reasonable period

TABLE 1
COMPARISON OF NUCLEAR PROPERTIES

	Range of Chemical Shifts at 23.4 kG	Natural Abundance (%)	Relative Sensitivity
^{1}H	1000 Hz (10 ppm)	99.98	1
^{13}C	5000 Hz (250 ppm)	1.1	1.59×10^{-2}
^{15}N	10000 Hz (1000 ppm)	0.37	1.04×10^{-3}

of time with a reasonable amount of sample. It is a function of the spectrometer, the intrinsic sensitivity of the nucleon studied, and the nature of the experiment. In addition, where labels are needed the relative ease or difficulty of the synthesis can be an important consideration.

The NMR properties which relate to resolution and sensitivity for the three nucleons, H-1, C-13, and N-15, are compared in Table 1. The following brief discussion will summarize the advantages and disadvantages of each nucleon, especially as they relate to enzyme studies.

H-1: High sensitivity is the major advantage. This feature is generally offset by the large number of protons in a molecule the size of a protein. In addition, the narrow range of chemical shifts combined with the presence of spin coupling results in further overlap of the resonances. Finally the large gyromagnetic ratio leads to strong dipolar interactions hence the potential for broad lines.

C-13: The main advantage is the larger range of chemical shifts and the corresponding decrease in the number of resonances per unit area of spectrum. The absence of homonuclear spin coupling and the ability to eliminate the heteronuclear scalar coupling to protons leaves carbon resonances as narrow lines. This is not necessarily the case for C-13 spectra of proteins. Carbons with directly bonded protons are severely broadened by the proton-carbon dipolar interaction. Unprotonated carbons are not so severely effected, and hence sharp resonances can be observed and many of these types of carbons can be resolved even for proteins of molecular weight of 20,000[2]. The main disadvantages are the low sensitivity of carbon-13 combined with its low natural abundance. This twin problem requires the use of large amounts of sample and long data acquisition time.

Syntheses of C-13 labels can be tedious and expensive when the molecular backbone itself must be generated. Furthermore labeling does not overcome the problem of broad lines associated with protonated carbons. Neither can this problem be alleviated by replacement of the proton with deuterium since the scalar coupling of deuterium to carbon can cause substantial line broadening[3].

N-15:   This nucleon has an even greater range of chemical shifts than C-13. However, its intrinsic sensitivity is an order of magnitude less than C-13. This low sensitivity makes meaningful N-15 NMR studies of dehydrogenases at natural abundance impossible,   for   all   practical   purposes, at this time. Therefore N-15 studies will require labeled compounds and are dependent on the relative ease of synthesis.

By setting as a goal the study of the activation of the pyridine ring, constraints are placed on the nature and location of the labels that can be used as probes.   Although a C-13 label of the carboxamide group has been synthesized[4] and an N-15 label could be easily synthesized, this group remains peripheral to the pyridine ring.   The changes in chemical shift that might be expected to result from activation of the pyridine ring would be comparable to effects that could arise from possible torsional changes or alterations in hydrogen bonding.   Therefore experimental results would be difficult to interpret. It is unfortunate, but other C-13 labels do not look promising at this time either. Carbon-13 labels of any of the ring positions except N3 are protonated carbons and hence will have linewidths in excess of 40 Hz even when bound to proteins of molecular weight as low as 18,000[2].   Thus C-13 labels even at the important N4 position may give resonances which are too broad to interpret.   In addition these labels would be difficult to synthesize.   An   example of such a synthesis is that for the N6 labeled nicotinamide[5].

The positions in the nicotinamide ring most intimately involved with the redox reaction of the coenzyme are the N1 and N4 (see Figure 1).   The atoms at both these positions rehybridize and hence they will be highly sensitive to the perturbations

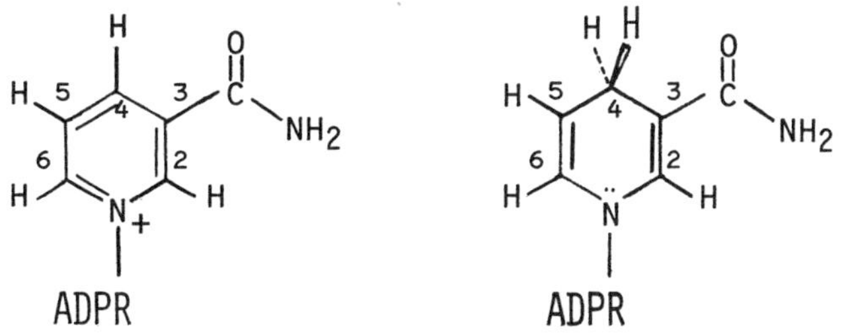

Figure 1.   The numbering of the pyridine ring.   In the text, $\underline{N}$ will refer to the nicotinamide ring and $\underline{A}$ will refer to the adenine ring.

which occur in the course of activation for hydride transfer. The N1 nitrogen and its role as an electron source or sink is a natural position for an N-15 label. In addition as an unprotonated atom its linewidth is expected to be only a few Hz even when tightly bound to dehydrogenases. This is in analogy to the case for unprotonated carbons. The major deterent for N-15 NMR studies, namely the synthesis of the label, has now been overcome with the developement of a facile, one step synthesis of N-15 nicotinamide[6].

The high intrinsic sensitivity of protons makes their role as a probe attractive especially considering that the redox reaction involves hydride transfer. Therefore the potential exists to study the very protons which are to be transferred. This makes it imperative to develop the methodology to facilitate the observation of the N4 protons. This primarily involves eliminating the strong intramolecular dipolar interactions which contribute greatly to the bound linewidths.

I will discuss work being conducted in my laboratory on two topics: 1). The synthesis and potential use of N-15 labeled NAD as a probe of the chemical nature of tightly bound NAD/dehydrogenase complexes. 2). The development of methods designed to enhance the observeability of proton resonances in order to study coenzyme interactions with dehydrogenases.

EXPERIMENTAL

N-15 Nicotinamide was prepared according to the literature procedure[6] from (99% N-15) ammonium chloride, (Korr Isotopes). [d_6]Nicotinamide was obtained from Merck Isotopes. Incorporation of these labeled nicotinamides into NAD was by pig brain NAD-glycohydrolase-catalyzed exchange, a method that has been employed for the synthesis of other NAD analogs[7]. The N-15 labeled model compound N-methyl nicotinamide was synthesized by the reaction of N-15 nicotinamide with methyl iodide and subsequent dithionite reduction gave the corresponding N-15 dihydronicotinamide[8]. Pig heart lactate dehydrogenase was obtained from Sigma Biochemicals and used without further purification. Polyacrylamide gels and specific activity indicated the protein to be more than 90% pure H4 type isozyme.

N-15 NMR spectra were obtained on a Varian XL-100 Fourier transform NMR spectrometer equipped with a Nicolet Technologies multinuclei probe operating at 10.14 MHz. Samples were run in 12 mm NMR tubes and D_2O was used as solvent and to provide the lock signal. Proton spectra were obtained at both 100 MHz and at 220 MHz.

N-15 NMR Studies of Pyridine Coenzymes: In the course of the synthesis of nicotinamide nucleosides, we have used the reaction of sugar amines[9] with the compound 1-N-(2,4-dinitrobenzene)nicotinamide[10], a Zincke salt[11]. Based on the mechanism proposed by Zincke[11]: a nucleophilic addition of the amine followed by ring opening, ring closing and finally elimination of dinitroaniline, it was apparent that this reaction had potential as a facile and general method for the synthesis of N1 labeled pyridines (see Figure 2). This reaction has lived up to expectations, giving 80-85% recovered yields based on initial N-15 ammonia[6].

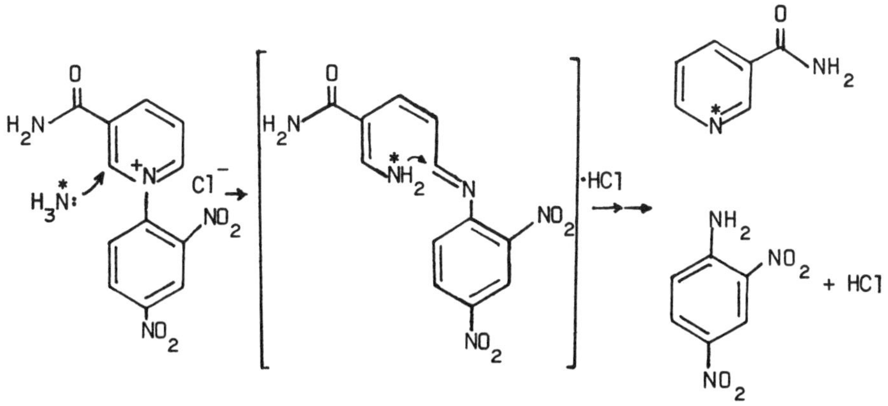

Figure 2. Synthesis of ^{15}N-1 nicotinamide via the Zincke Reaction. It is not clear whether the ammonia attacks the N2 position as drawn or at N6. Either case will yield the labeled nicotinamide.

N-15 NMR spectra were obtained for the model compounds and N-15 labeled NAD. The data are listed in Table 2. As can be seen the N1 nitrogen of nicotinamide is very sensitive to protonation, shifting 96.2 ppm upfield. In both the model compounds and the coenzyme, oxidation/reduction causes large shifts of 106.1 and 109.2 ppm respectively. The N1 position appears to be quite insensitive to ring current effects as indicated by a 0.2 ppm shift, 2 Hz, observed for NAD on lowering the pH to 2. Protonation of the adenine ring at low pH is known to cause destacking of the bases and comparable 0.2 ppm shifts have been observed for the proton resonances[12] and C-13 resonances[13]. The small shift observed for the N1 resonance, however, is within the range of the estimated error for the chemical shifts. It is interesting that the N1 appears to be also rather insensitive to the charge on the 5' phosphate. Neither cleaving the pyrophosphate backbone, nor protonating the 5' phosphate in the mononucleotide caused a greater than 1 ppm shift. This is in contrast to C-13 where the N2 and N6 carbons show significant shifts[13]. It is not

TABLE 2

NITROGEN CHEMICAL SHIFTS OF ^{15}N-NAD$^+$ AND MODEL COMPOUNDS

Compound[a]	Chemical Shift in ppm from NH$_4^+$ (external)
Nicotinamide pD = 7	278.2
Nicotinamide pD = 2	182.0
1-Methyl Nicotinamide	181.7
1-Methyl Dihydronicotinamide	75.6
NAD$^+$ pD = 7	197.9
NAD$^+$ pD = 2	198.1
NMN$^+$ pD = 7	199.0
NMN$^+$ pD = 2	198.8
NADH pD = 9	88.7
NMNH pD = 9	88.8

[a]Samples were all 0.1 M.   Temperature = 31° C.

clear because of the paucity of data whether N-15 heteroatoms in nucleotides will, in general, be insensitive to the charge of the 5' phosphate.

The sensitivity we have observed for the 99% N-15 label is approximately three times that for a comparable, unprotonated C-13 at natural abundance. Therefore N-15 NMR studies of [N-15]NAD tightly bound to dehydrogenases are feasible. The ready observation of single, unprotonated C-13 resonances for proteins of molecular weight 15-20,000[2] is encouraging for the observation of a 99% N-15 label on a typical dehydrogenase subunit of molecular weight 35,000. The low natural abundance of N-15 will also minimize the background absorption due to the nitrogens of the protein.

It should be noted that because of the low sensitivity of even an N-15 label the studies must be limited to tightly bound complexes. Any exchange broadening of the resonance will greatly increase the amount of time needed to obtain a useable spectrum.

Proton Isolation: Protons, as the second most sensitive nucleon, hold the potential for conducting studies at lower concentrations and with more modest requirements for the amount of enzyme needed. However, the abundance of proton resonances and the strong dipolar interactions of the proton nuclei represent a considerable problem in their use as a probe of the mechanism of dehydrogenases. By removing the interproton interactions a significant improvement in resolution can be obtained. The most direct method to eliminate these interactions is by deuterium labeling. To this end we have synthesized NAD in which all the protons of the nicotinamide ring have been replaced by deuterium except the N4 proton. The resulting N4 resonance is a sharp singlet as shown in Figure 3. Note that the absence of scalar coupling to the other ring protons results in about a five-fold increase in the signal to noise

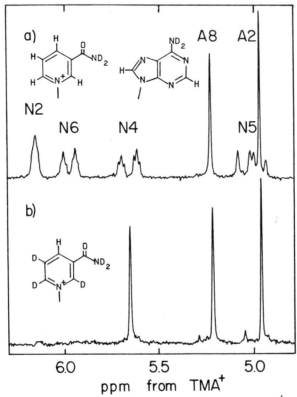

Figure 3. Portion of the ^{1}H-NMR spectra for 10 mM NAD$^+$ (top) and [2,5,6-d$_3$]NAD$^+$ taken using identical spectrometer parameters. Note the increased observeability of the N4 "proton label" in the bottom spectrum. The internal standard tetramethylammonium, TMA, was used.

ratio for the N4 proton. The N4 proton has now been isolated from dipolar interactions to a greater extent than is the physically isolated A2 proton as indicated by the longer value of the longitudinal relaxation time T_1 for the N4 proton.

The relaxation time of the proton resonances of a molecule the size of NAD is dominated primarily by the dipolar interaction between the proton nuclei[1]. In the extreme narrowing limit the longitudinal relaxation time T_1 is given by the following equation

$$\left(\frac{1}{T_1}\right)_{INTRA} = \left(\frac{1}{T_2}\right)_{INTRA} = \frac{\frac{3}{2}\gamma_H^4\hbar^2\tau_c}{r^6} \qquad \text{Eqn. 1}$$

following equation where r is the distance between the interacting dipoles, γ_H is the gyromagnetic ratio, $\hbar$ is Planck's constant and τ_c is the correlation time of the molecule. Note that the dipolar relaxation is primarily a function of the distance between the nuclei, and their tumbling rate. With deuterium substitution, the parameters stay the same except that γ_D is 1/6th that of a proton. Therefore its dipolar contribution to the relaxation of a proton is about 1/40th that of the proton it replaced [14]. Not only does the smaller gyromagnetic ratio for deuterium isolate a proton but the consequences of deuterium replacement can provide conformational information.

The distance between any two protons can be calculated if they are dipolar-coupled by measuring the effect on the relaxation time, T_1, of the proton of interest when the other proton is replaced with deuterium[14]. The distance r is calculated by recombining equation 1 and introducing the change in T_1 as shown in

$$<r> = \left[\frac{\frac{3}{2}\gamma^4\hbar^2\tau_c}{\Delta(T_1^{-1})/(0.94)}\right]^{\frac{1}{6}} \qquad \text{Eqn. 2}$$

equation 2. Some results for NADH are shown in Table 3. Note that perdeuteration of the dihydronicotinamide ring causes a significant increase in T_1 for the adenine A2 proton. Methanol, a known destacking agent, eliminates this effect. The observed change in T_1 corresponds to an average distance of 4.2 Å. However, use of a model for the solution conformation of NADH where there is a 30-40% population of a base-stacked folded form leads to a calculated distance for that conformation of about

3.8 Å. The observation of any effect of perdeuteration on the T_1 of the A2 proton in NADH indicates that the protons of the dihydronicotinamide ring make a contribution to the dipolar relaxation of the A2 proton. Therefore, these protons must be spatially juxtaposed for a period of time comparable to the molecular reorientation time of the dinucleotide. The distance values which have been calculated can not be considered definitive since it is not known, at this time, whether the dipolar interaction is due to one or more protons on the dihydronicotinamide ring. However, the results do give clear indication of a close association of the bases. This is consistent with the presence of a significant population of a folded conformation. No effect has been observed on the T_1 of the

using selective deuteration of the nicotinamide ring has also shown no effect on the T_1 of the A2 proton[15]. This absence of an effect in the oxidized coenzyme may be a consequence of its smaller population of folded conformation[12]. Estimates based on the distance measurement for NADH indicate that the smaller percentage of folded form for NAD would cause a change in T_1 too small to be observed.

TABLE 3

RELAXATION DATA FOR THE ADENINE A2 PROTON[a]
Calculated Interbase Distance in NADH

	T_1 (sec.)	ΔT_1^{-1}	$<r>$ (Å)
NADH (10° C)	5.06		
d_5-NADH (10° C)	6.49	0.044	4.2
NADH (60% methanol, 10° C)	4.08		
d_5-NADH (60% methanol, 10° C)	4.15		

[a]Samples were all 3 mM and 12 mm NMR tubes were used. Data was acquired at 100 MHz and represents an average of three separate determinations. The reproducibility was within 5%. The smaller value in methanol may reflect the greater viscosity of the solution.

Enzyme Binding: Since the relaxation time determines the linewidth of the proton resonances and since dipolar interactions dominate proton relaxation, it follows that the linewidth for a coenzyme bound to a dehydrogenase will be a function of the interproton distance as well as the tumbling time of the protein. This latter parameter is in turn a function of the molecular weight of the protein and to a lesser extent of its shape[1]. The dependence of the linewidth on the correlation time of the protein and the interproton distance is listed in Table 4. I have used the molecular weights for two extremes of dehydrogenases, dihydrofolate reductase at molecular weight 18,000 and a typical, tetrameric dehydrogenase with a molecular weight of 140,000. The calculations show that the closer two protons are, the broader will be their resonances when bound to a protein, even one as small as dihydrofolate reductase. It is clear, however, that decreasing the intramolecular dipolar interaction should greatly narrow the expected bound linewidths. These calculated values should only be taken as estimations of the relative order of linewidths. Other factors, primarily kinetic effects, can make a considerable contribution to linewidths in the absence of domination by dipolar relaxation. An example of the resolution improvement resulting from deuterium replacement is given by our earlier studies with NADH binding to dehydrogenases. We have shown that just removal of the geminal interaction between the N4 methylene protons caused sufficient decrease in the bound linewidth to allow observation of the interaction of the separate N4 resonances[16].

In preliminary studies of the binding of [2,5,6-d$_3$]NAD to pig heart lactate dehydrogenase a considerable improvement in the observeability of the resonances has been found. The plot of the chemical shift of the NAD protons with increasing NAD/enzyme ratio clearly shows the large enzyme-induced shifts (see Figure 4). The results are in substantial agreement with those observed by Lee et al.[17] for other lactate dehydrogenases. The downfield shift of the N4 and A8 proton is indicative

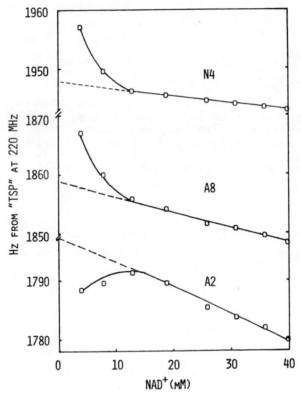

Figure 4. Plot of the chemical shift of the aromatic protons for [2,5,6-d$_3$] NAD$^+$ at 23° C as a function of concentration at a constant concentration of pig heart lactate dehydrogenase at 35 mg/ml, which corresponds to a 1 mM solution of binding sites. The upfield shift observed for all the protons with increasing concentration is caused by self-association of NAD$^+$.[12] The internal standard tetramethylammonium was used, TMA$^+$ = 3.2 ppm.

of the unfolding of the coenzyme into a region of the protein devoid of aromatic
residues. The X-ray structure of pig heart lactate dehydrogenase[18] indicates that
the coenzyme is unfolded, and that aromatic residues are relatively distant from
these protons. In addition a close association is observed between the tyrosine-85
residue and the A2 portion of the adenine ring. This proximity to an aromatic amino
acid residue can explain its upfield shift.

TABLE 4

CALCULATED LINEWIDTHS[a]

Dependence of the Bound ^{1}H Linewidths on both Interproton
Distance and Molecular Weight of the Dehydrogenase

Interproton Distance	18,000 mw Enzyme (Hz)	140,000 mw Enzyme (Hz)
CH_2^b (r = 1.7 Å)	68	338
$CH-CH^c$ (r = 2.4 Å)	9	43
$CH\cdots CH^d$ (r = 3.6 Å)	0.8	3.7

[a]The values are calculated for 100 MHz and assume a τ_c of 20 ns. for an 18,000 mw protein and 100 ns. for a 140,000 mw protein[2]. All values assume only a two proton interaction.
[b]Intramolecular distance between geminal protons.
[c]Intramolecular distance between vicinal protons.
[d]Intermolecular distance between closely spaced neighboring residues.

Double resonance experiments of NAD bound to lactate dehydrogenase show that
the relaxation of the N4 and A2 protons are dominated by dipolar interaction with
residues of the protein. This result clearly indicates that the goal of isolating
the N4 proton from intramolecular dipolar interactions has been achieved. Further
study is now in progress on the magnitude of the bound linewidth and its dependence
on the intermolecular proton distances in the active site. The decrease in bound
linewidths also should allow analysis of the dependence of the linewidth and
chemical shift of the coenzyme resonances on the concentration of enzyme or the
magnetic field strength used, a topic I will not discuss. From these experiments
should come detailed information on the dynamics of binding and the chemical
environment of the active site.

Proton studies are primarily limited to experiments where the coenzyme is in either fast or intermediate exchange. This is necessitated by the difficulty inherent in trying to observe a single proton of a tightly bound complex on a protein in which there may be hundreds of protons in the same region. Small enzymes like dihydrofolate reductase may still allow direct observation of single protons of tightly bound complexes since the resonances of the coenzyme should still be narrow and there will be fewer protein resonances with which to overlap.

In summary, the two techniques that I have described should provide complementary information. The N-15 label should be a probe for tightly bound complexes, giving information about the hybridization state at the N1 position for given binary and ternary complexes. Proton isolation, on the other hand, provides an enhanced resolution of the important N4 resonances of the pyridine moiety which can be used to study the dynamics of the interaction with dehydrogenases. The possibility exists for direct observation of the N4 protons of the coenzyme when bound to small dehydrogenases like dihydrofolate reductase.

ACKNOWLEDGEMENTS

I would like to thank Dr. Nathan O. Kaplan and Dr. Susan S. Taylor for technical advice and stimulating discussions. This work was in part supported by NIH grant GM-22982 and the grants NIH-RR-00892, UCSF NMR Research Resource Center and NIH-RR-00708, UCSD NMR/MS Research Resource Center.

REFERENCES

1). James, T.L. (1975) Nuclear Magnetic Resonance in Biochemistry, Academic Press, New York.

2). Oldfield, E., Norton, R.S. and Allerhand, A. (1975) J. Biol. Chem. 250, 6381-6402.

3). Browne, D.T., Kenyon, G.L., Packer, E.L., Sternlicht, H. and Wilson, D. M. (1973) J. Am. Chem. Soc. 95, 1316-1323.

4). Blumenstein, M. and Raftery, M.A. (1972) Biochemistry 11, 1643-1648.

5). Byrson, T.A., Wisowaty, J.C., Dunlap, R.B., Fisher, R.R. and Ellis, P.D. (1974) J. Org. Chem. 39, 1158-1160.

6). Oppenheimer, N.J., Matsunaga, T.O. and Kam, B.L. submitted.

7). Williams, T.J., Zens, A.P., Wisowaty, J.C., Fisher, R.R., Dunlap, R.B.,Byrson,

T.A. and Ellis, P.D. (1976) Arch. Biochem. Biophys. 172, 490-501.

8). Karrer, P., Schwarzenbach, G., Benz, F. and Solmssen, U. (1936) Helv. Chim. Acta. 19, 811-828.

9). Atkinson, M.R., Morton, R.K. and Naylor, R. (1965) J. Chem. Soc. (London) 610-615.

10). Lettre, V.H., Haede, W. and Ruhbaum, E. (1953) Annalen 579, 123-132.

11). Zincke, T. (1905) Annalen 341, 365-379.

12). McDonald, G., Brown, B., Hollis, D.P. and Walter, C.F. (1972) Biochemistry 11, 1920-1930.

13). Blumenstein, M. and Raftery, M.A. (1973) Biochemistry 12, 3585-3590.

14). Akasaka, K., Imoto, T. and Hatano, H. (1973) Chem. Phys. Lett. 21, 398-400.

15). Zens, A.P., Byrson, T.A., Dunlap, R.B., Fisher, R.R. and Ellis, P.D. (1976) J. Am. Chem. Soc. 98, 7559-7564.

16). Lee, C.-Y., Oppenheimer, N.J. and Kaplan, N.O. (1974) Biochem. Biophys. Res. Commun. 60, 838-843.

17). Lee, C.-Y., Eichner, R.D. and Kaplan, N.O. (1973) Proc. Nat. Acad. Sci. USA. 70, 1593-1597.

18). Eventoff, W., Rossmann, M.G., Taylor, S.S., Torff, H.-J., Meyer, H., Keil, W. and Kiltz, H.-H. (1977) Proc. Nat. Acad. Sci. USA. 74, 2677-2681.

THE MECHANISM AND ENERGETICS OF THE GLUTAMATE DEHYDROGENASE REACTION

HARVEY F. FISHER and ALAN H. COLEN

Laboratory of Molecular Biochemistry, Veterans Administration Hospital, Kansas City, Missouri 64128, and the University of Kansas School of Medicine

ABSTRACT

We outline an approach to the study of mechanisms of enzyme catalyzed reactions which involves the study of properties of principal complexes, the resolution of the time course of the catalytic reaction by various means, and the correlation of the occurrence of specific complexes to resolved features of the time course. An example of the detailed study of one cooperative ternary complex is provided. The transient-state time course of the catalytic reaction and its resolution into discrete phases is described, and chemical characteristics are assigned to such resolved steps. The development and interconversion of the major complexes are then correlated with the kinetically resolved phases.

A thermodynamic study of the reaction is described. As examples, the thermo-dynamic parameters for the formation of two ternary complexes (one positively cooperative, the other negatively cooperative) are discussed. The concept of thermodynamic cooperativity parameters is developed and applied to the two complexes. It is noted that because of extreme enthalpy-entropy compensation, all parameters for both systems are similar except for ΔG_C° itself, which is a small resultant. Thermodynamic profiles based on both temperature dependence of individual transient rate constants and direct calorimetric measurements are presented. Two notable features of these profiles are the deep trough in the middle of the reaction sequence and the large proportion of the total energetic changes that are accomplished by α-ketoglutarate binding. Mechanistic alternatives are considered and a mechanism involving a transimination followed by carbinolamine formation is proposed.

Finally, we cite evidence that the homologous coenzyme binding site common to all pyridine-nucleotide dehydrogenases, has, in the case of bovine glutamate dehydrogenase, undergone an evolutionary split such that the adenylate specific portion of that site no longer functions in coenzyme binding, but has in fact become the allosteric ADP binding site.

INTRODUCTION

We can best describe the results of our investigation of the mechanism of the glutamate dehydrogenase reaction in terms of the rationale we have used to guide those studies. We have first studied the various binary, ternary, and quaternary complexes formed between the enzyme and its substrates, products, coenzyme and

95

allosteric modifiers, determining their stoichiometry, optical and other physical properties, and the kinetics of their formation. We then attempt to resolve the time course of the reaction in as many different ways as possible, so that discrete steps or at least small groups of steps are distinguishable. To provide such resolutions we have used transient state spectrophotometry, transient state calorimetry, partial reactions (in which one reactant is omitted), order of addition effects and some other rather specialized techniques. More recently we have started using the cryoenzymologic technique pioneered by Fink[1] in which the use of organic cosolvents permits the study of transient features at temperatures so low (-50°C) that some steps of the reaction are slowed by factors of as much as 10^7. The next step in the approach is to use our knowledge of the properties of the various complexes to identify their occurrence in the various phenomenologically resolved phases of the reaction.

All of this is a necessary prelude to the determination of the detailed chemical and energetic mechanisms of the interconversion of one complex into another. The first steps of this approach have now been essentially completed, at least to the currently attainable level of detail; we are now attacking the last and most important step. We will briefly summarize here some examples of these various initial steps of our investigation and sketch the most important results of the correlation of the occurrence of complexes with the phases of the reaction. We will then describe some new preliminary findings on the nature of the interconversion of these complexes.

AN EXAMPLE OF THE DETAILED STUDY OF A TERNARY COMPLEX

The interaction of ADP with glutamate dehydrogenase (E) has some very favorable properties and we have frequently used ADP complexes to develop the techniques necessary for the study of more reactive (and more elusive) complexes[2]. The "static" difference spectrum of the E-ADP complex (obtained by measuring the difference between the absorbance of an enzyme-ADP mixture and that of the sum of absorbances of the separate components) is shown as the solid line in Figure 1. When ADP and enzyme are mixed in a stopped flow experiment a relatively slow first order change in absorbance is observed (Figure 2). The amplitudes of these time dependent absorbance changes are plotted as open circles in Figure 1, and it is obvious that we are measuring the kinetics of formation of the E-ADP complex. However, the measured amplitudes of the kinetic signal become increasingly smaller than the statically measured signal as ADP concentration is increased, indicating that there is a much faster ADP-concentration-dependent process producing the same red-shifted signal preceding the slow step we observe, and indeed the non-linear dependence of the rate constant of the observed process on the sum of enzyme and ADP concentrations indicates that we are actually observing a slow isomerization step following a rapid (but weaker) initial binding step which must involve the adenine ring of

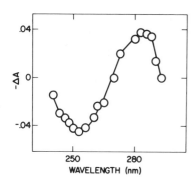

Fig. 1 (——) Static difference spectrum of E-ADP complex[2]o: amplitudes from
experiments as in Fig. 2.

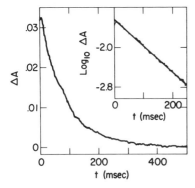

Fig. 2 ΔA at 253 nm on mixing E + ADP. Inset: first
order plot of same data[2].

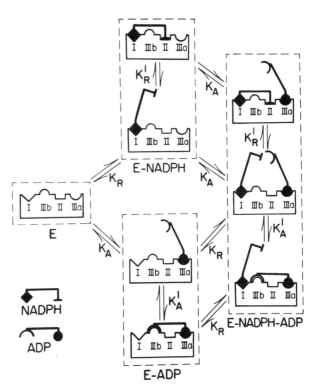

Fig 3. Ligand exclusion model of binding of NADPH
and ADP to E. ◆ represents nicotinamide ring of
NADPH. ● represents adenine ring of ADP[2].

ADP. Gel filtration binding studies in the simultaneous presence of enzyme ADP,
and NADPH proved the existence of an E-ADP-NADPH complex with a stoichiometry of
one ADP and one NADPH per active site, and in which the dissociation constant of
each nucleotide was raised seven-fold in the presence of the other. The kinetics
and concentration dependence of binding of the two nucleotides, together with consideration of the spectroscopic properties of the complex, leads to the scheme for
the formation of a negatively cooperative heterotropic complex shown in Figure 3.
In the ternary complex, each nucleotide is bound through an interaction with its
chromophoric ring while the isomerizations of the ribose diphosphate and ribose
adenylate tails of the two ligands compete for separate but mutually exclusive
secondary binding sites.

In a quite separate study we have found that hydrogen exchange of the amide
hydrogen atoms of NADPH with D_2O could be followed by mixing the two in a stopped
flow spectrophometer and following the time dependent perturbation of the reduced
nicotinamide by the incorporated deuterium atoms. We found that in the presence
of glutamate dehydrogenase (or any pyridine nucleotide dehydrogenase) the deuterium

97

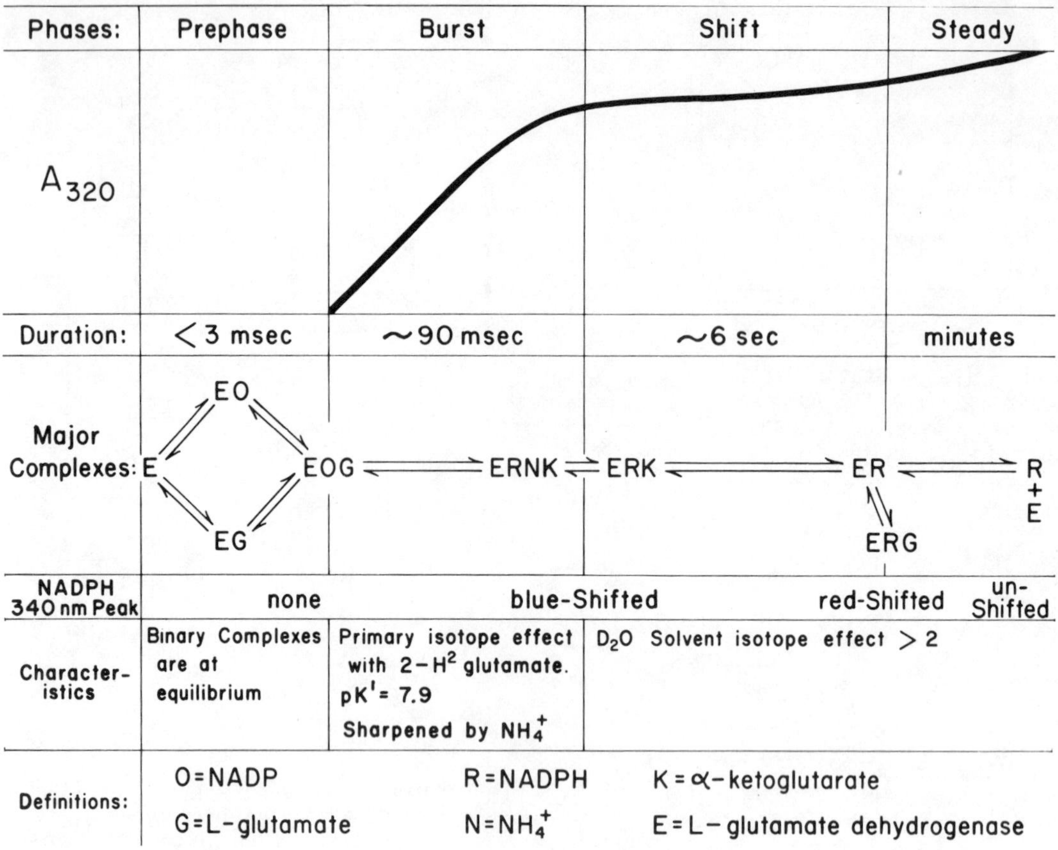

Phases:	Prephase	Burst	Shift	Steady
A_{320}				
Duration:	< 3 msec	~ 90 msec	~ 6 sec	minutes

Major Complexes:

$$E \rightleftharpoons \begin{matrix} EO \\ EG \end{matrix} \rightleftharpoons EOG \rightleftharpoons ERNK \rightleftharpoons ERK \rightleftharpoons ER \rightleftharpoons R + E$$
$$ER \rightleftharpoons ERG$$

NADPH 340 nm Peak	none	blue-Shifted	red-Shifted	un-Shifted
Characteristics	Binary Complexes are at equilibrium	Primary isotope effect with 2-H^2 glutamate. pK$'$ = 7.9 Sharpened by NH$_4^+$	D$_2$O Solvent isotope effect > 2	

Definitions:

O = NADP	R = NADPH	K = α-ketoglutarate
G = L-glutamate	N = NH$_4^+$	E = L-glutamate dehydrogenase

Fig. 4 Transient state intermediates in the glutamate dehydrogenase reaction.

exchange was abolished[3]. A recent experiment indicates that in the E-ADP-NADPH complex we have just described the hydrogen exchange of the amide is not restored, a result anticipated by the scheme of Figure 3.

THE CORRELATION OF THE EXISTENCE OF SPECIFIC COMPLEXES WITH THE KINETICS OF THE REACTION

When a solution of glutamate dehydrogenase is mixed rapidly with a solution containing NADP and L-glutamate in a stopped flow spectrophotometer, there is a rapid burst of absorbance in the 320 to 360 nm range followed by much slower changes in absorbance[4,5]. While the precise shape of the time course is dependent on wavelength and relative concentrations of reactants, the solid curve shown in the upper panel of Figure 4 shows the general nature of the spectroscopic phenomena. We have available to us four distinguishable "colors" of the 340 nm band to aid us in resolving such curves: NADP is, of course, colorless, free NADPH has a normal maximum at 340 nm, E-NADPH (ER) and E-NADPH-L-glutamate (ERG) are both red shifted with maxima at 344 nm, while E-NADPH-ketoglutarate (ERK) is blue shifted, with a maximum at 333 nm[6]. These spectral discriminents permit us to separate the time course of the reaction into four phases, and using our knowledge of the properties of the complexes and a number of other more specific experimental probes we can make assignments of major complexes to the occurrence of these phases, as shown in Figure 4. Initial rate studies of the burst phase prove that the two binary complexes enzyme-NADP (EO) and enzyme-L-glutamate (EG) and the ternary complex EOG all equilibrate rapidly on the millisecond time scale of the overall reaction[7]. The first observed phase, the "burst" is due to the production of a blue-shifted peak as defined above and thus must involve the ERK complex. The amplitude of the burst is roughly proportional to enzyme concentration. Using the known extinction coefficient for ERK, the burst height accounts for approximately 85% of the active sites of the enzyme. The remainder is largely EOG. A 2.2-fold isotope effect on the burst rate proves that a hydrogen transfer does in fact occur during this phase[8]. In the presence of added ammonia the burst is "sharpened" and the burst amplitude diminished, indicating clearly the point in the reaction in which the carbon-nitrogen bond of glutamate is cleaved. The dependence of amplitude and rate effects on ammonia concentration indicate the existence of an ammonia-containing complex ERKN which is in equilibrium with ERK and has the same spectrum. The rate constant for the burst phase has a pK' = 8.1 and could thus involve lysine 126 which has been shown to have such a pK and to be involved in the catalytic process.

At this point in the time course, the overall reaction rate is slowed by a rather unusual and apparently obligatory process involving the conversion of blue-shifted species to red-shifted ones. This process takes a full six seconds and is subject to a two-fold solvent isotope effect in D_2O[9]. The process can be mimicked quantitatively in all detail by the partial reaction: ERK + G $\rightleftharpoons$ ERG + K. The product of this phase is clearly a mixture of ER and ERG, both of which are red-shifted.

As this process nears completion, we observe the final phase of the reaction characterized by the release of free NADPH, identified by its unshifted spectrum, while the concentration of red-shifted complexes remains unchanged. This final phase is in fact the "steady-state initial rate" observed in conventional kinetic studies using catalytic concentrations of enzyme.

There are, of course, reactions with other coenzymes, other substrates, and a great variety of modifiers, all of which provide further details of enzyme binding sites and functional group interactions; but this sketch of the basic reaction (all features of which are independent of the state of protein association) provides a sufficient background of the aspects of the work which we are going to discuss now.

It was recognized by Iwatsubo[4] in his pioneer work on the transient state kinetics that the steady state rate of this enzyme reaction (with dicarboxylic substrate) is limited by the rate of release of NADPH from the various tight ternary complexes

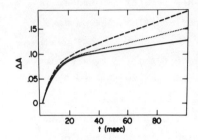

Fig. 5 The effect of the order of addition of ADP on the catalytic reaction kinetics. The conditions are: no ADP (———); enzyme preincubated with ADP (---); reactants and ADP added to enzyme simultaneously (···).

in which it is involved. Iwatsubo showed, furthermore, that the activation of the steady state rate of ADP and its inhibition by GTP both resulted from effects of those modifiers on post-burst stages. Returning to the E-R-ADP complex whose properties we discussed in some detail earlier in this paper, we now describe the use of order-of-addition stopped flow experiments in determining the details of how ADP and its enzyme complexes actually cause its effects on the catalytic reaction.

In Figure 5, the solid curve represents the time course of the reaction in the absence of ADP. The dotted and dashed curves represent the same reaction in the presence of ADP but with different orders of addition of components. When ADP is preincubated with enzyme the time course (dashed line) shows an immediate activation of the post burst release of free NADPH. In contrast, when ADP, coenzyme and substrate are mixed with the enzyme simultaneously, (dotted line) the activation lags behind, building up in a time commensurate with that required for the isomerization step measured in the ADP binding experiment shown in Figure 2. Referring to the ligand binding scheme of Figure 3, it is clear that the initial binding of ADP by its adenine ring is without effect on the kinetics, but the subsequent interaction of the ribose pyrophosphate tail of ADP interferes with the interaction of the ribose-diphosphate-ribose interaction of NADPH, thus loosening its binding.

THE ENERGETICS OF THE GLUTAMATE DEHYDROGENASE REACTION AND OF THE COOPERATIVE COMPLEXES FORMED IN THAT REACTION

There remains one set of properties in which our knowledge of the complexes of this system, or indeed of any system, is far from complete -- that property is the energetics. We do know the overall thermodynamic parameters of the chemical reaction catalyzed by glutamate dehydrogenase. A knowledge of those parameters for the formation of the various complexes and for the processes connecting those complexes would not only provide an additional means of relating individual complexes to reaction kinetics, but may provide important information on the nature of the bonding in these complexes, and may thereby point the way to specific mechanisms of interconversion. The recent availability of commercial flow microcalorimeters of sufficient sensitivity provides a means of doing this, and we now summarize the most striking conclusions of our studies thus far.

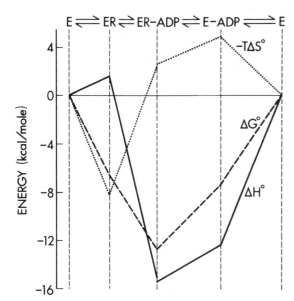

$$E \rightleftharpoons ER \rightleftharpoons ER-ADP \rightleftharpoons E-ADP \rightleftharpoons E$$

Fig. 6 Thermodynamic parameters of the formation of the E-NADPH-ADP complex. $\Delta H°$'s are from direct calorimetric measurements. $\Delta G°$'s are from most accurate methods of measurement for that particular process under the same conditions.

In Figure 6 we portray the thermodynamic parameters for the formation of the E-R-ADP complex and for its related binary complexes. (In this and in following figures the parameters are plotted in such a way that forces favoring binding are negative and forces opposing binding appear positive.) It can be seen that ER formation is entropy driven and enthalpy opposed, while E-ADP and E-R-ADP are enthalpy driven and entropy opposed. Such enthalpy-entropy compensation is a familiar pattern in many physical systems. If one subtracts the sum of any of the binary parameters from the corresponding ternary parameter, one obtains a parameter of cooperativity which we designate as $\Delta G_C°$, $\Delta H_C°$ and $\Delta S_C°$. These parameters are plotted for the E-R-ADP system in Figure 7A and for the dead-end inhibitor complex E-R-L-glutamate in Figure 7B. It is seen that in both cases there is a large exothermic driving force opposed (or compensated) by a more-or-less equivalent entropic force. The ΔG_C which is positive for E-R-ADP (negative cooperativity) and is negative for E-R-L-glutamate (positive cooperativity) appears to be a smaller, almost trivial resultant of the two larger opposing forces. In the cases of E-R ternary complexes with D-glutamate and with glutarate, the $\Delta H_C°$'s and $\Delta S_C°$'s show patterns very similar to the data portrayed in Figure 6, with $\Delta G_C°$'s again being smaller in magnitude, negative for D-glutamate but slightly positive for glutarate. The $(\Delta C_p)_C$ for the E-R-ADP system is -100 cal. $deg^{-1}M^{-1}$ while that for all three E-R-G systems is identical: -400 cal. $deg^{-1}M^{-1}$, a quite large value and striking in itself. It is as if some rather general, energetically large interaction occurs when NADPH and any one of a number of dissimilar second ligands are bound at the same time on the enzyme surface. The free energy change that accompanies this general interaction however is quite variable in both magnitude and direction, and apparently depends on very specific details of precise molecular structure. In other words it does not appear to be a direct result of the general interaction. The nature

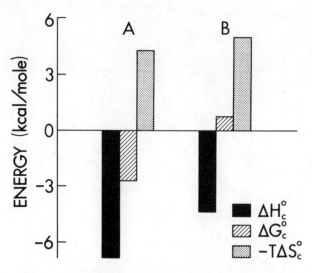

Fig. 7 Cooperativity parameters for A: E-R-G complex and B: ER-ADP complex. A cooperativity parameter is defined as the difference between that parameter for a given ternary complex and the sum of those parameters for the binary complexes.

of the general interaction itself, and particularly the large negative ΔC_p involved is the object of current study in our laboratory as well as that of Julian Sturtevant who has observed similar large negative ΔC_p's in the formation of binary complexes of related pyridine nucleotide dehydrogenases[10].

THE THERMODYNAMIC PROFILE OF THE ENZYME-CATALYZED REACTION

We have obtained the profile of the major features of the reaction from two kinds of information. The first kind of measurement is simply the classic approach of measuring the temperature dependence of individual rate constants obtained from our various transient state kinetic studies. For example, we can follow the formation of the E-R complex by protein fluorescence quenching as shown in Figure 8A. The concentration dependence of the first order rate constants follows the equation: $k_{app} = k_+[R] + k_-$; so that we obtain k_+ from the slope and k_- from the intercept. The slopes of the temperature dependence of these two k's (Figure 8B) gives us $\Delta H^{\ddagger}$ from the transition state *down* to E-R in one direction and *down* to E + R in the other, providing the maxima for that step in the profile.

The second type of information is obtained from direct calorimetric measurement of $\Delta H°$ of formation of the various complexes to provide the minima of the enthalpy profiles and of the various partial reactions required to hook up those minima. Examples of this type of measurement have been provided in Figures 6 and 7.

From a large number of these two kinds of experiments we have assembled the energetic profiles shown in Figure 9. The total free energy hill to be climbed

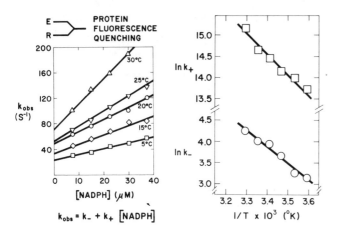

$k_{obs} = k_- + k_+ [NADPH]$

Fig. 8A Kinetics of E-NADPH complex formation.

Fig. 8B Arrhenius plots of forward and reverse rate constants obtained from slopes and intercepts of data in Fig. 8A.

in the stoichiometric reaction is 20 Kcal. The extra displacements from the data to known complexes in the figure represent the driving forces from H_2O on one side and from H^+ on the other. We know their magnitude but not their precise location, so we have simply included them at the ends of the reaction. The overall enthalpic rise to be overcome in the forward direction is 15.7 Kcal.

We now proceed to correlate the features of the thermodynamic profile of Figure 9 with the transient-state kinetic features of Figure 4. The first "prephase" steps proceed over low activation barriers to actually decrease both free energy and enthalpy in forming the E-O-G complex. The next step, which involves hydrogen transfer from glutamate to coenzyme goes over high activation barriers, dropping even lower in free energy but attaining some of the increase in enthalpy required for the overall reaction as E-NADPH-α-ketoglutarate-NH_3 (ERKN) is formed. In the next step NH_3 leaves without much energetic effect producing the first of two successive ERK complexes. The system now goes over the highest free energy barrier. This is the very slow blue-to-red shift phase. The product is a second ERK complex which is thermodynamically very similar to the first but which is kinetically much more reactive. Finally α-ketoglutarate leaves in a step involving a large entropy compensated increase in enthalpy. There are two particularly striking features of the overall profiles. The first is the very pronounced free-energy trough in the middle of the reaction sequence whose bottom actually is lower than either reactants or products; the second is the large number of steps involving α-ketoglutarate complexes and the large fraction of the overall $\Delta G°$ and $\Delta H°$ of the reaction involved in forming or in breaking up those complexes.

MECHANISTIC IMPLICATIONS OF THE THERMODYNAMIC PROFILES

Some general mechanistic possibilities are sketched in Figure 10. We have already demonstrated that the α carbon-hydrogen bond of L-glutamate is broken during the burst phase and that the amine moiety of L-glutamate leaves at a point later in time. These two facts are sufficient to establish that the first observable step of the reaction must produce an enzyme-bound product having the stoichiometry of α-iminoglutarate. (It does not at all follow, however, that the product has a bond structure anything like such an imine.) As shown in Figure 10, there are two general mechanistic routes that the reaction could take: Path I involves an attack by water to form a carbinolamine which then simply loses ammonia to form E-α-ketoglutarate; and Path II in which an attack on the α-iminoglutarate by an enzyme amino group to form a quaternary diamino compound is followed by loss of the original α-amino nitrogen as ammonia to yield an enzyme-bound imine. After this transimination sequence, the enzyme bound imine accepts a water molecule to form a carbinolamine, which in turn dissociates to produce enzyme bound α-ketoglutarate.

We may now compare each of these two general pathways with the thermodynamic profiles of Figure 9. Viewing the reaction in the reverse direction (from right to left) we note that the ERK complex is in fact defined in our work as whatever entity which is formed on the addition of α-ketoglutarate to the enzyme in the absence of ammonium. Its formation, therefore, must include all steps to the right of the dotted line in Figure 10 for either of the possible pathways. We have mentioned the fact that the breakdown of ERK is characterized by a two-fold primary isotope effect in D_2O. Such an effect would be expected at the point of entry of

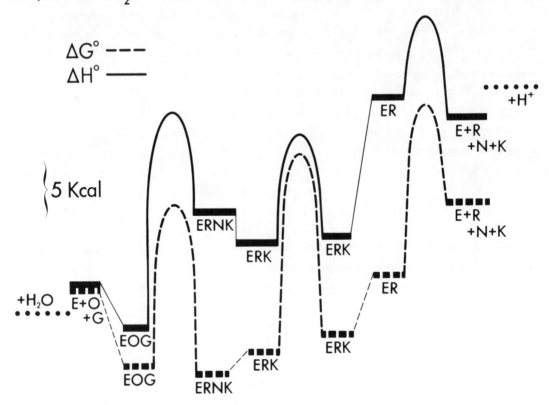

Fig. 9   Enthalpy and free energy profiles for major features of the glutamate dehydrogenase reaction.

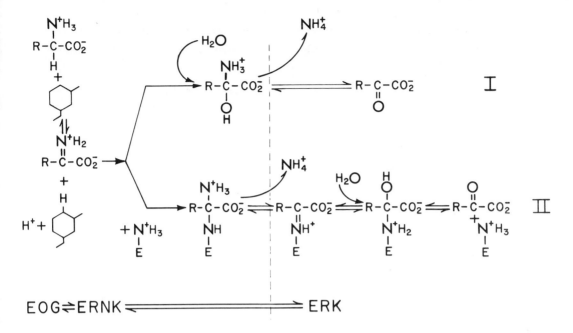

Fig. 10 Mechanistic alternatives for post-hydrogen-transfer events in the catalytic reaction.

the required water molecule. In pathway I this point of entry is not included in the steps leading to the formation of ERK, but pathway II does include that entry point in the predicted part of the sequence. Furthermore, it can be seen from Figure 9, the formation of ERK (measured directly in a flow calorimeter by mixing ER and K) is a highly exothermic reaction with a $\Delta H°$ of -10.5 Kcal. It would be surprising if the simple binding of α-ketoglutarate as such to the enzyme without extensive changes in covalent bonds as envisioned in Pathway I were to evolve that much heat. To evaluate the $\Delta H°$'s of the two steps postulated as representing ERK formation in Pathway II we have used very rapid continuous flow calorimetry to investigate the reaction: $CH_3CHO + H_2NOH$ $CH_3CH = NOH + H_2O$, which is very similar to the reverse of the last two steps of Pathway II[11]. The results of such an experiment, in which the flow rate is progressively increased so that a decreasing residence time of the reaction after mixing occurs is shown in Figure 11. It can be seen that while the overall reaction has a $\Delta H°$ of 15.7 Kcal., the initial step, that of carbinolamine formation has a $\Delta H°$ of 10 Kcal. This is precisely the $\Delta H°$ observed for that portion of pathway II which the model mixture is intended to mimic. Thus, if we make the single assumption that the carbinolamine complex is stabilized through hydrogen bonding of the hydroxyl group to the enzyme, the following assignments of structure to the various kinetically identified complexes are in quite reasonable agreement with the thermodynamic parameters of Figure 9: following the reaction in the forward direction (left to right) ERKN would be the quaternary diamino complex of pathway II. The two kinetically distinguishable ERK complexes would then be related in some way to the enzyme-bound imino form and the

105

carbinolamine form. Since we have not as yet established the relationship between the kinetically discernible events and the specific mixture of complexes formed in the partial reactions on which we carry out our direct calorimetric measurements, these assignments must be regarded as quite speculative at this time. The scheme of pathway II, however, may explain one fact which has puzzled many investigators over the years. Despite the inclusion of enzyme-bound iminoglutarate in almost every mechanism ever drawn for this reaction, the reaction mixture and partial reaction mixtures have been very refractory to borohydride reduction. While this could be explained by assuming inaccessibility of the C=N group to borohydride, it is also possible that both imino forms (if indeed they exist) are very transitory, and that in the steady state the enzyme-substrate complexes exist predomininately as a mixture of quaternary diamino and carbinolamine forms.

EVOLUTIONARY ASPECTS OF THE GLUTAMATE DEHYDROGENASE BINDING SITE

Based on a large number of high resolution x-ray crystallographic structures of enzymes, Rossman et al. have pointed out the existence of a tertiary structure common to NAD and ATP utilizing enzymes called the "NAD binding domain"[12]. This nucleotide fold typically is composed of two rather distinct sub-domains, one specific for the nicotinamide moiety of the coenzyme, and the other specific for the adenylate moiety. A rather stylized diagram of one such nucleotide fold is shown in Figure 12. Using secondary structure predictions and primary sequence information,

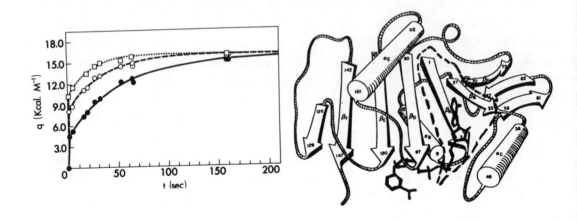

Fig. 11 Time dependence of heat evolution on mixing 1 mM sodium pyruvate with 16 mM (•-•), 57 mM (-o-o-), and 290 mM (.▫-▫.) total hydroxylamine at pH 6.0, 25° C. Flow rates ranged from 0.14 to 23 ml min^{-1}.

Fig. 12 Diagrammatic representation of NAD binding domain of glyceraldehyde phosphate dehydrogenase.

Wooten has located two sequences in bovine glutamate dehydrogenase which appear to have some structural homology with the nucleotide binding domain of other pyridine-nucleotide dehydrogenases[13]. For some time now we have had the notion that this homologous feature has in the case of glutamate dehydrogenase evolved to take on a somewhat different form, one in which the two subdomains have become slightly separated both in space and therefore in function. We were led to this thought by a number of striking differences between the binding of NADPH to this enzyme and the binding of it and related coenzymes to other dehydrogenases. For example, the reduced coenzyme binds more weakly, has an endothermic $\Delta H°$, and a small ΔC_p, and shows no red-shift in the 260 nm region of the spectrum, whereas typical dehydrogenases show tighter binding, a large exothermic $\Delta H°$ and a large negative ΔC_p, and are red-shifted in the 260 nm region. On the other hand, ADP binding to bovine glutamate dehydrogenase shows the same properties as does coenzyme binding to most dehydrogenases. As we have shown earlier in this paper, ADP is bound at a site which is partly separated, yet apparently sufficiently close to bound NADPH that mutual steric hindrance can occur. Putting these two lines of thought together, it occurred to us that possibly the adenylate domain of the common fold had, in the case of this enzyme alone, moved some distance away from the nicotinamide binding domain, generating an ADP binding regulatory site and leaving a rather weakened nucleotide binding site. The recent report by Stellwagen[14] that a dye, known as Cibacron Blue F3GA, or "reactive blue 2" forms rather tight specific 1:1 complexes with active sites having the common fold conformation provided us with an opportunity to test our idea. While our results are preliminary at this time, and have some points of considerable complexity, we can at least be reasonably sure of the following conclusions. Ciba blue appears to bind at least two molecules per active site rather than one as is the case with other dehydrogenases. The dye is a competitive inhibitor against NADPH. About one half of the binding (as measured by equilibrium dialysis, difference spectroscopy, or calorimetry) is abolished by the presence of ADP. The difference spectra change in form as a function of dye concentration, suggesting two slightly different modes of binding. At least one of the binding modes becomes increasingly weaker as enzyme concentration is increased, suggesting interference by subunit association. This was confirmed by light scatter studies which showed the dye does indeed promote the dissociation of the enzyme. Therefore the notion that the nucleotide fold has undergone an evolutionary structural split in glutamate dehydrogenase and, in so doing, has by that fact generated an allosteric ADP binding site can now be seriously considered.

ACKNOWLEDGEMENTS

This work was supported in part by grants from the National Science Foundation (GS33868X) and from the General Medicine Institute of the National Institutes of Health (GM15188).

REFERENCES

1. Fink, A.L. (1976) J. Theor. Biol., 61, 419.

2. Colen, A.H., Cross, D.G. and Fisher, H.F. (1974) Biochemistry, 13, 2341.

3. Cross, D.G., Brown, A. and Fisher, H.F. (1976) J. Biol. Chem., 251, 1785-1788.

4. Iwatsubo, M. and Pantaloni, D. (1967) Bull. Soc. Chim. Biol., 49, 1563-1572.

5. Fisher, H.F. (1973) Advances in Enzymology, 39, 369.

6. Cross, D.G. (1972) J. Biol. Chem., 247, 784-789.

7. Colen, A.H., Prough, R.A. and Fisher, H.F. (1972) J. Biol. Chem., 247, 7905-7909.

8. Fisher, H.F., Bard, J.R. and Prough, R.A. (1970) Biochem. Biophys. Res. Commun., 41, 601.

9. Colen, A.H., Wilkinson, R.R. and Fisher, H.F. (1975) J. Biol. Chem., 250, 5243-5246.

10. Sturtevant, J. (1977) Proc. Natl. Acad. Sci. U.S.A., 74, 2236.

11. Fisher, H.F., Stickel, D.C., Brown, A. and Cerretti, D.J., Amer. Chem. Soc. (In press).

12. Rossman, M.G., Liljans, A., Branden, C.I. and Banaszak, L.J. (1975) Enzymes, 3rd Ed. 11, Chapter 2.

13. Wooten, J. (1974) Nature, 252, 542.

14. Stellwagen, E. (1977) Accts. Chem. Res., 10, 92.

CONFORMATIONAL CHANGES IN DEHYDROGENASE REACTION MECHANISMS

JOSEPH D. SHORE, KAREN D. LUCAST AND WILLIAM R. LAWS
Edsel B. Ford Institute, 2799 West Grand Blvd., Detroit, Mich.,
48202

ABSTRACT

Horse liver alcohol dehydrogenase exists in two pH-linked conformational states, which can readily be observed by the quenching
of protein fluorescence. Collisional quenching studies with KI
demonstrated that 50% of the protein fluorescence is due to TRP-15,
at the surface of the molecule, with the remainder due to Trp-314,
a buried residue. The residue quenched at alkaline pH and by ternary complex formation with NAD^+ and Trifluoroethanol is Trp-15.
The alkaline quenching curve of LADH showed the same pK_a, 9.8,
when the bound water of the active center zinc atom was replaced
by imidazole, indicating that the water molecule was not the ionizing group linked to the conformational change. Replacement of
the water by imidazole, however, resulted in a 5-fold greater rate
constant for dissociation of NADH from the enzyme due to a more
rapid conversion of enzyme to the native conformation.

INTRODUCTION

Horse liver alcohol dehydrogenase, by virtue of the fortui-
tous property of pH-linked conformational changes measured by
the quenching of protein fluorescence, provides a useful system
for relating conformational changes to dehydrogenase reaction
mechanisms. Previous studies indicated that the enzyme exists in
two pH-linked conformational states with a pK_a of 9.8 and alkaline
quenching of 35% of the protein fluorescence (1). NAD^+ selectively
binds to the protonated form of the enzyme with subsequent perturbation of the pK_a to 7.6, while ethanol and trifluoroethanol bind
to the unprotonated form of the binary complex and convert it to
the maximally quenched alkaline conformation (1,2).

Liver alcohol dehydrogenase contains two tryptophan residues
per subunit, Trp-314 in a hydrophobic millieu at the center of the
molecule near the subunit interface, and Trp-15 at the surface
with one face of the indole ring exposed to solvent (3). Both
tryptophan residues are quite far from the binding site of the
nicotinamide ring of oxidized coenzyme, with distances of 17 and

27 Å for residues 314 and 15, respectively. It is probable that the fluorescence of LADH is almost totally due to these residues, with the quenching upon ternary complex formation resulting from a conformational change.

Imidazole, an inhibitor of the LADH reaction, has been studied extensively in the past (4,5). It abolishes the pH-dependence of NAD^+ binding, and causes a 7-fold weaker binding of NADH to the enzyme and a 5-fold increase in turnover number for ethanol oxidation. Recent crystallographic studies (6) indicate that it replaces the bound water molecule of the zinc atom at the active center of the enzyme, and that the ternary enzyme-NADH-imidazole complex forms orthorhombic crystals similar to apoenzyme, as compared to the triclinic crystals obtained with binary enzyme-NADH complex (7). This strongly implies that the conformational change due to coenzyme binding is prevented by removal of the zinc-bound water and replacement by imidazole.

The current studies were performed to determine which tryptophan residue reflects the conformational change, to ascertain whether the ionizing group linked to the conformational change is the water of hydration of the active center zinc, and to delineate the involvement of the water molecule in the conformational change related to the binding of NADH.

MATERIALS AND METHODS

Alcohol dehydrogenase was prepared from horse livers by methods previously described and its concentration was determined both by assay and by titration with NADH in the presence of isobutyramide (2). Coenzymes were purchased from Sigma Chemical Company and trifluoroethanol was a product of Aldrich Chemical Company. Fluorescence titrations were performed using Farrand spectrofluorimeter and coenzyme binding and dissociation rates were determined on a Durrum-Gibson stopped-flow spectrophotometer with a fluorescence attachment. Stern-Volmer plots were performed at a constant salt concentration of 0.5 M through the use of sodium chloride. The stock KI solution contained 10^{-4} M sodium thiosulfate to prevent formation of I_3^-. The reciprocal Stern-Volmer plot was based on the method of Lehrer (8), with the intercept providing the maximal protein fluorescence fraction quenched by a collisional process. The fluorescence levels for the NAD^+ titration were corrected for both dilution and for inner-filter effects by using the antilog

of the absorbance at 290 nm. Determination of the rate constant
for coenzyme dissociation was performed by using another dehydrogenase and its substrate as previously reported (9).

RESULTS

  The quenching of the protein fluorescence of LADH by KI was
substantial, and is expressed as a reciprocal Stern-Volmer plot in
Figure 1. From the extrapolated intercept of this plot, it can be
seen that approximately 50% of the protein fluorescence could be
quenched by a collisional process.
Since there are two tryptophan
residues per subunit, with only one,
Trp-15, exposed to solvent, it is
reasonable to assume that this
residue is contributing approximately 50% of the LADH protein
fluorescence.

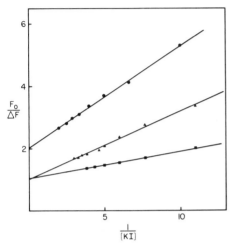

Figure 1. Reciprocal Stern-
Volmer plot of LADH protein
fluorescence quenching by KI.
●, 10 μN enzyme; ▲, 10 μN enzyme in 8 M urea; ■, 10 μM
tryptophan. Excitation 290 nm,
emission 335 nm, in 0.1 μ phosphate at pH 7.2 and 25° C.

  In order to determine which of the two tryptophan residues is
responsible for the 35% quench of protein fluorescence which occurs
when a ternary complex is formed, NAD$^+$ was added to enzyme in
the presence of trifluoroethanol. The results of these experiments
are presented in Figure 2. In the case of enzyme in the absence
of salt, a maximal quenching of protein fluorescence was observed
as previously reported (1). The sample containing 0.5 M NaCl also
showed the same final level but a weaker binding due to competition
of chloride for the binding site of the pyrophosphate group of coenzyme (3). At a concentration of 0.5 M KI, the protein fluorescence is already 40% quenched due to a collisional process affecting

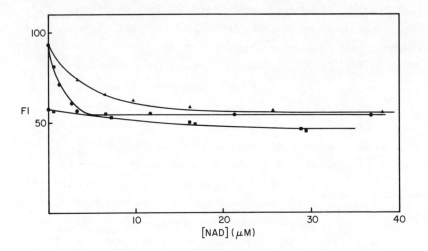

Figure 2. Quenching of LADH fluorescence by formation of a ternary complex with NAD⁺ and trifluoroethanol. LADH, 2.8 μN and trifluoroethanol, 2.7 mM in all experiments. ●, LADH in 0.1 μ phosphate, pH 7.2; ▲, in the presence of 0.5 M NaCl; ■, in the presence of 0.5 M KI. Excitation 290 nm, emission 335 nm.

Trp-15, and formation of the ternary complex caused less than 10% additional quenching. The results of this experiment indicate that the tryptophan residue at the surface of the LADH molecule, Trp-15, is the residue whose quantum yield is markedly diminished by ternary complex formation. Since this residue is 27 Å from the nicotinamide binding site on the same subunit, and 60 Å from the nicotinamide binding site of the adjacent subunit, the conformational change due to ternary complex formation is being transferred over a considerable distance. Thus, an event at the active center is causing substantial change in the millieu of the tryptophan residue at the enzyme surface. Whether this change results from further exposure of the residue to solvent, or interaction with adjacent charged functional groups, is still to be determined.

Since it has been established that imidazole replaces the water of hydration of the active center zinc atom (6), the effect of increasing pH on protein fluorescence of LADH was determined in the presence of saturating (10 mM) imidazole. The plotted results yielded a curve identical with that obtained previously for the alkaline quench in the absence of imidazole (1), with a pK_a of 9.8. This indicates that the ionizing group with an apparent pK_a of 9.8 linked to the conformational state of the enzyme is not the water molecule liganded to zinc. These results do not, however, preclude a role for the water in the conformational change resulting from the binding of oxidized or reduced coenzyme. It is feasible that the water molecule could be essential for conformational changes due to coenzyme binding without changing its state of ionization.

Recent studies, showing that the enzyme-NADH-imidazole complex crystallizes in the same form as apoenzyme (7), imply that the zinc-bound water is essential for the conformational change resulting from NADH binding. To determine the rate of this conformational change and relate it to the reaction mechanism, we have determined the rate constants for NADH binding and dissociation in the presence and absence of imidazole. The bimolecular rate constant for NADH binding was the same, $1 \times 10^7 \ M^{-1} \ sec^{-1}$, in the presence and absence of saturating imidazole. The rate constants for dissociation of NADH are substantially different in the presence and absence of imidazole, as demonstrated in Figure 3. Dissociation of

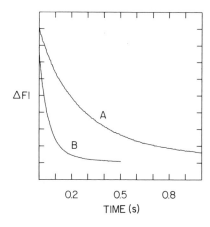

Figure 3. Rate constants for dissociation of NADH from binary and ternary complexes. Syringe 1, 185 μN Lactate dehydrogenase and 10 mM pyruvate in both experiments; Syringe 2, curve A, 40 μN LADH and 36 μM NADH; Syringe 2, curve B, 40 μN LADH, 36 μM NADH and 10 mM Imidazole. Excitation, 340 nm, emission, >410 nm, 0.1 μ phosphate, pH 7.0, 25° C.

NADH from the binary complex with enzyme occurred at 3 sec.$^{-1}$ while the rate constant for dissociation of NADH from the ternary LADH-

NADH-imidazole comples was 14.2 sec.$^{-1}$. These results correlated well with past reports that the dissociation constant for NADH was 7-fold higher in the presence of imidazole (4) and that the turnover number for ethanol oxidation, which is limited by NADH dissociation, was 5 times as fast (5). The identical binding rate constant in the presence of imidazole, with an enhanced dissociation rate, indicates that the conformational change occurs subsequent to the binding of the reduced nicotinamide ring which provides the fluorescent signal. The 3 sec.$^{-1}$ rate constant for binary complex dissociation is probably limited by the rate of the conformational change while the 14 sec.$^{-1}$ rate constant is an inherent value for dissociation from the ternary complex without the involvement of a conformational change. The previously reported increase in turnover number at maximum velocity in the presence of imidazole (5) to 14 sec.$^{-1}$ indicates that the replacement of zinc-bound water or aldehyde of the binary enzyme-NADH complex, and associated return of the enzyme to its native conformation, must be substantially more rapid than 14 sec.$^{-1}$.

SUMMARY

Protein fluorescence quenching studies indicate that 50% of the fluorescence of LADH is due to Trp-15, at the surface of the molecule, and that this residue is 70% quenched by formation of a ternary complex with NAD^{+} and trifluoroethanol. The mechanism by which a change at the active site results in alteration of the millieu of a residue 27 Å away awaits further structural and dynamic studies. Replacement of zinc-bound water at the active site with imidazole did not alter the alkaline protein quenching curve, indicating that the water molecule is not the ionizing group linked to the pH-dependent conformational states of LADH. The water is required, however, for the conformational change resulting from the binding of the nicotinamide ring of NADH since replacement of the water with imidazole prevents this change. This is manifested by a 5-fold increase in the rate constant for dissociation of NADH from the ternary complex as compared with binary complex. The 3 sec.$^{-1}$ rate constant for coenzyme dissociation and turnover number at maximum velocity in the absence of imidazole is thus limited by the rate of a conformational change returning the enzyme to its native state. Replacement of the water or aldehyde molecules liganded to the active site zinc results in an enhanced rate for this conformational change, substantially greater than 14 sec.$^{-1}$.

ACKNOWLEDGEMENT

This work was supported by National Science Foundation Grant PCM 74-04112. The technical help of Mr. Otto Urschel and secretarial assistance of Ms. B. Miles is gratefully acknowledged.

REFERENCES

1. Wolfe, J. K., Weidig, C. F., Halvorson, H. R., Shore, J. D., Parker, D. M. and Holbrook, J. J. (1977) J. Biol. Chem., 252, 433-436.

2. Shore, J. D., Gutfreund, H., Brooks, R. L., Santiago, D. and Santiago, P. (1974) Biochem., 13, 4185-4190.

3. Branden, C.-I., Jornvall, H., Eklund, H. and Furugren, B.(1975) in The Enzymes, Vol. XI, P. D. Boyer ed., Academic, N. Y., pp. 103-190.

4. Theorell, H. and McKinley-McKee, J. S. (1961) Acta Chem Scand., 15, 1811-1833.

5. Theorell, H. and McKinley-McKee, J. S. (1961) Acta Chem. Scand., 15, 1834-1865.

6. Boiwe, T. and Branden, C.-I. (1977) Eur. J. Biochem., 77, 173-179.

7. Branden, C.-I. (1977) in 2nd Int. Symposium on Pyridine Nucleotide-Dependent Dehydrogenases, Sund, H., ed., de Gruyter, Berlin, In Press.

8. Lehrer, S. S. (1971) Biochem., 10, 3254-3263.

9. Weidig, C. F., Halvorson, H. R. and Shore, J. D. (1977) Biochem., 16, 2916-2921.

IRON-SULFUR ENZYMES

HYDROGENASE OF *CLOSTRIDIUM PASTEURIANUM*

LEONARD E. MORTENSON

Department of Biological Sciences, Purdue University, West Lafayette,
Indiana 47907 (U.S.A.)

ABSTRACT

Hydrogenase of *Clostridium pasteurianum* catalyzes the reversible reaction $H_2 +$
Fd_{ox} $Fd_{red} + 2 H^+$. In this paper the properties of this enzyme are presented
that show this enzyme to be an iron sulfur protein with three tetrameric FeS centers.
The possible role of these three centers in catalysis by the enzyme and the regulation of hydrogenase activity by the E_h in the cell are discussed. Evidence is
presented that suggests two of the centers are concerned with the role of hydrogenase in maintaining the proper electron flow during metabolism whereas the
third center is concerned with catalysis.

INTRODUCTION

Purified hydrogenase (H_2ase) from *C. pasteurianum* has a mol. wt. of 60,500 as
determined by electrophoresis on polyacrylamide gel in the presence of sodium dodecyl sulfate (SDS). Since a similar value was obtained with the native enzyme by gel
filtration, this hydrogenase seems to lack dissociable subunits[1,2].

Hydrogenase prepared by our present procedure[3] has a specific activity of about
550 units per mg protein in the H_2 production assay (1 mM dithionite-reduced methyl
viologen at pH 8 in 50 mM Tris·Cl) and exhibits only one sharp protein band when
subjected to SDS-gel electrophoresis. It contains 12 gatoms each of iron and acid-
labile sulfide per mole[1,3]. Earlier results showed a lower sulfide content but
this problem was eliminated[4] when it was found that the *anaerobic* decomposition
product(s) of the dithionite that is routinely included in the buffers for H_2ase
purification strongly inhibited the formation of methylene blue from sulfide. This
problem was circumvented by using small volumes of concentrated H_2ase samples, and
by the use of higher Fe^{+++} in the reagent.

The iron and sulfide content of clostridial H_2ase preparations might be regarded
as unique because other H_2ase preparations have been reported to contain less iron
and sulfide. Recently however, a hydrogenase from a *Desulfovibrio gigas* (Le Gall
personal communication) has been found to contain twelve iron atoms. The clostridial H_2ase purified by another laboratory[2] from the same organism was reported to
contain 4.0-4.5 and 4.7-4.9 gatoms of iron and sulfide, respectively, whereas H_2ase
from *Chromatium* was reported to have been purified to homogeneity and to contain ~4
Fe and ~4 sulfide groups per ~100,000 daltons[5,6]. Preparations of H_2ase from *Desulfovibrio vulgaris* with iron contents of 1-8 gatoms have been described[7,8,9] but

recently Fe and sulfide contents of 7–9 gatoms were reported for H_2ase of *D. vulgaris*[10]. It will be interesting to compare the electron paramagnetic resonance spectra of these other H_2ases with the structure of their Fe·S centers determined by the recently developed ligand exchange reaction[11].

The Structure of the Iron–Sulfur Clusters

The Fe·S clusters of H_2ase with about 11.2 gatoms of Fe and sulfide each per 60,500 daltons have been extruded intact from hydrogenase by the ligand exchange method. The total Fe and sulfide of this H_2ase were recovered quantitatively in the form of the extruded product, $[Fe_4S_4(SR)_4]^{2-}$ where R = Ph. An average of 2.8 Fe·S tetramers for this H_2ase, containing 11.2 gatoms of Fe, were found. Control experiments carried out with spinach and bacterial Fd also showed that the dimeric centers of spinach Fd, $[Fe_2S_2(SR)_4]^{2-}$, did not react to form tetrameric centers $[Fe_4S_4(SR)_4]^{2-}$ under the extrusion conditions used. Based on these studies one can conclude that fully active H_2ase from *C. pasteurianum* contains three tetrameric iron sulfur centers. Earlier the Fe sulfur center of a partially purified H_2ase preparation was shown to be tetrameric but no quantitation was reported[2].

Optical Absorption Spectra

The UV–visible absorption spectrum of H_2–reduced H_2ase[1] increases in the 400 nm range when H_2ase is subjected to controlled oxidation by O_2 or methylene blue. The molar extinction coefficients of H_2–reduced and controlled O_2–oxidized H_2ase at 400 nm are 2.9×10^4 M^{-1} cm^{-1} and 3.6×10^4 M^{-1} cm^{-1}, respectively. Prolonged exposure of H_2ase to O_2 eventually inactivates it and it loses adsorption in the 300–500 nm range. These absorption spectra of H_2ase are of the bacterial Fd type which, of course, also suggests the presence of tetrameric Fe·S cluster(s) (this suggestion was confirmed by the extrusion method described above).

Carbon monoxide, a strong field ligand for iron atoms and a potent inhibitor of H_2ase was examined for its possible effect on the adsorption spectrum of H_2ase in the UV–visible region. Other than a slight general increase in absorption, no other distinct change was observed in the wave length region between 350 and 700 nm.

Circular Dichroism

In the visible region, oxidized clostridial Fd (23 and 53 µM) shows distinct CD bands at the following wave lengths and signs: 570 nm (+), 415 nm (+), and 370 nm (−). Reduction of Fd with dithionite abolishes or greatly diminishes the distinct CD bands between 400 and 600 nm, a result different from that obtained with the 8 Fe Fd from *Peptostreptococcus elsdenii*[12]. In contrast to Fd purified H_2ase shows weak optical activity in the visible region and distinct CD bands are observed only at protein concentrations higher than 50 µM (1 cm cell). Hydrogen reduced H_2ase whows the CD bands, 540 nm (−) 460 nm (+), 375 nm (−), 340 (−), ~300 nm (+) which do not change upon controlled oxidation with O_2. This clearly shows that H_2ase and the 8 Fe Fds are different although these two proteins show similar properties in optical absorption, EPR spectra and their tetrameric Fe·S structure. This difference in CD shows that the configuration of the Fe·S chromophores of H_2ase is unique and distinguishes hydrogenase from the more general electron transferring Fe·S proteins. CD bands resulting from the polypeptide of H_2ase (below 250 nm) show an helical content of about 10% with H_2-reduced H_2ase. A previous report on the CD of hydrogenase was in error because of a contaminating 60,000 dalton component[13].

Electron Paramagnetic Resonance Spectra of H_2ase at Different Redox States

A "g = 1.94" type EPR spectrum was reported with partially purified H_2-reduced, H_2ase of *C. pasteurianum*[14]. Oxidation of this H_2ase with oxygen resulted in a loss of this signal but incubation of the oxidized sample under H_2 restored it[15]. Purified H_2ase samples have been titrated with the oxidant, ferricyanide, and with the reductant dithionite and their EPR spectra measured[16]. This was performed in the presence of a mediator mixture of dyes so the redox potential (E_h) of the solution could be directly measured by a platinum electrode. H_2ase could not be maintained at an E_h more negative than -400 mV because active H_2 formation always took place when the E_h was more negative than -400 mV and the E_h which initially was much more negative than -400 mV, rapidly increased to -400 mV. The E_h was maintained more negative than -400 mV by a combination of different H_2 partial pressures and proton concentrations. Three EPR spectra of purified H_2ase were seen under the latter conditions. The spectrum in the presence of H_2 or $S_2O_4^-$ was similar to the spectrum previously reported for H_2 reduced, partially purified H_2ase[14], with the exception that some fine features, probably resulting from contaminants, were missing. A similar spectrum also was seen with a partially purified H_2ase from the same organism[2]. The same g = 1.94 type spectrum seen at -400 mV was also seen with H_2ase at E_h values more negative than -400 mV, and also with dithionite-reduced H_2ase in the presence or absence of a catalytic amount of either Fd or MV. But, when H_2ase was titrated with oxidant, the spectrum changed first into a complex family of weak signals at an E_h of about -360 mV. On further oxidation to -300 mV this spectrum changed into a distinct rhombic type spectrum. The rhombic spectrum of H_2ase at -330 mV was readily converted to the "1.94" type spectrum by reacting it with H_2 or dithionite in the presence or absence of catalytic amounts of either Fd or MV. Further oxidation of the form of H_2ase with the rhombic spectrum caused a reversible loss of the EPR signal.

The EPR spectrum of H_2ase under H_2 or in the presence of dithionite is much more complex than the spectra seen in simple $s = \frac{1}{2}$ paramagnetic centers and is similar to the EPR spectrum of the Fd of *Micrococcus lactilyticus*[17]. There is evidence that in the latter Fd there are two $s = \frac{1}{2}$ centers each arising from a tetrameric Fe·S cluster and that coupling most likely dipolar in origin occurs between its two tetrameric centers. Integration of spin intensity of clostridial hydrogenase, with spinach Fd and CuEDTA as standards, showed the presence of 1.66–1.80 spins per molecule for either H_2 or dithionite-reduced H_2ase. This suggests that two of the tetrameric Fe·S centers of H_2ase are in the trianionic state, $[Fe_4S_4(SR)_4]^{3-}$, when the enzyme is under H_2 or in the presence of dithionite.

The complex spectra at –360 mV show features of both the "reduced" species (under H_2) and the "oxidized" species (E_h more positive than –300 mV). In addition lines are seen that are caused by a second rhombic species with g-values of about 2.077, ~1.98, 1.908. This appears to be like a single-center Fd resonance and could come from those H_2ase molecules in which only one cluster is present as the trianion[16].

The EPR spectrum at – 330 mV (rhombic) is of the high potential FeS protein type since all its g-values are greater than 2.0 and even though the line shape is not typical of the high potential type, it has a well-defined set of three g-values. Integration of the rhombic spectrum gave a spin concentration of 0.69 spin/mole for H_2ase at E_h values from –154 mV to –330 which suggests that only one of the tetrameric Fe·S centers, probably the one that appears to be silent at –400 mV, is oxidized to the monoanionic state, $[Fe_4S_4(SR)_4]^{-1}$.

When H_2ase with the H_2-reduced EPR spectrum is incubated under carbon monoxide, a complex spectrum similar to that previously reported[2] is observed. This confirms that this potent inhibitor of H_2ase interacts with the Fe·S center(s) or with site(s) in the immediate vicinity of the Fe·S centers, again suggesting an involvement of tetrameric centers in catalysis by H_2ase.

Function of the Electron Carriers in the Production of H_2 by H_2ase

In anaerobic oxidative metabolism by *Clostridium*, excess reducing power is generated inside the cell and the electrons are transmitted via electron carriers, such as ferredoxin, to hydrogenase which reduces H^+ to H_2[18]. This poses the question, what role during the formation of H_2 by H_2ase does the electron carrier play? In the absence of any added electron carrier, purified H_2ase oxidizes $Na_2S_2O_4$ as measured by a decrease in $A_{315\ nm}$ of the dithionite solution and a dilution of 4x did not affect the specific rate of dithionite oxidation. It was demonstrated manometrically that H_2 was produced by the latter oxidation since the gas produced served as the electron donor for the H_2ase-catalyzed reduction of methylene blue in the side-arm of the Warburg flask used. The maximum rate of H_2 production from dithionite in the absence of an electron carrier by highly purified H_2ase was about 4 units/mg which is 10^{-3} times the V_{max} of H_2ase-catalyzed H_2 production with reduced Fd plus 1 mM MV. Dilution of the H_2ase did not affect the specific rate (~4 units/mg) of this H_2-producing activity which suggests that a dissociable carrier was not complexed to the hydrogenase. This is in contrast to the response shown for maximum H_2ase activity (H_2 production) where reduced Fd, with an apparent K_m of 49.5 μM, was required[1,18,19]. Hydrogenase produces H_2 from reduced Fd or reduced MV according to Michaelis-Menten type kinetics, and in this reaction apparent K_m values of 49.5 μM and 6.25 mM were estimated for reduced Fd and MV, respectively[1]. Although the K_m for reduced Fd is about 1% of the K_m for reduced MV, the maximal rate of H_2 production from dithionite-reduced Fd, $[V_{max(Fd)}]$, is only about 40% of that from dithionite-reduced MV, $[V_{max(MV)}]$. Addition of 1 mM MV to a reaction mixture containing H_2ase, Fd, and dithionite greatly enhances the rate of H_2 production and the $[V_{max(Fd + 1 \text{ mM MV})}]$ is about four times higher than $[V_{max(Fd)}]$. At a fixed level of Fd, (18.5 and 157 μM were used), the enhancing effect of MV decreased rapidly when the concentration of MV was increased and the enhancement by MV became insignificant when the MV concentration was increase to 4 mM. If the reduction of *free* Fd by dithionite is rate-limiting, then the initiation of H_2 production by H_2ase should be accompanied by an accumulation of oxidized Fd and an increase in $A_{400 \text{ nm}}$ should result [at 400 nm (A oxidized Fd)/(A reduced Fd = 2.21]. Since no significant change in $A_{400 \text{ nm}}$ was observed when H_2ase was added to this dithionite-reduced Fd solution and since H_2 production proceeded linearly with time during the same period, the reduction of Fd by dithionite appeared not to be rate limiting.

Several explanations for this affect of MV on hydrogenase activity can be proposed. First when dithionite-reduced Fd serves as the sole electron carrier for the H_2ase-catalyzed production of H_2, the reaction is limited by the rate by which reduced Fd can donate electrons to H_2ase. This rate limiting step could be the rate of the dissociation of oxidized Fd from H_2ase and the association of another reduced Fd with H_2ase or the rate at which electrons are exchanged between exogenous reduced Fd and the bound oxidized Fd. In the presence of about 1 mM MV and excess dithionite, H_2ase-bound oxidized Fd could be reduced rapidly thereby enhancing the rate of H_2 production. At higher MV concentrations, one would have to propose that reduced MV competes with reduced Fd for sites on H_2ase and that the enhancement is decreased.

A second interpretation for the enhancement of H_2 evolution by low concentrations of MV is that a more negative redox potential is maintained in the presence of MV. In fact the E_h of the standard H_2 production assay mixture containing 1 mM MV was measured and found to be -550 mV. Since the E_h of the H_2ase-reduced Fd system is more positive than -550 mV, the inclusion of reduced MV into the latter system (with dithionite) would bring the E_h to the level of the MV/reduced MV half cell. By maintaining the Fd half cell at a more negative redox potential, all Fd would be reduced and a more favorable driving force or a lowered energy barrier for H_2 production would result.

The major function of Fd, the electron donor for hydrogenase is to pass electrons to H_2ase when the redox potential inside the cell (reflected by the ratio of reduced Fd to oxidized Fd) reaches a level which limits the rate of the energy-yielding oxidative metabolism because of a lack of reducible electron carriers. The key enzyme in glucose catalysis, pyruvate:Fd oxidoreductase, is coupled to the reduction of oxidized Fd[20,21]. Although other reductive biosynthetic reactions require reduced Fd, the amount of reduced Fd needed is much less than the amount produced during oxidative metabolism[22]. Thus it appears that the H_2ase of *C. pasteurianum* functions to adjust the intracellular redox potential so that a constant ratio of oxidized to reduced Fd is maintained. Dihydrogen production by H_2ase at a high rate occurs only at redox potentials more negative than -400 mV. Because the midpoint potential (E_o') of Fd half cell is -410 mV at pH 7.0, at this pH H_2ase maintains nearly equal concentrations of oxidized and reduced Fd to satisfy both the oxidative and the reductive metabolic reactions. Accordingly, H_2ase will rarely function at its maximal turnover rate. The data also suggest that once H_2ase is reduced by reduced Fd the enzyme reduces protons without the need of a bound carrier.

This discussion assumes that no compartmental arrangement is required inside the cell for the coupled reactions of H_2ase, Fd and pyruvate:Fd oxidoreductase. The E_h for the best balance of the hydrogenase-Fd system may not be ideal for the function of the cells NADH:Fd oxidoreductase, which thermodynamics predict should function best at an E_h more positive than -400 mV so a proper ratio of NAD^+ to NADH can be maintained. This suggests that a mechanism, perhaps some sort of energy-dependent compartmentalization of cellular contents or coupling to 3-phosphoglyceraldehyde dehydrogenase, the major enzyme producing NADH during glucose catalysis, may be necessary so that reactions functioning best at different E_h environments can proceed in the required directions and at the required rates.

Inhibitors of H_2ase
_____

Inhibition of H_2ase activity in whole cells and cell-free extracts by metal ligands such as carbon monoxide and cyanide has been observed and these inhibitions were used to support the contention that H_2ase is a metalloenzyme[18]. In work with purified H_2ase from *C. pasteurianum*, the effect of CO, CN^-, F^-, and N_3^-, all potential ligands for the iron atoms, has been examined. It is concluded that at 50 mM, F^- and N_3^- do not inhibit H_2 production catalyzed by H_2ase with dithionite-reduced methyl viologen as the electron donor.

Carbon monoxide, similar to results with cell-free preparations, is a potent inhibitor of purified H_2ase. At 6×10^{-4} atm H_2 evolution was inhibited 50% when H_2ase was assayed with dithionite as the reductant in the absence of any added electron carrier. The oxidation of H_2 by purified H_2ase was also inhibited by CO when either MV, BV, or methylene blue was the electron acceptor. The inhibition by CO was reversed when the CO was removed from the hydrogenase by evacuating and flushing several times with an inert gas. With purified H_2ase, cyanide was found to cause both inhibition and inactivation of H_2ase. Inhibition of H_2ase by CN^- was observed when the enzyme was assayed in the direction of H_2 production using dithionite-reduced MV as the electron carrier under a H_2 atmosphere.

ACKNOWLEDGEMENTS

The research from our laboratory reported here was supported by Grant PCM 75-04336 from the National Science Foundation and Grant AI 04865 from the National Institute of Health.

REFERENCES

1. Chen, J.S. and Mortenson, L.E. (1974) Biochim. Biophys. Acta, 371, 283-298.
2. Erbes, L., Burris, R.H. and Orme-Johnson, W.H. (1975) Proc. Natl. Acad. Sci. USA, 72, 4795-4799.
3. Mortenson, L. In "Methods in Enzymology" (in press 1978).
4. Chen, J.S. and Mortenson, L.E. (1977) Anal. Biochem., 79, 157-165.
5. Gitliz, P.H. and Krasna, A.I. (1975) Biochemistry 14, 2561-2568.
6. Kakuno, T., Kaplan, N.O. and Kamen, M.D. (1977) Proc. Natl. Acad. Sci. USA, 74, 861-863.
7. Haschke, R.H. and Campbell, L.L. (1971) J. Bacteriol., 105, 249-258.
8. Le Gall, J., Dervartanian, D.V., Spilka, E., Lee, J.P. and Peck, H.O. Jr. (1971) Biochim. Biophys. Acta, 234, 525-
9. Yagi, T. (1970) J. Biochem (Tokyo), 68, 649-657.
10. Yagi, T., Kimura, K., Daidoji, H., Sakai, F., Tamura, S. and Inokuchi, H. (1976) J. Biochem (Tokyo), 79, 661-671.
11. Gillum, W.O., Mortenson, L.E., Chen, J.S. and Holm, R.H. (1977) J. Amer. Chem. Soc., 99, 584-595.
12. Atherton, N.M., Garbett, K., Gillard, R. D., Mason, R., Mayhew, S.J., Peel, J.L. and Stangroom, J.E. (1966) Nature, 212, 590-593.
13. Multani, J.S. and Mortenson, L.E. (1972) Biochim. Biophys. Acta, 256, 66-70.
14. Nakos, G. and Mortenosn, L.E. (1971) Biochim. Biophys. Acta, 227, 576-583.
15. Mortenson, L.E. and Nakos, G. (1973) In "Iron-Sulfur Proteins" (W.M. Lovenberg, ed.), Academic Press, N.Y., pp. 37-64.
16. Chen, J.S., Mortenson, L.E. and Palmer, G. (1976) In "Iron and Copper Proteins", (Yasunobu, K.T., Mower, H.F. and Hayaishi, O., eds.), Plenum Pub. Corp., N.Y., pp. 68-82.
117. Mathews, R., Charlton, S., Sands, R.H. and Palmer, G. (1974) J. Biol. Chem., 249, 4326-4328.
18. Mortenson, L.E. and Chen, J.S. (1974) In "Microbial Iron Metabolism" (J.B. Neilands, ed.), Academic Press, Inc., N.Y., pp. 231-276.
19. Mortenson, L.E. and Chen, J.S. (1975) In "Microbial Production and Utilization of Gases (H_2, CH_4, CO)". Akademie der Wissenschaften zu Göttingen, pp. 97-108.
20. Mortenson, L.E., Valentine, R.C., and Carnahan, J.E. (1963) J. Biol. Chem., 238, 794-800.
21. Uyeda, K. and Rabinowitz, J.D. (1971) J. Biol. Chem., 246, 3111-3119.
22. Mortenson, L.E. (1968) In "Survey of Progress in Chemistry" (A.R. Scott, ed.), Academic Press, New York, Vol. 4, pp. 127-163.

REACTION MECHANISM OF TRIMETHYLAMINE DEHYDROGENASE

DANIEL J. STEENKAMP, HELMUT BEINERT, WILLIAM McINTIRE, AND THOMAS P. SINGER

Department of Biochemistry and Biophysics, University of California, San Francisco, Molecular Biology Division, Veterans Administration Hospital, San Francisco, CA. and Institute for Enzyme Research, University of Wisconsin, Madison, WISC., (USA).

ABSTRACT

Trimethylamine dehydrogenase, a bacterial enzyme which catalyzes the oxidative demethylation of trimethylamine, contains a novel flavocoenzyme, 6-cysteinyl-FMN, and a single Fe-S cluster as redox active groups. Reduction of the enzyme with dithionite gives rise to a typical ferredoxin-type reduced iron-sulfur EPR signal at g=1.94, while reduction with excess substrate results in a complex pattern of EPR signals which is attributed to the interaction of two paramagnetic species. This "spin-coupled" form of the enzyme is characterized by a three-banded absorbance spectrum with maxima at 365, 435, and 510 nm. Spectrophotometric titration data indicate that a substantial excess of substrate is required to obtain the spin-coupled form at neutral pH, although maximal bleaching of the 443 nm absorption band, attributable to reduction of the flavin moiety, is accomplished by addition of one mole or less of substrate at all pH values. Further experiments by the stopped-flow method indicated that both the extent and rate of appearance of the spin-coupled form of the enzyme is regulated by the binding of substrate or non-oxidizable substrate analogues such as tetramethylamonium chloride.

INTRODUCTION

Trimethylamine dehydrogenase (TMADH; EC 1.5.99.7), an enzyme from facultative methylotrophic bacteria, catalyzes the oxidative demethylation of trimethylamine to dimethylamine and formaldehyde in the presence of phenazine methosulfate (PMS) as electron acceptor[1]

$$(CH_3)_3N \ + \ H_2O + PMS \rightleftharpoons (CH_3)_2NH + CH_2O + PMSH_2$$

TMADH from bacterium W3A1 has a molecular weight of 147,000 daltons and is comprised of two subunits of 70,000-80,000 daltons molecular weight, as judged by SDS-gel electrophoresis[2,3] Evidence was recently obtained for the presence of an unusual covalently bound flavin, 6-cysteinyl-FMN, at the active site[4], as well as of a single Fe-S cluster containing the integral core unit, Fe_4S_4, in the enzyme[5,6]. Pronounced differences have been observed between dithionite and substrate reduced TMADH, both by EPR and optical spectroscopy. The EPR-spectrum of the substrate reduced enzyme was of particular interest since it showed a complex pattern of signals indicative of spin-coupling between two paramagnetic species, presumably a flavosemiquinone and a reduced Fe-S center, which must be in close proximity in the interacting form of the enzyme[6].

Steady state kinetic data on the reaction mechanism of TMADH were not readily interpretable and it was necessary to propose conformational isomeration of both stable forms of the enzyme in a double displacement mechanism to account for the observed product inhibition pattern[2]. Moreover, substrate inhibition by both PMS and the alternate substrate, diethylamine, suggested that both these substrates can combine with the wrong stable form of the enzyme.

This paper summarizes recent equilibrium binding and stopped-flow kinetic studies on the reaction mechanism of TMADH. Two distinct intermediates have been identified during the reductive phase of the catalytic cycle. Initial reduction of the flavin moiety by the substrate is very fast, while the formation of a second, spin-coupled species is a slower reaction. The rate and extent of appearance of this spin-coupled form seems to be regulated by trimethylamine and other amines capable of combining with the reduced enzyme. These studies also imply a conformational flexibility in the enzyme which may be of importance in catalysis.

METHODS

Trimethylamine dehydrogenase was purified from bacterium W3A1[2]. The enzyme was stored in 20% ethylene glycol at -20°C and was transferred into 0.1 M sodium pyrophosphate buffer, pH 7.7 either by dialysis or gel chromatography on Sephadex G-25 prior to EPR or stopped-flow experiments.

For comparison with EPR and stopped-flow analyses the catalytic center activity of the enzyme was obtained from double reciprocal plots of steady state kinetic assays at a fixed saturating concentration of trimethylamine (3.33 mM) and variable PMS concentrations.

Stopped-flow kinetic measurements, absorption , and EPR spectra were measured and evaluated as described previously[7,8,9]. Details are given in the figure legends. Solutions of volatile amines used in anaerobic titrations of the enzyme were made anaerobic under an atmosphere of argon by incubation with 5.6 pmoles/ml of glucose oxidase, 10 μmoles/ml of glucose, 10 μmoles/ml of sodium acetate, pH 5.2 and sufficient catalase to prevent accumulation of peroxide.

The amino acyl coenzyme of TMADH and apoflavodoxin was prepared as described[10,11] and binding of the latter to the amino acyl coenzyme was monitored by optical spectroscopy.

RESULTS AND DISCUSSION

The visible absorption of TMADH, as shown in Fig. 1a and 1b is bleached maximally by the anaerobic addition of about 1.5 moles dithionite per mole enzyme, in agreement with the presence of one flavin moiety and an iron-sulfur center in the enzyme. The final featureless spectrum obtained at the end of the dithionite titration differed appreciably from the spectrum of the reduced enzyme

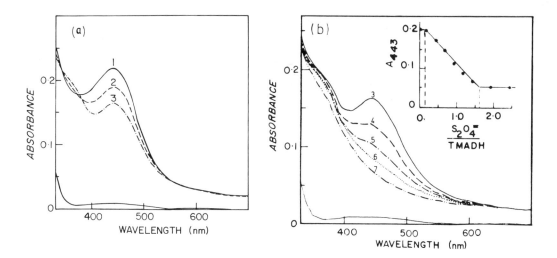

Fig. 1. Anaerobic dithionite titration of trimethylamine dehydrogenase: (a) Quasi-isosbestic course of titration. The curves shown are 1, oxidized enzyme, 2 and 3 after the addition of 0.42 and 0.66 molar equivalents of dithionite respectively. (b) Further non-isosbestic titration. Curves 3,4,5,6 and 7 represent the addition of 0.66, 0.93, 1.17, 1.44 and 1.93 molar equivalents of dithionite. The buffer was 0.1 M Na-pyrophosphate. The dithionite solution was standardized against riboflavin.

Fig. 2. Comparison between the spectral properties of trimethylamine dehydrogenase reduced with an excess of substrate and that obtained at the end of dithionite titration such as in Figure 1.

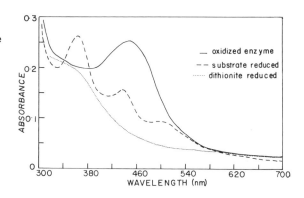

species obtained after the addition of an excess of trimethylamine (Figure 2).

The difference between substrate reduced and dithionite reduced trimethylamine dehydrogenase is even more dramatic when examined by EPR spectroscopy (Figure 3). When the enzyme is reduced anaerobically by an excess of dithionite, a rhombic signal (Fig. 3 bottom curve) with g values of 2.035, 1.925, and 1.85, typical of a reduced Fe-S protein, is observed. When the enzyme is reacted with trimethylamine, however, a different, complicated set of signals appears (Fig. 3, top and center curves) which is seen at 34 GHz (top curve, Fig. 3) and 9.2 GHz (center curve, Fig. 3). By integration this set of signals accounts for 1.5 electron equivalents per mole of enzyme. The relationship of the principal lines at these two frequencies is indicated in the figure by broken lines. The center spectrum at 9.2 GHz was recorded at 20 times the amplification of the bottom

Fig. 3. EPR spectra of trimethylamine dehydrogenase reduced with an excess of trimethylamine (top and center curve) or dithionite (bottom curve). Enzyme, 32 μM, in 0.1 M pyrophosphate buffer, pH 7.7, was mixed anaerobically with trimethylamine hydrochloride (final concentration 5 mM) for the center spectrum and at approximately 10 times the enzyme concentration for the top spectrum and was frozen approximately one minute after mixing. The bottom spectrum was obtained after addition of a 10-fold excess of dithionite to 32 μM enzyme. The conditions of EPR spectroscopy were: top spectrum, 34 GHz, 3 mwatt, 5 G modulation amplitude and 13°K; center and bottom spectrum, 9.2 GHz, 2.7 mwatt, 8 G modulation amplitude and 13°K. The amplification of the center spectrum was 20 times that of the bottom spectrum. The corresponding resonances at 34 and 9.2 GHz are connected by broken lines, to show the large spread of the (sharp) lines of the Fe-S center, as compared to the broad lines, indicating interaction. Two new, sharp lines in the center in the top and

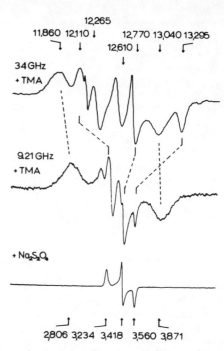

center spectra, which are not seen in the bottom spectrum (reduced with dithionite), are also due to interaction. The numbers next to the arrows designate magnetic field strength in gauss, at the indicated frequencies.

spectrum for equivalent amounts of enzyme. Comparison of the intensity of the line at 3,560 G, in the center and bottom curves, which differ by a factor of 20 in amplification, reveals that only about 5% of the Fe-S cluster of the former yields the same signal (i.e., is in the same state) as the dithionite-reduced enzyme. The substrate-reduced form, similarly, shows only weak free radical signals (e.g., the sharp line in the top spectrum between 12,110 and 12,265 G). As indicated by the broken lines, the resonances of the reduced cluster are spread out at 34 GHz, as expected for the anisotropic features of a paramagnetic species. On comparing the 34 GHz and 9.2 GHz spectra, however, it may also be seen that other lines do not spread out to the same extent [cf. the broad ones at 2,800 and 3,871 (9.2 GHz) and the two more narrow lines in the center]. The failure of these lines to spread out on increasing the field strength is indicative of spin-spin interaction. The spin-spin interaction, moreover, manifests itself in an unusually strong half-field signal around $g = 4$ (Figure 4). The most likely explanation of these observations is formation in the enzymatic reaction of a flavin semiquinone and a reduced Fe-S cluster, which are sufficiently close to interact, as indicated by the EPR spectra.

During anaerobic titration with dithionite the principal signals emerging are a free radical signal (flavin semiquinone) and the signal of the reduced Fe-S center, with only weak signals of the kind observed with trimethylamine as reductant superimposed (not shown). This may be interpreted to indicate that the two paramagnetic sites are not in the same favorable position for interaction when

Fig. 4. Half field signal of trimethylamine dehydrogenase after reaction with
trimethylamine. Only the signal at
g = 4.3 is present before addition of the
substrate. The sample preparation and
conditions of spectroscopy at 9.2 GHz
are those of Figure 3. The numbers
represent g values.

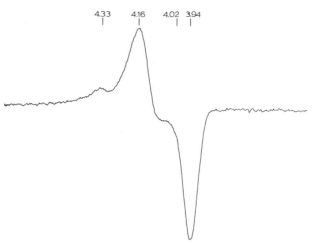

dithionite is the reductant, as they are in the presence of trimethylamine. This would be com-

patible with a substrate induced conformation change in the enzyme.

When TMADH was titrated with trimethylamine under anaerobic conditions at pH 7.7 the 443 nm

absorption band of the enzyme was maximally bleached by the addition of a stoichiometric amount

of trimethylamine. However, a considerable excess of substrate was required to maximize the

appearance of the spin-coupled form of the enzyme (Figure 5). This enhancement of the g = 4

Fig. 5. Titration of trimethylamine
dehydrogenase with trimethylamine,
followed by spectrophotometry and
EPR spectroscopy. Individual samples
of 50 μM enzyme in 0.1 M NaPP$_i$,
pH 7.7 were reduced anaerobically
with different amounts of trimethylamine. Absorption spectra were
recorded at 20°C. After 20 min
the samples were frozen and EPR
spectra recorded at 30°K. Absorbance at 445 nm and 365 nm
and development of the half-field
signal at g = 4.0, characteristic
of spin-spin interaction, as a percentage of maximal signal observed
with an excess of trimethylamine,

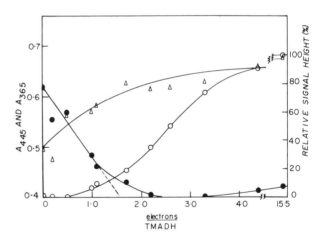

are plotted as functions of a number of reducing equivalents added as trimethylamine per mole of
the enzyme (1 mole of trimethylamine is assumed equivalent to 2 e$^-$). With excess trimethylamine
the number of e$^-$ recovered by double integration of the signal centered at g = 2.0, characteristic
of spin-spin interaction, was 1.4 e$^-$/mole of enzyme.

half field interaction signal by the addition of excess substrate shown in Figure 5 was accompanied

by relatively minor spectral changes, notably an increase in absorbance at 365 and 510 nm (see

also Figure 9).

Further investigation showed that the transfer of electrons from the reduced flavin moiety to the Fe–S center is pH-dependent, as might have been expected for a reaction in which a proton is liberated. A qualitative insight into this aspect of TMADH was obtained by the anaerobic titration of the enzyme with trimethylamine at three different pH values, i.e., at 7.0, 7.7, and 8.5. In order to facilitate interpretation of the data a complication resulting from the release of dimethyl-amine during the reduction of TMADH by trimethylamine should be emphasized at this stage. The reduction of TMADH at pH 8.5 by trimethylamine using sub-stoichiometric amounts of substrate is biphasic (Figure 6). If sufficient time was allowed for the slow phase to reach completion, maxi-

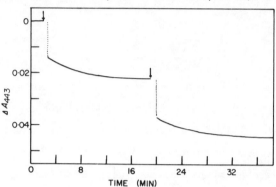

Fig. 6. Biphasic reduction of trimethyl-amine dehydrogenase by sub-stoichio-metric amounts of trimethylamine under anaerobic conditions. The arrows indi-cate the addition of 2 nmole quantities of trimethylamine to 30 nmoles of the enzyme in 0.1 M Na-pyrophosphate, pH 8.5 at room temperature.

mal bleaching of 443 nm absorption band of the enzyme was achieved at this pH by the addition of 0.5 mole of substrate/mole of enzyme. This could readily be accounted for by release of dimethyl-amine from the reduced form of the enzyme, as predicted from steady state kinetics[2], since di-methylamine, apart from being a product, is also a slow substrate. The anaerobic titration of TMADH with dimethylamine is spectrally indistinguishable from that obtained with trimethylamine (Figure 7). The rate of reduction of the enzyme is much slower, however, and maximal bleaching

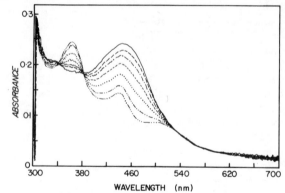

Fig. 7. Anaerobic titration of trimethyl-amine dehydrogenase with dimethylamine at pH 8.5 in 0.1 M Na-pyrophosphate. Spectra were recorded after the addition of 0 (——), 0.1 (– ––), 0.2 (–·–·), 0.4 (- - - -), 0.6 (·····), 0.9 (–··–··) and 4.15 moles dimethylamine per mole of enzyme.

of the absorption band at 443 nm requires a mole of dimethylamine/mole of TMADH (Figure 8). The rate of reaction of dimethylamine with TMADH is considerably slower at pH 7.7 than at 8.5 and be-comes imperceptible at pH 7.0. This pH-effect on the rate of reaction of the enzyme with dimethyl-amine most probably accounts for a previously reported discrepancy in the literature regarding the

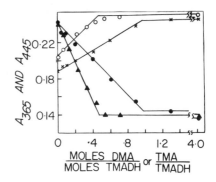

Fig. 8. Replot of the data from the anaerobic titration shown in Figure 7. The curves represent absorbance values at 365 nm (x—x) and 445 nm (●—●) as a function of the dimethylamine (DMA) to enzyme ratio. Also shown are the absorbance values at 365 nm (o—o) and 445 nm (▲—▲) in an anaerobic titration of the enzyme with trimethylamine (TMA) at the same pH. For purposes of comparison the data in the two experiments were normalized to the same initial absorbance at 445 nm.

utilization of dimethylamine as a substrate by the enzyme[1,2].

Further evidence that the enzyme releases both dimethylamine and formaldehyde during the reductive half reaction came from reduction of TMADH with ^{14}C-TMA, and separation of the products on Sephadex G-25 under anaerobic conditions. When the reaction and chromatography were carried out at pH 7.7, the reduced enzyme contained in the effluent had only about 1/10th of the label required if one of the methyl groups from trimethylamine had remained enzyme bound, whereas after reaction at pH 7.0 no label was associated with the enzyme protein.

Reduction of the flavin moiety and generation of the spin-coupled form during anaerobic titration of TMADH with trimethylamine was most clearly separated into two distinct phases at pH 7.0, where only 26% of the maximal EPR signal height at g = 4 was obtained with a stoichiometric amount of substrate. Such a spectrophotometric titration is shown in Figures 9a and 9b. A ∿ 10-fold excess

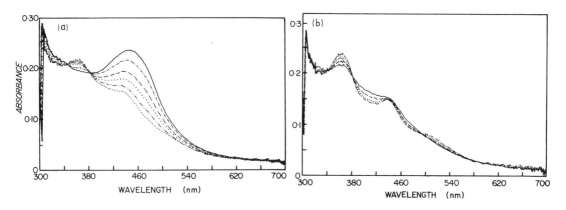

Fig. 9. Anaerobic titration of trimethylamine dehydrogenase with trimethylamine at pH 7.0 in 0.1 M K-phosphate. (a) Titration of the enzyme with a stoichiometric amount of substrate. Spectra were recorded after the addition of 0 (——), 0.254 (-·-·), 0.355 (– ––), 0.51 (····), 0.66 (—··—··) and 1.117 moles trimethylamine per mole of enzyme. (b) Further titration with excess trimethylamine. The spectra shown were recorded after the addition of 1.117 (——), 2.13 (– ––), 8.22 (····) and 20.40 moles of trimethylamine per mole of enzyme.

of trimethylamine was required at pH 7.0 to maximize the absorbance change at 365 nm, indicative of the spin-coupled form (Figure 10). At pH 8.5 on the other hand, maximal absorbance change at

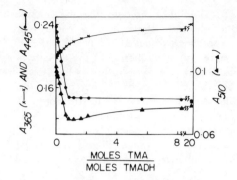

Fig. 10. Replot of the data from the experiment shown in Figure 9. The curves represent absorbance values at 365 nm (x—x), 445 nm (•—•) and 510 nm (▲—▲) as functions of the trimethylamine (TMA) to enzyme ratio.

365 nm is observed upon addition of somewhat less than a stoichiometric amount of trimethylamine per mole of TMADH (Figure 8). Moreover, the reversal of the absorbance changes at 510 nm, which characterizes the titration at the lower pH, is almost imperceptible at pH 8.5 and 64% of the maximal EPR signal height at g = 4 was obtained in an independent experiment when adding a stoichiometric amount of trimethylamine to the enzyme. The spectrophotometric titration data obtained at pH 7.7 presented a situation intermediate between those observed at pH 7.0 and 8.5, 38% of the maximal EPR interaction signal being generated by addition of one mole of trimethylamine.

Since the spin-coupled form of the enzyme was also detected by EPR spectroscopy after reduction of the enzyme with alternate substrates, such as dimethylpropynylamine and diethylamine, the question arose whether amines which are not oxidized by the enzyme but nonetheless combine with it, would also be effective in enhancing the appearance of the spin-coupled form by binding to the substrate reduced enzyme. The enzyme was therefore reduced at pH 7.7 with 0.85 moles of trimethylamine. Subsequent addition of tetramethylammonium chloride, a quaternary amine which protects the enzyme against inhibition by monoamine oxidase type 'suicide' inhibitors[1] but is not a substrate, resulted in spectral changes typical of formation of the spin-coupled form (Figure 11).

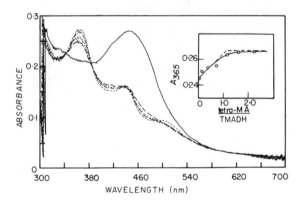

Fig. 11. Effect of tetramethylammonium chloride on substrate-reduced trimethylamine dehydrogenase. The absorbance of the enzyme at 443 nm was maximally bleached by the addition of 0.84 moles of trimethylamine. Subsequent addition of tetramethylammonium chloride resulted in the spectral changes shown. The curves represent the spectrum of the oxidized enzyme (———), after the addition of 0.84 moles of trimethylamine (– – –), and after the addition of 0.35 (· · · ·), 0.70 (–·–·–) and 1.59 moles of tetramethylammonium chloride per mole of enzyme. The inset shows the progress of the titration of reduced enzyme with tetramethylammonium chloride.

Since maximal absorbance change at 365 nm was obtained by using less than 2 moles of tetramethylammonium chloride per mole of enzyme, it is clear that the appearance of the spin-coupled form is modulated by combination of the amine with a single, unique binding site.

The enhancement of the spin-coupled form of the enzyme by tetramethylammonium chloride could also be observed by EPR spectroscopy. Thus the addition of a 90-fold molar excess to a sample of the enzyme previously reduced with a stoichiometric amount of trimethylamine enhanced the $g = 4$ signal height from 38.5% to 90% of the maximal obtained in the presence of excess trimethylamine.

One further observation in particular emphasized the conformational flexibility of the enzyme in response to occupancy of the binding site which modulates appearance of the spin-coupled form. Whereas the addition of 90 moles of tetramethylammonium chloride per mole of enzyme resulted in the appearance of no detectable interaction signal, subsequent reduction of the enzyme with dithionite resulted in the development not of the rhombic signal centered at $g = 1.94$, typical of *B. polymyxa* ferredoxin   but of the interaction signal usually observed after reduction of TMADH with substrate. Anaerobic spectrophotometric titration of the enzyme with dithionite in the presence of tetramethylammonium chloride, moreover, showed that the enzyme now consumed only one mole, as opposed to 1.5 moles of dithionite for complete reduction of the redox active groups in the enzyme (Figure 12). This observation implies either that the electrode potential for the half reaction

$$FlH^{\cdot} + e^- + H^+ \rightleftharpoons FlH_2$$

has been considerably altered by the binding of tetramethylammonium chloride or, alternatively, that the flavin moiety in the spin-coupled form is inaccessible to dithionite.

Fig. 12. Anaerobic titration of trimethyl-amine dehydrogenase with dithionite in the presence of tetramethylammonium chloride. The curves represent absorption spectra of the oxidized enzyme and after the addition of 10 moles tetramethylammonium chloride (o···o) followed by 0.323 (– – –), 0.648 (–·–·), 0.968 (····) and 1.36 moles of dithionite per mole of enzyme. The inset shows the progress of the titration at 443 and 365 nm.

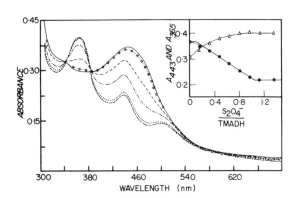

Stopped-flow kinetics of TMADH

The data reported above cannot be relied upon to assess whether the spin-coupled form of the enzyme is an obligatory intermediate in the catalytic turnover of the enzyme with PMS as electron acceptor. It was, therefore, of interest to determine the rate of formation of the spin-coupled form of the enzyme by stopped-flow kinetics, since the spectrophotometric titration data suggested that the appearance of the spin-coupled form may be associated with absorbance increases at 510 nm and 365 nm. As might have been expected in the case of an enzyme containing more than one chromo-phoric group, spectral changes observed in the stopped-flow apparatus were complex and suggested the formation and decay of several intermediates.

Bleaching of the absorption band at 445 nm by trimethylamine was extremely rapid and depended on substrate concentration[6]. Because of minor, slow subsequent absorbance changes the pseudo-first order nature of this reaction could not be conveniently established beyond reasonable doubt. However, when analyzed by the Guggenheim procedure at least 80 percent of the absorbance change associated with the fast bleaching yielded linear semilogarithmic plots. The data were analyzed in terms of the partial reaction sequence, assuming first a rapid equilibrium for the sub-

$$E_{ox} + S \underset{k_2}{\overset{k_1}{\rightleftharpoons}} ES_{ox} \xrightarrow{k_3} F$$

strate binding step and, secondly, that this step does not entail any significant spectral change. The latter supposition is based on the fact that about 90% of the absorbance difference at 450 nm between the oxidized and substrate-reduced enzyme was associated with the rapid reaction and also on the observation that a substrate analog, such as tetramethylammonium chloride does not alter the spectrum of the oxidized enzyme significantly, although the protection of the enzyme against 'suicide' inhibitors suggest that it binds at the active site. Double reciprocal plots of the dependence of the observed kinetic constants on substrate concentration are shown in Figure 13 . The near zero Y-axis intercepts of these plots do not allow for an accurate estimation of the rate constant k_3 and, moreover, suggests that the assumption of rapid equilibrium in the substrate binding step usually employed in the analysis of such data may be valid in this case. Since the catalytic center activities of TMADH at pH 7.0 and 7.7 and 10°C are only 33.2 and 58.8 per min., predicting first-order rate constants of at least 0.554 and 0.98 sec^{-1}, respectively, for the rate-limiting step in catalysis, the bleaching of the enzyme by substrate is obviously much too fast to be rate-limiting. Figure 13 also clearly shows that the rate decreases as the pH is lowered. In fact, at pH 8.5 the reaction became too fast to be accurately measured under pseudo first order conditions.

The rapid initial bleaching of the enzyme by substrate was followed by at least two slower reactions which, fortunately, seemed to show relatively little spectral overlap. The absorption spectral changes of these reactions, therefore, could conveniently be extrapolated back to zero time at several different wavelengths, in order to determine the spectrum of the rapidly formed intermediate, as described[12,13]. Comparison of the difference spectrum between the latter and the oxidized enzyme with the difference spectrum between the oxidized and reduced amino acyl coenzyme after binding to *Azotobacter* apoflavodoxin (Figure 14), indicate that this reaction must involve reduction of the flavin moiety in the enzyme. Also shown in Figure 14 are the relative magnitudes at several different wavelengths of the two slower reactions which under the conditions of the experiment had half lives of about 80 ms and 200 ms respectively. It is evident that the data of Figure 14 are essentially in agreement with the equilibrium binding data of Figures 5 and 8 and indicate that the formation of the spin-coupled form of the enzyme can be observed by the stopped-flow method,

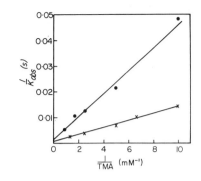

Fig. 13. Double reciprocal plots of the first order rate constants for the rapid bleaching of trimethylamine dehydrogenase by trimethylamine at pH 7.0 (0.1 M K-phosphate)(●—●) and at pH 7.7 (0.1 M Na-pyrophosphate) (x—x). The reaction was followed at 450 nm in the stopped-flow apparatus at 10°C.

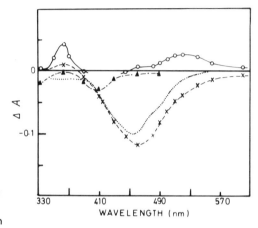

Fig. 14. Absorption spectral changes reflecting the formation of different intermediates during reduction of trimethylamine dehydrogenase by trimethylamine. Stopped-flow experiments were performed at 18°C in 0.1 M Na-pyrophosphate, pH 7.7 at a fixed final concentration of trimethylamine of 0.5 mM. For purposes of comparison the data from different experiments were all normalized to an enzyme concentration of 9 μM. The curves represent the spectral changes involved in the rapid bleaching of the enzyme ($t_{1/2} \sim 1.5$-2.0 ms)(x —x), in the formation of the spin-coupled form of the enzyme ($t_{1/2} \sim 80$ ms)(o—o), and in a slower reaction with $t_{1/2} \sim 200$ ms (△—△). The difference spectrum between the oxidized and reduced amino acyl coenzyme bound to *Azotobacter* apoflavodoxin is also shown (·····).

particularly at 365 nm and at 510 nm, as a reaction proceeding with a half life of about 80 ms under these conditions.

The slow reaction with $t_{1/2} \sim 200$ ms at 18°C and pH 7.7 showed the least spectral overlap with the faster reaction($t_{1/2} \sim 80$ ms) at 332 nm and gave linear semilogarithmic plots at this wavelength (Figure 15). This reaction was, within experimental error, invariant with substrate concentration. However, comparison of the observed rate constants with the rate constants predicted from the catalytic center activity of the enzyme (Table I) suggest that this reaction is most probably not involved in catalysis because of a 2-fold discrepancy between the expected and observed rates at pH 8.5. It should be noted that the minimal mechanism which was proposed for TMADH from steady state kinetic data involved the retention of the aldehyde product on the reduced form of the enzyme, in contrast to the results from ^{14}C-labelling experiments described above and, therefore, it is not entirely improbable that this slow reaction may be due to the release of formaldehyde in a step which is not on the main pathway during catalytic turnover with PMS as electron acceptor.

Examination of the spectral changes observed at 365 nm and 510 nm, which, in accordance with

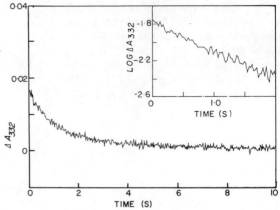

Fig. 15. Slow reaction in the reduction of trimethylamine dehydrogenase by trimethylamine. The enzyme and substrate concentrations were 9.5 μM and 2 mM, respectively, in 0.1 M Na-pyrophosphate, pH 7.7 and 10°C. The inset shows a semilogarithmic plot of the data. The reaction was observed at 332 nm in the stopped-flow apparatus.

the binding equilibrium experiments discussed above, must be correlated with the appearance of the unique spin-coupled form of the enzyme, revealed an unusual substrate dependence, the complete quantitative analysis of which has not yet been attempted. Several qualitative aspects are, however, of interest. At pH 7.7 the total absorbance change at 510 nm associated with the generation of the spin-coupled form appeared to increase as the substrate concentration was increased (Figure 16). Moreover, the reaction was biphasic at low substrate concentrations, but followed a simple logarithmic decay at higher substrate levels. Comparison of the spectral changes at 365 nm (Figure 17) with those at 510 nm indicated, however, that much of the faster phase in the biphasic increase in absorbance at low substrate concentrations was obscured at 510 nm by the rapid decrease in absorbance due to reduction of the flavin moiety which precedes formation of the spin-coupled form. The observed increase in the absolute magnitude of the reaction at 510 nm with increasing trimethylamine concentration is, therefore, artifactual. In accordance with the data of Figure 5, the faster phase, which is partly obscured at 510 nm, must be associated with the substrate independent equilibrium,

$$FlH_2 \cdot Fe_4S_4 \ ox \ \rightleftharpoons \ FlH \cdot Fe_4S_4 \ red + H^+,$$

since at pH 7.7 about a third and at pH 8.5 about two-thirds of the interaction signal is generated in equilibrium binding studies upon the addition of a stoichiometric amount of substrate, apparently without any requirement for the binding of an additional substrate molecule to the reduced enzyme. This supposition is supported by the fact that the faster of the two phases disappear, both at 365 nm and at 510 nm, at elevated substrate levels, suggesting that the formation of the spin-coupled form can proceed by alternate pathways. These considerations may be summarized in the partial reaction sequence presented in Scheme I.

It should be noted that in Scheme I the substrate binding steps cannot be in rapid equilibrium, since in that case a simple logarithmic appearance of the spin-coupled form might have been expected over a wide range of trimethylamine concentrations.

Fig. 16. Dependence of the reaction observed at 510 nm in the stopped-flow apparatus when mixing trimethylamine dehydrogenase with trimethylamine, on trimethylamine concentration. The buffer was 0.1 M Na-pyrophosphate, pH 7.7, the temperature 10°C and the enzyme concentration 9.5 µM.

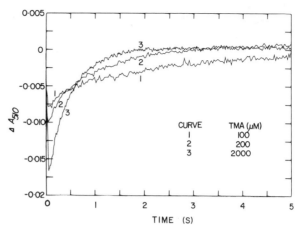

CURVE	TMA (µM)
1	100
2	200
3	2000

The reaction sequence depicted in Scheme I is supported by the behavior of the system at pH 8.5 (Figure 17b). At the latter pH increasing concentrations of substrate actually decreased the rate of formation of the spin-coupled form of the enzyme. Semilogarithmic plots of the data appeared linear at the lowest substrate concentrations, became non-linear at intermediate levels of substrate and was again linear at the highest substrate concentration, suggesting a changeover to an alternate reaction pathway.

$$FlH_2 \cdot Fe_4S_{4ox} \underset{TMA}{\rightleftharpoons} TMA \cdot FlH_2 \cdot Fe_4S_{4ox}$$

$$-H^+ \big\Updownarrow H^+ \qquad\qquad H^+ \big\Updownarrow -H^+$$

$$FlH^\cdot \cdot Fe_4S_{4red} \underset{TMA}{\rightleftharpoons} TMA \cdot FlH^\cdot \cdot Fe_4 S_4{}_{red}$$

SCHEME I

TABLE I

COMPARISONS OF RATE CONSTANTS FROM STOPPED-FLOW DATA WITH PREDICTED RATE CONSTANTS FROM CATALYTIC CENTER ACTIVITY

pH	Observed Rate Constant Wavelength		Theoretical Rate Constant from V_{max} (PMS)
	332 nm	365 nm	
7.0	0.38	0.296	0.554
7.7	0.689	2.31	0.98
8.5	0.64	5.78	1.54

The catalytic center activity of the enzyme was determined from double reciprocal plots in which PMS was varied at a fixed saturating concentration of trimethylamine (3.33 mM) at 10°C. The buffers used both in stopped-flow and steady state kinetic experiments were 0.1 M sodium pyrophosphate adjusted to pH 7.7 and 8.5 with HCl and 0.1 M potassium phosphate pH 7.0.

The observed rate constants were obtained at 2 mM trimethylamine. The reaction at 332 nm, however, was essentially invariant with substrate concentration.

Further evidence that the binding of the substrate to the reduced form of the enzyme, or a subsequent isomerization, is a slow step, was obtained by following the appearance of the spin-coupled form at pH 7.0 at 365 nm. The absorbance change recorded upon reaction of the enzyme with 0.2 mM trimethylamine was characterized both by a lag in the overall reaction and by a section which was almost zero order (Figure 17b). These features are characteristic of a sequential equilibrium in which an intermediate species approaches steady state for part of the reaction, when the formation of the final product is monitored.

Whereas the rate constants for the appearance of the spin-coupled form of the enzyme was fast enough at pH 7.7 and 8.5 for this step to be involved in catalysis, this was not the case at pH 7.0. Even at a concentration of 5 mM trimethylamine where the data gave linear semilogarithmic plots, the pseudo-first order rate constant for the reaction was only 0.31 sec^{-1}, in contrast to 0.55 sec^{-1} predicted from the catalytic center activity at 10°C. Moreover, the apparent K_m for trimethylamine at a near saturating level of PMS (2 mM) was only 10 μM. (It should be noted that the apparent first order rate constants obtained at elevated substrate levels cannot conveniently be extrapolated to infinite substrate concentration in a double reciprocal plot. Semilog-

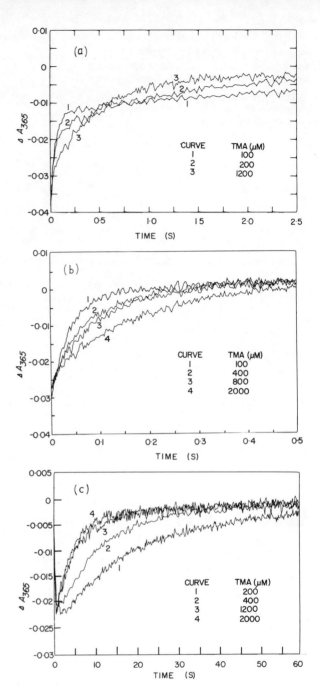

Fig. 17. Dependence of the reaction observed at 365 nm in the stopped-flow apparatus when mixing trimethylamine dehydrogenase with trimethylamine, on trimethylamine concentration at different pH values. The temperature was 10°C and enzyme concentration 7 μM. (a) In 0.1 M Na-pyrophosphate, pH 7.7. (b) In 0.1 M Na-pyrophosphate, pH 8.5. (c) In 0.1 M K-phosphate, pH 7.0.

arithmic plots of the data only appear linear at the higher substrate concentrations and the apparent rate constants which are then obtained give a curved double reciprocal plot.) The available data, therefore, suggest that the spin-coupled form of the enzyme is not involved in catalysis at pH 7.0 when PMS is the electron acceptor. However, it is unlikely that this is also the case with the physiological oxidant, since non-participation of the spin-coupled form of the enzyme in catalysis would obviate the requirement for an Fe-S center. Moreover, pronounced discrepancies between steady state kinetic data and results from rapid reaction kinetics which are performed at much higher enzyme concentrations have been noticed in the past[14] and could perhaps account for the difference between the steady state and stopped-flow kinetic data presented here.

ACKNOWLEDGEMENTS

This research was supported by the National Institutes of Health (HL 16251 and GM 12394). Helmut Beinert is the recipient of Research Career Award 5-K06-GM-18442 from the National Institute of General Medical Sciences.

REFERENCES

1. Colby, J. and Zatman, L.J. (1974) Biochem. J. 104, 555-567.
2. Steenkamp, D.J. and Mallinson, J. (1976) Biochim. Biophys. Acta 429, 705-719.
3. Colby, J. and Zatman, L.J. (1971) Biochem. J. 121, 9P-10P.
4. Steenkamp, D.J., Kenney, W.C. and McIntire, W. (1977) submitted for publication.
5. Hill, C.L., Steenkamp, D.J., Holm, R.H. and Singer, T.P. (1977) Proc. Natl. Acad. Sci. U.S. 74, 547-551.
6. Steenkamp, D.J., Beinert, H. and Singer, T.P. (1977) Biochem. J.   in press.
7. Beinert, H., Hansen, R.E. and Hartzell, C.R. (1976) Biochim. Biophys. Acta 423, 339-355.
8. Orme-Johnson, N.R., Hansen, R.E. and Beinert, H. (1974) J. Biol. Chem. 249, 1922-1927.
9. Orme-Johnson, N.R., Hansen, R.E. and Beinert, H. (1974) J. Biol. Chem. 249, 1928-1939.
10. Steenkamp, D.J., Kenney, W.C. and Singer, T.P. (1977) submitted for publication.
11. Edmondson, D.E. and Tollin, G. (1972) Biochemistry 10, 124-132.
12. Cattalini, L., Ugo, R. and Orio, A. (1968) J. Am. Chem. Soc. 90, 4800-4803.
13. Hiyama, T. and Ke, B. (1971) Proc. Natl. Acad. Sci. U.S. 68, 1010-1013.
14. Beinert, H., Palmer, G., Cremona, T. and Singer, T.P. (1965) J. Biol. Chem. 240, 475-480.

INTERACTIONS BETWEEN SUCCINATE DEHYDROGENASE AND ITS MEMBRANE ENVIRONMENT

BRIAN A.C. ACKRELL   AND EDNA B. KEARNEY

Department of Biochemistry and Biophysics, University of California, San Francisco, and Molecular
Biology Division, Veterans Administration Hospital, San Francisco, CA., U.S.A.

and

ANGELO MERLI, BERND LUDWIG AND RODERICK A. CAPALDI*
Institute of Molecular Biology, University of Oregon, Eugene, Oregon, U.S.A.

ABSTRACT

The micro-environment of succinate dehydrogenase in the inner mitochondrial membrane (i) potentiates the activity of the enzyme in the phenazine methosulfate assay, (ii) protects the HiPIP-type non-heme iron-center of the enzyme from the deleterious effects of oxygen, and (iii) sensitizes the enzyme to specific inhibitors such as thenoyltrifluoroacetone and the carboxamides. To delineate further these membrane interactions we have initiated an approach to probe the immediate environment and the orientation of the enzyme using antibodies generated to pure succinate dehydrogenase. The antibody precipitates the enzyme, the isolated subunits, Complex II and any membrane fragment in which the enzyme is exposed. Immunoprecipitates obtained from detergent-dispersed Complex II or Complex III, isolated with Complex II as a contaminant, always contained the subunits of the enzyme plus two peptides of molecular weight 13,000 and 7,000, respectively. These peptides must therefore be closely associated with the enzyme in the membrane; they are not found in Complexes IV (cyt. oxidase) or V (ATPase). The antibody inhibits the various reductase activities of the enzyme and succinoxidase but not NADH oxidase activity of the sub-mitochondrial particle, ETP. The antibody does not inhibit succinoxidase of intact mitochondria. Treatment of ETP, where the matrix side is exposed, with ^{35}S-Diazobenzenesulfonate caused labelling of the 70,000 and 30,000 molecular weight subunits of the enzyme. This indicates that the enzyme is aligned along the inner surface of the mitochondrial membrane rather than in perpendicular orientation.

INTRODUCTION

The inner mitochondrial membrane, to which mammalian succinate dehydrogenase is firmly bound, provides an environment which protects the enzyme from the deleterious effects of oxygen, permits the action of the specific inhibitors thenoyltrifluoroacetone (TTF) and the carboxamides, potentiates the phenazine methosulfate (PMS) reductase activity of the enzyme, and facilitates electron transfer

*R.A.C. is an Established Investigator of the American Heart Association

143

to the respiratory chain.

When succinate dehydrogenase is removed from its protective membrane environment, its turnover number at 38° as measured by the PMS reductase assay[1] decreases by approximately 40%, a significant loss. Only a small part of this activity loss can be attributed to inactivation, since the major part of the enzyme can be reincorporated into the membrane with restoration of the original high activity[2]. If exposed to air, however, the extracted enzyme rapidly loses this reconstitutive ability, together with the EPR signal (g = 2.01) of its tetranuclear 4Fe-4S non-heme iron cluster (HiPIP center) and a succinate-ferricyanide reductase activity characterized by its low K_m for ferricyanide (250 μM)[3]. Since these three properties of the enzyme are lost at the same rate (half life ~ 50 min)[4], they are likely to be intimately related. Perhaps the integrity of the HiPIP center is vital both to the "low K_m" ferricyanide reductase activity and to reconstitutive ability, being required for electron flow to the respiratory chain.

The membrane-bound enzyme suffers a loss in activity similar to that resulting from solubilization when TTF or carboxamides are added to the particles[5,6], or when the ubiquinone in the particle is extracted; the latter is reversible on restoration of the ubiquinone content[5]. These treatments not only decrease the PMS-reductase activity of the enzyme but also interrupt electron transfer from the enzyme to oxygen via the resiratory chain. The site of action of the inhibitors is known from rapid-freeze EPR experiments to be in the region of the enzyme-ubiquinone interaction; thus in their presence reduction of the succinate-reducible non-heme iron centers of the enzyme is permitted but not their reoxidation by ubiquinone[7]. These inhibitors do not affect the PMS-reductase activity of the soluble enzyme. This difference suggests that a membrane component(s) in close proximity to the enzyme contributes to the binding sites for the inhibitors. Possibly the same component(s) allows the ubiquinone-enzyme interaction[5] since ubiquinone cannot by itself react with the soluble form of the enzyme. The questions that arise are what are these membrane components and how are they organized around the enzyme so that these interactions are possible. To try to answer these and related questions we have been using as probes antibodies generated against the enzyme.

RESULTS AND DISCUSSION

For the studies concerned with the structure-function relationships of the unmodified, soluble enzyme, it is necessary to use a pure preparation with full reconstitutive capacity. The best preparations available until now have been either reconstitutively active but only 30% pure[8] or pure but lacking full reconstitution activity[9]. An enzyme of the desired quality can be isolated, however, by treating Complex II with butanol under strictly anaerobic conditions and in the presence of succinate[10]. The major steps in this purification procedure are summarized in Table I. The resulting enzyme is essentially pure (~ 0.9 moles histidyl flavin/100,000 g protein) and has a high turnover number (12-14,000 at 38°) in both the PMS and low K_m ferricyanide reductase assays; its HiPIP

TABLE I

1-BUTANOL EXTRACTION OF SUCCINATE DEHYDROGENASE FROM COMPLEX II

Stage	Specific activity[a]	Total activity (units)	Yield (%)	Turnover number[b]
Complex II	45.5	5,600		11,700
After 1-butanol treatment at pH 9				
Aqueous phase	80.6	5,300	95.0	13,300
Residue	2.7	152	2.7	2,700
Adsorption on calcium phosphate gel				
Not adsorbed	3.2	47	< 1.0	
Gel wash	0	0	0	
Gel eluate	120.0	4,420	79.0	14,600

[a] Micromoles of succinate oxidized/min/mg of protein at 38° in the succinate/PMS/DCIP assay at V_{max}.

[b] Based on histidyl flavin content of enzyme fraction

center is intact as judged by EPR[10], and it is 85–90% reconstitutively active.

This type of preparation was used for immunizing rabbits. The antibody generated immunoprecipitates the soluble enzyme, Complex II, and any membrane fragment or sub-mitochondrial particle containing exposed succinate dehydrogenase, but does not react with Complex IV (cytochrome oxidase) or Complex V (ATPase). Complex II is a particulate preparation of succinate dehydrogenase containing some residual membrane components but behaves more like the soluble enzyme in terms of PMS reductase activity. The antibody also immunoprecipitates both the isolated 70,000 and 30,000 molecular weight subunits of the enzyme, which indicates that both subunits possess antigenic determinants and suggests the possibility of generating antibodies specific for each subunit. Only a single precipitin line was developed against pure enzyme in immunodiffusion (Figure 1), but more than one line was apparent when the antibody was reacted against inner mitochondrial membrane preparations (ETP) dispersed in 1% Triton X-100. Although the extra lines probably represent immunoprecipitation of various sized aggregates containing succinate dehydrogenase, for studies relating to the membrane environment it was important to establish that the antibody preparation was indeed specific for the enzyme and contained no additional antibodies generated against trace contaminants in the enzyme used as antigen.

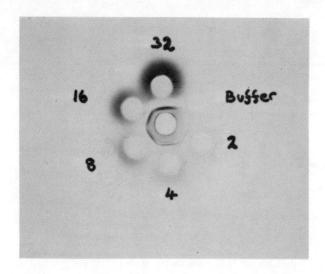

Fig. 1. Immunodiffusion of antibody preparation (as a γ-globulin fraction) against soluble succinate dehydrogenase. The antibody (50 µl) was placed in the center well and buffer and µg of enzyme in the outer wells, as indicated.

A comparison of the band patterns of Complex II (Figure 2A) and pure Complex III (ubiquinone-cyt. c reductase) (Figure 2B) obtained by SDS-polyacrylamide electrophoresis[11] showed that in addition to the large and small subunits of succinate dehydrogenase, designated in Figure 2A as C_{II-1} and C_{II-2}, respectively, Complex II contained two other peptides, not found in Complex III, with molecular weights of 13,000 and 7,000, respectively, which are designated in the figure as peptides C_{II-3} and C_{II-4}. A comparison of the respective band patterns also showed that Complex II preparations characteristically contain the Complex III peptides as contaminants. When this preparation of Complex II, which had been pre-labelled with ^{35}S-Diazobenzene sulfonate (DABS) to facilitate subsequent detection of the peptides, was treated with antibody in the presence of 1% Triton X-100, an immunoprecipitate was obtained which, as shown by SDS-polyacrylamide electrophoresis[13], contained succinate dehydrogenase (peptides C_{II-1} and C_{II-2}) plus the two peptides C_{II-3} and C_{II-4} and showed reduced amounts of the contaminating Complex III peptides (Figure 3A). In a more severe test of specificity, the succinate dehydrogenase antibody was reacted in the presence of Triton X-100 (1%) with a preparation of DABS-labelled Complex III which, as isolated, contained some Complex II as a contaminant. The immunoprecipitate which formed was washed, mixed with excess unlabelled Complex III, and the mixture subjected to SDS-polyacrylamide electrophoresis[13]. The resultant band patterns, determined by staining for protein and as peaks of radioactivity, are shown together in Figure 3B. The peaks of radioactivity (baragraph) represent only the peptides of the immunoprecipitate whereas the pattern elicited by staining (dashed line) includes

146

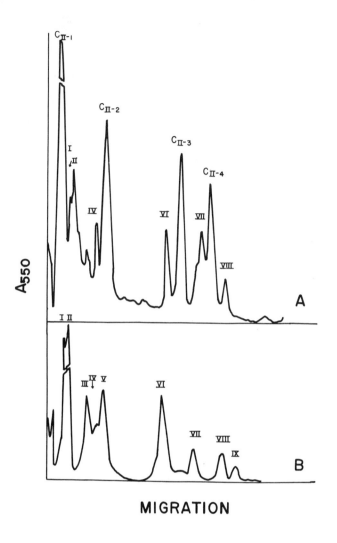

C_{II-1}
C_{II-2}
C_{II-3}
C_{II-4}

I
II
IV
VI
VII
VIII

A

A_{550}

I II
III IV V
VI
VII VIII
IX

B

MIGRATION

Fig. 2.   Peptide band patterns of Complex II(A) and Complex III (B) obtained by SDS-polyacrylamide electrophoresis [11] according to Capaldi et al [12].   (Reproduced by kind permission of the American Chemical Society.)

the peptides of both the immunoprecipitate and the additional Complex III included as marker.   It
can be seen that the immunoprecipitate contained predominantly the four bands of Complex II, but
relatively little of the Complex III; thus, the major peaks of radioactivity co-migrate with the four
peptides identified from Figure 3A as being the large and small subunits of succinate dehydrogenase
and the peptides C_{II-3} and C_{II-4}.   These experiments established the specificity of the antibody
preparation for succinate dehydrogenase, and, in addition, showed that succinate dehydrogenase is
closely associated with the two peptides, C_{II-3} and C_{II-4}, in particulate preparations of the enzyme, a more important point for the structural organization of the enzyme in the membrane.   The

147

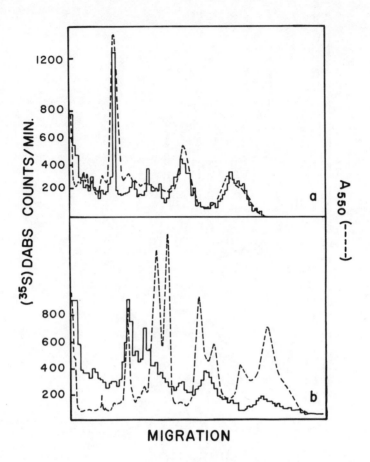

Fig. 3. Composition of immunoprecipitates obtained by reacting antibody with Complex II and a Complex III preparation containing Complex II as a contaminant. Band patterns were established by SDS-polyacrylamide electrophoresis according to Weber and Osborn[13]; baragraph, radioactive counts; dashed line, peptides stained with Coomassie Blue. Upper figure: immunoprecipitate from Complex II labelled with ^{35}S-Diazobenzene-sulfonate (DABS). Lower figure: the immunoprecipitate from the DABS-labelled Complex III preparation was mixed with unlabelled Complex III prior to electrophoresis.

relevance of these peptides to the effects of the mitochondrial membrane environment on succinate dehydrogenase and their possible function in binding the enzyme to the membrane will be studied once they have been isolated in native form.

The ability of the antibody to inhibit the various catalytic activities of both membrane-bound and soluble forms of the enzyme was investigated with the intention of using the antibody to probe the orientation of the enzyme in the membrane. Our antibody preparation was found to be a potent inhibitor of succinate dehydrogenase activity, the inhibition being characteristically rapid but never complete. Saturation curves for the inhibition of the PMS and low K_m ferricyanide reductases of the reconstitutively active soluble enzyme obtained under strictly anaerobic conditions, and of the PMS reductase activity of Complex II are given in Figures 4 and 5, respectively.

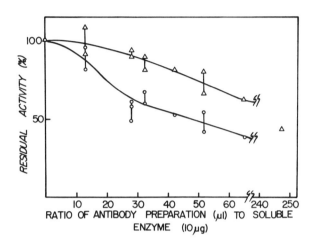

Fig. 4. The inhibition of the PMS ($\triangle$) and low K_m ferricyanide reductase (0) activities of reconstitutively active succinate dehydrogenase. The antibody preparation (as a γ-globulin fraction in 0.9% NaCl) was incubated with the enzyme at 22° in the presence of 20 mM succinate, 100 µM EDTA and 20 mM Tris-SO$_4$ buffer, pH 7.5, under an atmosphere of argon, with glucose, glucose oxidase and catalase added to ensure anaerobiosis. Residual activity is expressed as a percentage of the activity of a control sample of enzyme treated similarly but with 0.9% NaCl or control γ-globulin fraction instead of antibody. Activity was determined at 38° at a single PMS (1.08 mM) or ferricyanide (200 µM) concentration, as in previous work[4].

Double reciprocal plots of the inhibition of Complex II by the optimum amount of antibody showed that the inhibition is accompanied by a decrease in the apparent K_m for PMS (Figure 6A) from 0.53 to 0.36 mM but no change in the apparent K_m for succinate (0.4 mM) (Figure 6B). The maximum inhibition of the succinate dehydrogenase of ETP in the PMS assay was $\sim$ 65%, a little more than observed with the soluble enzyme and Complex II, and characterized by a change in the apparent K_m for PMS from 0.56 to 0.18 mM (Figure 7). The major part of this inhibition may arise in all preparations because the binding of the antibody at certain sites on the enzyme prevents the binding of succinate. The following evidence is consistent with this interpretation: (i) the maximum inhibition observed in ETP was the same both when monitored in the PMS assay and by succinoxidase (Table II), although the rate-limiting steps are different in the two assays, and the succinoxidase activity of ETP is only 50% of the PMS reductase activity; (ii) the inhibition is not associated with

149

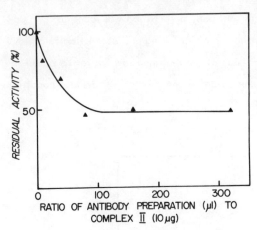

Fig. 5. Inhibition of the PMS reductase activity of Complex II by antibody. Complex II, prepared according to Baginsky and Hatefi[14] but without removing the deoxycholate, was incubated with the antibody preparation at 22° in the presence of 20 mM succinate- 50 mM Tris-SO$_4$,buffer, pH 7.5. Activity was determined at 38° at a single concentration of PMS (1.08 mM), and is expressed as percentage of the activity of a control sample incubated without antibody.

a change in K$_m$ for succinate (Figure 6B); and (iii) the maximum inhibition observed is very similar for the soluble enzyme (54%), Complex II (52%) and ETP (64%) (Figures 4,5 and 7). If this interpretation of the data be correct, this inhibition must be caused by antibody directed to the 70,000 molecular weight subunit wherein the active site is located.

It is clear that further binding of the antibody must take place, however, since multiple antigenic sites are expected to be present on the enzyme and the antibody has been shown to precipitate the 30,000 molecular weight subunit of the enzyme. The changes in K$_m$ for PMS observed in

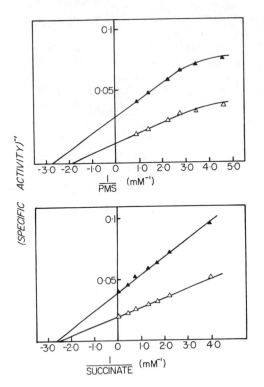

Fig. 6. The effect of antibody on the apparent K$_m$ for PMS and for succinate. Complex II was incubated with antibody at a ratio of 80 μl/10 μg enzyme as in Figure 5. Upper figure: assays were conducted at 38° with varying PMS at constant succinate concentration (20 mM); Complex II (△), Complex II + antibody (▲). Lower figure: assays at 38° with varying succinate but constant PMS concentration (1.08 mM); Complex II (△), Complex II + antibody (▲).

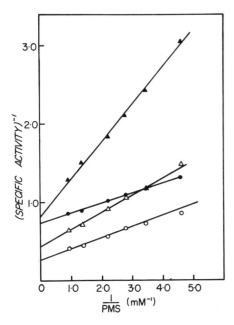

Fig. 7. The effect of antibody on the inhibition of ETP by TTF. Antibody (128 µl) and ETP (0.1 mg) were incubated at 22° in 20 mM succinate, 250 mM sucrose-50 mM Hepes buffer, pH 7.5, for 10 min before the PMS reductase activity was measured at 38° in the presence and absence of TTF (100 µM). Open symbols, control ETP; closed symbols, ETP plus antibody; circles, no addition; triangles, plus TTF.

the remaining active enzyme may result from binding of antibody to these alternate sites. It appeared possible that antibodies reacting with alternate sites might provide information on the site of action of the inhibitors TTF and carboxamides, by interfering with their inhibitory action in submitochondrial particles. The usual effect of TTF is one of inhibiting succinoxidase and halving the PMS reductase activity of ETP (Table II and lower two curves of Figure 7). No significant change in V_{max} occurred, however, when TTF was added to antibody-treated ETP; only a slope change in

TABLE II

THE EFFECT OF ANTIBODY ON THE ACTIVITY OF SUCCINATE DEHYDROGENASE OF ETP

	PMS Reductase* (µmoles succ. oxidized/min/mg)	Inhibition %	Succinoxidase* (µg atom 0/min/mg)	Inhibition %
ETP	3.8	-	1.77	-
ETP + TTF (100 µM)	2.3	40	0	100
ETP + antibody	1.3	66	0.77	57
ETP + antibody + TTF	1.1	71	0	100

*Activities measured at 38°

the double reciprocal plot (upper two curves of Figure 7) was seen. The residual succinoxidase activity of the ETP remained completely sensitive to TTF (Table II). It appears, therefore, that the antibody bound to sites on the enzyme not involved in succinate binding modified the effect of TTF

151

but did not prevent its binding. These results are obviously complex and do not permit localizing different antigenic sites on the enzyme, or relating the site of action of TTF to a particular region of the enzyme. Clarification may be possible with the use of subunit-specific antibodies.

We attempted next to see if the enzyme spanned the membrane: an inhibiting effect of the antibody on the succinoxidase activity of intact mitochondria would indicate both the accessibility of the enzyme on the outer surface of the inner membrane and its essential nature for electron transport to the cytochrome system. The criterion for intactness of the mitochondria was the lack of oxidation of externally added NADH. In beef heart mitochondria prepared by the Nagarse method [15] the rate of NADH oxidation was about 7% of that observed after the mitochondria had been sonicated. The results given in Table III show that antibody at the level used did not inhibit succinoxidase activity, although it caused 50% inhibition of succinoxidase in sonicated mitochondria.

TABLE III

THE EFFECT OF ANTIBODY PREPARATION ON THE OXIDASE ACTIVITIES OF MITOCHONDRIA

	NADH	Malate/Pyruvate	RCI	Succinate	RCI
	μg atom 0/min/mg*	μg atom 0/min/mg*		μg atom 0/min/mg*	
Mitochondria	0.03	0.33	2.2	0.27	1.6
Mitochondria + antibody	0.03	0.41	1.6	0.43	1.3
Mitochondria + control Serum	0.04	0.41	1.6	0.43	1.4
Sonicated mitochondria + antibody	0.45	-	-	0.15	-
Sonicated mitochondria + control serum	0.45	-	-	0.3	-

*Activities at 30°.

It can also be seen that the antibody preparation caused no breakage of the mitochondria, although some uncoupling of respiratory control was observed, and that the antibody apparently did not inhibit transport of substrates into the mitochondria. Partial inhibition, however, would probably not have been noticed if electron transport is rate-limiting. This point is of interest in view of possible similarities in binding sites of transport carriers and succinate dehydrogenase for dicarboxylate substrates. These results would indicate that the antigenic determinants of succinate dehydrogenase are exposed only on the matrix side of the mitochondrial membrane, consistent with the idea that the enzyme does not span the membrane.

Experiments relevant to this point have recently been carried out involving DABS-labelling of ETP. After labelling, the ETP were treated with F_1 (ATPase)-specific antibody to select only those particles certain to be in the everted configuration. These were then dispersed with Triton X-100 and treated with the succinate dehydrogenase-specific antibody to precipitate those fragments containing the enzyme. Both the 70,000 and 30,000 molecular weight subunits of the enzyme proved to be labelled by the DABS, which suggests that the enzyme is aligned along the surface of the membrane, rather than in perpendicular orientation. Experiments in progress with labelling of intact mitochondria should provide a definite answer to this question.

In summary, the use of the succinate dehydrogenase-specific antibody to probe the micro-environment of the enzyme in the membrane has already provided much new information, showing that the enzyme exists in the membrane as at least a four peptide complex, which includes the two subunits of the enzyme plus the two peptides C_{II-3} and C_{II-4}, which may possibly be involved in binding to the membrane. The organization of these latter peptides around the enzyme and with which subunit they are associated is not known. DABS-labelling experiments indicate that at least two of the peptides of this tetra-peptide complex are exposed to the matrix side of the inner mitochondrial membrane, namely, the large subunit, which contains the flavin moiety[9] and substrate binding site[16] of the enzyme, and the small subunit, which is thought to contain the HiPIP center of the enzyme and to communicate with ubiquinone of the respiratory chain.

ACKNOWLEDGEMENTS

This investigation was supported by grants from the National Institutes of Health (HL 16251), and the National Science Foundation (PCM 76-11656 and PCM 74-00957 A02).

REFERENCES

1. Singer, T.P. (1974) Methods Biochem. Anal. 22, 123-175.

2. Ackrell, B.A.C., Kearney, E.B., and Singer, T.P. (1977) J. Biol. Chem. 252, 1582-1588.

3. Vinogradov, A.D., Gavrikova, E.V., and Goloveshkina, V.G. (1975) Biochem. Biophys. Res. Commun. 65, 1264-1269.

4. Beinert, H., Ackrell, B.A.C., Vinogradov, A.D., Kearney, E.B., and Singer, T.P. (1977) Arch. Biochem. Biophys. 182, 95-106.

5. Rossi, E., Norling, B., Persson, B., and Ernster, L. (1970) Eur. J. Biochem. 16, 508-513.

6. Mowery, P.C., Steenkamp, D.J., Ackrell, B.A.C., Singer, T.P., and White, G.A. (1977) Arch. Biochem. Biophys. 178, 495-506.

7. Ackrell, B.A.C., Kearney, E.B., Coles, C.J., Singer, T.P., Beinert, H., Wan, Y-P., and Folkers, K. (1977) Arch. Biochem. Biophys. 182, 107-117.

8. King, T.E. (1967) Methods Enzymol. 10, 322-331.

9. Davis, K.A., and Hatefi, Y. (1971) Biochemistry 10, 2509-2516.

10. Ackrell, B.A.C., Kearney, E.B., and Coles, C.J. (1977) J. Biol. Chem. 252, 6963-6965.

11. Swank, R.T., and Munkres, K.D. (1971) Anal. Biochem. 39, 462-477.

12. Capaldi, R.A., Sweetland, J., and Merli, A. (1977) Biochemistry, in press.

13. Weber, K., and Osborn, M. (1969) J. Biol. Chem. 244, 4406-4412.

14. Baginsky, M.L., and Hatefi, Y. (1969) J. Biol. Chem. 244, 5313-5319.

15. Smith, A.L. (1967) Methods Enzmol. 10, 81-86.

16. Kenney, W.C., Mowery, P.C., Seng, R.L., and Singer, T.P. (1976) J. Biol. Chem. 251, 2369-2373.

NITRATE REDUCTASE COMPLEX FROM THE CYTOPLASMIC MEMBRANE OF *Escherichia coli*

JOSE RUIZ-HERRERA

Departamento de Genética y Biología Molecular

Centro de Investigación y de Estudios Avanzados

del I.P.N. México, D.F. México.

ABSTRACT

Escherichia coli utilizes nitrate as final electron acceptor under anaerobic conditions by means of a membrane-bound multienzyme system. This system includes specific formate dehydrogenase, two forms of cytochrome b_1 and nitrate reductase.

Both formate dehydrogenase and nitrate reductase have been solubilized by different procedures. Both enzymes contain molybdenum and non-heme iron. Formate dehydrogenase in addition contains selenium. Purified nitrate reductase behaves as a dissociating-associating system which is converted by proteolysis to a non dissociating form.

The nitrate reductase complex is under the control of at least seven genes: Chl A-G. All the components of the complex are induced by nitrate and repressed by oxygen.

INTRODUCTION

Several facultative bacteria can utilize nitrate as an alternate electron acceptor under anaerobic conditions by the mechanism named "nitrate respiration" (for a review of this concept see Egami[1]). In *Escherichia coli* the process of nitrate reduction is carried out by a membrane-bound complex, the nitrate reductase complex which utilizes formate as the main reductant. This complex is constituted by formate dehydrogenase, two forms of cytochrome b_1 and nitrate reductase proper[2,3]. The study of this complex offers the possibility to explore the organization of respiratory enzymes into membranes. Nitrate reductase is a dispensable component of the cell and it is controlled by enviromental conditions which are easily manipulated. The genetic control of the complex has been explored and mutants lacking their different components have been isolated. Formate dehydrogenase and nitrate reductase have been extracted from the cell membrane and have been purified, thus their mutual interactions and their association within the cytoplasmic membrane are amenable to experimental manipulation.

COMPONENTS OF THE NITRATE REDUCTASE COMPLEX

Formate dehydrogenase. Several lines of evidence suggest that formate is the physiological electron donor for nitrate reduction in E.coli. Formate is a more efficient substrate than glucose for nitrate reduction by whole cells[2], mutants lacking formate dehydrogenase do not form nitrite when grown anaerobically in a medium containing nitrate[3,4]; NADH was an inefficient reductant of cytochrome b_1 in membranes isolated from cells grown with nitrate[2] and finally it is possible to reconstitute from the purified components a complex which reduces nitrate with formate as electron donor[5].

It has been suggested that E.coli contains two forms of formate dehydrogenase, one soluble and one particulate[6,7]. The former is involved in H_2 evolution by the hydrogenlyase complex and the second in nitrate reduction. Further studies revealed that both enzymes are membrane-bound[8]. The properties of both enzymes: Km, electron acceptor, and optimum pH and temperature are different[8]. The association of both enzymes with cytochromes, their regulation and their genetic control are also different (see below). The formate dehydrogenase component of the nitrate reductase complex has been solubilized from the bacterial membrane with deoxycholate[5,9-11], and with the non-ionic detergent Brij 36T[8,12]. The enzyme has been found associated with cytochrome b_1 which remains attached during the several steps of purification[9-11]. Purified formate dehydrogenase forms aggregates of different molecular weight[11,13]. From a particular preparation of the enzyme, a molecular weight of 590,000 was estimated[11]. It was reported[11] that the enzyme contains 3 polypeptides (α, β and γ) of molecular weights 110,000, 32,000 and 20,000 respectively in a ratio $\alpha_4 \beta_4 \gamma_2$ but since the enzyme is present in different aggregation stage, other structures are possible.

Formate dehydrogenase is one of the few enzymes that contain selenium, this appears bound covalently to the α subunit[11]. No acid-labile selenide was detected in the enzyme. Besides selenium the enzyme also contains molybdenum, non-heme iron, acid-labile sulfide and as mentioned above heme (cytochrome b_1) in the following relative molar amounts: 0.96, 0.95, 14, 13 and 1.0.

Formate dehydrogenase utilizes artificial two electron-acceptors, phenazine methosulfate (PMS) and methylene blue are the most efficient[2]. Km for PMS was 0.33 mM. Different values of Km for formate have been described, the lowest was 0.013 mM[8]. Formate dehydrogenase is very sensitive to sodium cyanide and sodium azide, iodoacetamide and p-hydroxy-mercuribenzoate are less inhibitory[11].

<u>Nitrate Reductase</u>. Nitrate reductase was originally extracted by Taniguchi and Itagaki[14] heating particulate fractions of E.coli at alkaline pH in the presence of nitrate. Since this report, the enzyme has been solubilized with BRIJ 36T[15], heating at alkaline pH[16], heating at neutral pH[17], by extraction with deoxycholate in the presence of ammonium sulfate[11,18] and by use of Triton X-100[19]. It is noteworth that nitrate reductase is solubilized in active form with sodium dodecyl sulfate (SDS)[13]. This form of the enzyme runs in SDS-containing polyacrylamide gels as a single polypeptide of apparent molecular weight 105,000, and it probably represents an incompletly unfolded molecule. Data obtained from the several preparations of nitrate reductase are conflicting. Reported molecular weights differ from 320,000[18] to 1,000,000[14]. Their polypeptide composition are also different but all the authors report a large subunit of molecular weight 142,000[16], 155,000[11] or 150,000[17]; and a small subunit of variable molecular weight about 58,000[16]; 63,000[11] or 55,000[17] (but see Clegg[19]) in different ratios. A third polypeptide of molecular weight about 19,000-20,000 associated with nitrate reductase was identified as cytochrome b_1[5,11,19]. A summary of these data is shown in Table 1.

TABLE 1

CHARACTERISTICS OF NITRATE REDUCTASE SOLUBILIZED BY DIFFERENT METHODS

Method of Solubilization	Molecular weight	Sedimentation value	Subunits and molecular weight	Reference
Heat , alkaline	1,000,000	25.0S	-	14
	720,000-773,600	23.0S	I-142,000 II-~58,000	16
Deoxycholate	320,000	-	-	18
and $(NH_4)_2SO_4$	448,000	16S	I-155,000 II- 63,000 III- 19,000	11
Heat, neutral	200-880,000	10S-24S	I-150,000 II- 55,000	17
Triton X-100	220,000-880,000	9.9S-22.4S	I-150,000 IIA- 67,000 IIB- 65,000 III- 20,000	19

Comparing three techniques of membrane solubilization: heat, BRIJ-36T and SDS, it was observed that the enzyme differed in size[13] which suggests that depending on the method of enzyme removal from the membrane, different stages of aggregation are obtained. Lund and DeMoss[17] clarified the problem when they observed that nitrate reductase behaved as an associating-dissociating system with molecular weights from 200,000 to 880,000 and sedimentation values of 10S-24S depending on the concentration of the enzyme. Similar results were reported by Clegg[19]. The variability in the molecular weight of the small subunit(s) has been attributed to the action of endogenous proteases[17,20]. By limited proteolysis with trypsin the heterogeneous forms of nitrate reductase were transformed into a non-associating homogeneous form of molecular weight 200,000[21]. In this molecular form the large subunit was unaffected but the small polypeptides were all converted to a subunit of molecular weight 43,000. According to DeMoss[21] the active subunit of nitrate reductase contains the large and small polypeptides in a ratio 1:1. With the exception of Taniguchi and Itagaki[14] who reported a very low value, the rest of nitrate reductase preparations have been reported to contain about 1 atom of molybdenum per 200,000 daltons[16-18]. The enzyme contains also non-heme iron (9-12 atoms per 200,000 daltons)[14,17,18] and about 12 acid-labile sulfides per 200,000 daltons[17,18]. The precise location of molybdenum and none-heme iron in the molecule has not been ascertained. It has been reported that electron paramagnetic spectra of purified nitrate reductase showed $Mo(V) - Mo(III)$ transitions and nitrogen hyperfine structure from an NO complex[22], but these results were not confirmed and it was concluded that the molybdenum centre of nitrate reductase was similar to that of other molybdo-enzymes[23]. As mentioned above, in some preparations a cytochrome b_1 component was found associated with nitrate reductase[11,19,20,24]. This polypeptide was released by heating[11].

Kinetic parameters of nitrate reductase solubilized by different techniques are very similar[13]: optimum pH is 7.3; Km for the electron donor, reduced methyl viologen is $4.3-6.3 \times 10^{-7}$M and Km for nitrate is $8.3-12.5 \times 10^{-5}$M.

Cytochrome b_1. As described previously preparations of formate dehydrogenase extracted from the cytoplasmic membrane of *E.coli* appeared associated with cytochrome b_1[9-11]. Addition of nitrate to anaerobic cultures produced an increase in the level of cytochrome b_1[2,25] (see below). These results strengthed the idea that cytochrome b_1 was an obligatory carrier of the electrons from formate to nitrate reductase. Spectral studies of cells grown under different cultural conditions and some nitrate reductase-less mutants revealed that the cytochrome b_1 component present in cells grown anaerobically with nitrate was qualitatively and geneticaly different from that present in cells grown under other conditions[2]. Cytochrome b_1 from cells grown anaerobically without nitrate had the α-band at 555 nm[2] (later on this value was re-calculated to 555.5 nm by use of the fourth-derivative spectra[26]); cells grown aerobically contained two cytochrome b components with α bands at 555 and 562 nm respectively[2]. Several mutants unable to re-duce nitrate lacked cytochrome b_1 (555 nm) when grown anaerobically with nitrate, but synthesized normal amounts of cytochrome b_{555} and b_{562} when grown aerobically[2].

Analysis of the kinetics of reduction of cytochrome b_1 component from induced cells by formate and its oxidation by nitrate was inter-preted to mean the existence of two and not one cytochromes with identi cal spectra but different redox potential[2]. This idea has been chal-lenged by Stouthamer[27] who suggested that biphasic kinetics was due to a change in the steady state of a sole cytochrome, but the effect of n-heptyl hydroxyquinoline-N-oxide and ascorbate[2] are against this in-terpretation. Moreover, both formate dehydrogenase and nitrate re-ductase have been isolated with bound cytochrome b_1 (see above). Since both enzymes do not share any polypeptide it was concluded that both cytochrome b_1 components are different[11]. From the purified compo-nents, an active complex was reconstituted "in vitro"[5].

SYNTHESIS AND REGULATION OF THE NITRATE REDUCTASE COMPLEX.

Nitrate reductase is induced by nitrate and repressed by oxygen[4]. Formate dehydrogenase and specific cytochrome b_1 components are also induced by nitrate and repressed by oxygen[3,28] (Fig. 1A). Induction and derepression of all components from the nitrate reductase complex show biphasic kinetics (Fig. 1A). For nitrate reductase this behavior was interpreted as the result of a change in the internal redox poten-tial[4]. Accordingly, a formate dehydrogenase-less mutant showed a linear rate of nitrate reductase synthesis because it was unable to raise the internal redox potential after nitrate reductase was synthe-sized.

Removal of nitrate originates a decline in the synthesis of nitrate reductase, formate dehydrogenase and cytochrome b_1 which eventually stops (Fig. 1B). Nitrate reductase is not degraded, but formate dehy-drogenase and cytochrome b_1 are inactivated or destroyed (Fig. 1B). This difference probably reflects variable degree of susceptibility to degrading proteases.

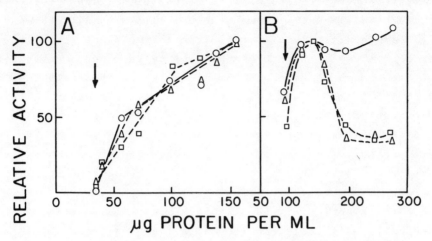

Fig. 1. Effect of nitrate addition (A) and nitrate removal (B) on the synthesis of nitrate reductase (o), formate dehydrogenase (□) and cytochrome b_1 (Δ), by *Escherichia coli*. Arrows indicate time of nitrate addition (A) or removal (B). Data are given as total activities.

Removal of oxygen in the presence of nitrate gave rise to an increase in the levels of formate dehydrogenase, cytochrome b_1 and nitrate reductase, but rate of synthesis of these components was not biphasic (Fig. 2A). Addition of oxygen repressed the synthesis of all components (Fig. 2B). Again nitrate reductase was stable, but formate dehydrogenase and cytochrome b_1 were degraded. The late synthesis of cytochrome b_1 observed corresponds to the aerobic component b_{555}, which is different from the cytochrome involved in nitrate reduction (see above).

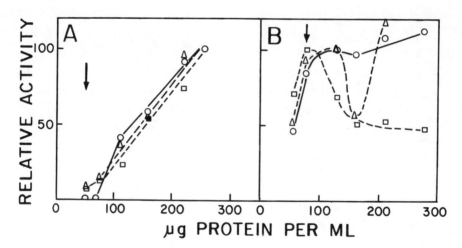

Fig. 2. Effect of transfer to anaerobiosis (A) and aerobiosis (B) in a nitrate-containing medium on the synthesis of nitrate reductase (o), formate dehydrogenase (□) and cytochrome b_1 (Δ) by *Escherichia coli*. Arrows indicate time of change in the atmosphere. Data are given as total activities.

It has been hypothesized that regulation of nitrate reductase operates through a redox-sensitive repressor[4]. Stouthamer[27] suggested that
the factor regulating the synthesis of nitrate reductase was the oxidation-reduction state of the components of the respiratory chain. However it has been found that synthesis of nitrate reductase in hem A mutants is repressed by oxygen independently whether the cells are grown
or not with δ-amino levulinic acid. It was therefore concluded that
cytochromes are not the sensors responsible for oxygen repression[29].

A detailed kinetic analysis showed that under aerobic conditions mRNA
for nitrate reductase was accumulated but it was not translated since
inactive forms of nitrate reductase were not detected[30]. It was concluded that oxygen inhibited both transcription and translation but its
exact mode of action was not ascertained[30].

Selenite is necessary for the synthesis of formate dehydrogenase
whereas molybdate is necessary for the synthesis of both, formate dehydrogenase and nitrate reductase[8,26,31]. ChlD mutants lack all the components of the nitrate reductase complex, but form normal levels when
the growth medium is supplemented with 1 mM molybdate[32]. The mutants
accumulate an inactive form of nitrate reductase which can be converted
to the active component in the absence of net protein synthesis when
molybdate is added to whole cells, but not to cell-free extracts[33]. It
was concluded that the product from ChlD gene was necessary for the
transformation of molybdate or its insertion into the inactive protein.
Addition of tungstate to wild type cells had similar phenotypic effects
than mutations in the ChlD gene[34]. Under these conditions, no inactive
formate dehydrogenase was detected, instead a new fast-migrating band,
not associated with cytochrome b_1 was observed in polyacrylamide gels
from membrane-solubilized proteins. Addition of molybdate to the cells
gave rise to normal formate dehydrogenase and nitrate reductase and the
association of cytochrome b_1 with the former enzyme. Insertion of functional cytochrome b_1 however is not necessary for the synthesis of nitrate reductase or formate dehydrogenase, since normal levels of both
enzymes are synthesized by hem A mutants grown in the absence of δ-amino
levulinic acid[35].

GENETIC CONTROL OF THE NITRATE REDUCTASE COMPLEX.

Mutants unable to reduce nitrate to nitrite "in vivo" have been isolated by several methods: resistance to chlorate, inability to grow
with non-fermentable substrates and inability to use formate to reduce
nitrate "in vivo". Mutants resistant to chlorate have been divided in
five classes: A, B, C, D and E. Mutants A, B, D and E are pleiotropic

and have lost not only nitrate reductase but also hydrogenlyase. 93% of the spontaneous chlorate-resistant mutants[36] and 98% of the mutants induced with nitrosoguanidine[37] belong to classes A, D and E linked to the gal-bio region. On the other hand selection of mutants unable to reduce nitrate with formate as electron donor yield 48% of the ChlC class[37] which are linked to the try region. ChlC mutants have alterations only in the nitrate reductase complex[38]. By the same techniques two new mutant classes were identified: ChlF closely linked to ChlC, and ChlG located at zero minutes of the genetic map of *E.coli*[37]. Map of Chl[r] mutants in the chromosome of *E.coli* is shown in Fig. 3.

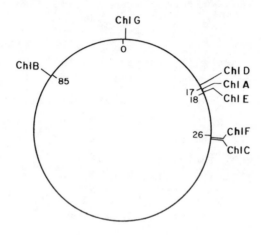

Fig. 3. Map of Chl[r] mutants in the chromosome of *Escherichia coli*.

Since most mutants are pleiotropic it has been difficult to stablish the specific role of the several loci. It has been suggested that ChlA is involved in the synthesis of a molybdenum factor, ChlB in the synthesis of an association factor (FA) necessary for the attachment of the Mo cofactor, and ChlD for the processing of molybdenum (see above). ChlC is apparently the structural gene of nitrate reductase and ChlF the structural gene of formate dehydrogenase. The possible roles of ChlE and ChlG loci are unknown (for reviews see Bachmann et al[39], Haddock and Jones[40]).

ACKNOWLEDGEMENTS
   This work was supported by subvención PNCB 071 from the Consejo Nacional de Ciencia y Tecnología (CONACYT), México.

REFERENCES

1. Egami, F. (1974) Origins of life, 5, 405-413.

2. Ruiz-Herrera, J. and DeMoss, J.A. (1969) J. Bacteriol., 99, 720-729.

3. Ruiz-Herrera, J., Showe, M.K. and DeMoss, J.A. (1969) J. Bacteriol. 97, 1291-1297.

4. Showe, M.K. and DeMoss, J.A. (1968) J. Bacteriol., 95, 1305-1313.

5. Enoch, H.G. and Lester, R.L. (1974) Biochem. Biophys. Res. Commun., 61, 1234-1241.

6. Peck, H.D. and Gest, H. (1957) J. Bacteriol., 73, 706-721.

7. Gray, C.T., Wimpenny, J., Hughes, D. and Mossman, M. (1966) Biochim. Biophys. Acta, 117, 22-31.

8. Ruiz-Herrera, J., Alvarez, A. and Figueroa, I. (1972) Biochim. Biophys. Acta, 289, 254-261.

9. Itagaki, E., Fujita, T. and Sato, R. (1961) Biochem. Biophys. Res. Commun., 5, 30-34.

10. Linnane, A.W. and Wrigley, C.W. (1963) Biochim. Biophys. Acta, 77, 408-418.

11. Enoch, H.G. and Lester, R.L. (1975) J. Biol. Chem., 250, 6693-6705.

12. Ruiz-Herrera, J. and Figueroa, I. (1972) Rev. lat-amer. Microbiol., 14, 99-102.

13. Ruiz-Herrera, J. and Villarreal-Moguel, E. (1975) in Isozymes, Vol. 2, Market, C.L., ed., Academic Press, New York, pp. 539-546.

14. Taniguchi, S. and Itagaki, E. (1960) Biochim. Biophys. Acta, 44, 263-279.

15. Villarreal-Moguel, E.I., Ibarra, V., Ruiz-Herrera, J. and Gitler, C. (1973) J. Bacteriol., 113, 1264-1267.

16. MacGregor, C.H., Schnaitman, C.A. and Normansell, D.E. (1974) J. Biol. Chem., 249, 5321-5327.

17. Lund, K. and DeMoss, J.A. (1976) J. Biol. Chem., 251, 2207-2216.

18. Forget, P. (1974) Eur. J. Biochem., 42, 325-332.

19. Clegg, R.A. (1976) Biochem. J., 153, 533-541.

20. MacGregor, C.H. (1975a) J. Bacteriol., 121, 1102-1110.

21. DeMoss, J.A. (1977) J. Biol. Chem., 252, 1696-1701.

22. Dervartanian, D.V. and Forget, P. (1975) Biochim. Biophys. Acta, 379, 74-80.

23. Bray, R.C., Vincent, S.P., Lowe, D.J., Clegg, R.A. and Garland, P.B. (1976) Biochem. J., 155, 201-203.

24. MacGregor, C.H. (1975b) J. Bacteriol., 121, 1111-1116.

25. Wimpenny, J. and Cole, J.A. (1967) Biochim. Biophys. Acta, 148, 233-242.

26. Lester, R.L. and DeMoss, J.A. (1971) J. Bacteriol., 105, 1006-1014.

27. Stouthamer, A.H. (1976) in Advances in Microbial Physiology Vol. 14, Rose, A.H. ed., Academic Press, London, pp. 315-375.

28. Ruiz-Herrera, J. and Alvarez, A. (1972) Ant. V. Leeuwenhoek J. Microbiol. Serol., 38, 479-491.

29. MacGregor, C.H. and Bishop, C.W. (1977) J. Bacteriol., 131, 372-373.

30. Ruiz-Herrera, J. and Salas-Vargas, I. (1976) Biochim. Biophys. Acta, 425, 492-501.

31. Pinsent, J. (1954) Biochem. J., 57, 10-16.

32. Glaser, J.H. and DeMoss, J.A. (1971) J. Bacteriol., 108, 854-860.

33. Sperl, G.T. and DeMoss, J.A. (1975) J. Bacteriol, 122, 1230-1238.

34. Scott, R.H. and DeMoss, J.A. (1976) J. Bacteriol., 126, 478-486.

35. Ramírez-Hernández, A., Ruiz-Herrera, J. and Villarreal-Moguel, E.I. (1976) Abstracts, X Mex. Congr. Microbiol., Monterrey, Mexico p. B1-3.

36. Casse, F. (1970) Biochem. Biophys. Res. Commun, 39, 429-436.

37. Glaser, J.H. and DeMoss, J.A. (1972) Molec. gen. Genetics, 116, 1-10.

38. Guest, J.R. (1969) Molec. gen. Genetics, 105, 285-297.

39. Bachman, B.J., Low, K.B. and Taylor, A.L. (1976) Bacteriol. Rev., 40, 116-167.

40. Haddock, B.A. and Jones, C.W. (1977) Bacteriol. Rev., 41, 47-99.

ON THE METAL CENTERS IN NITROGENASE

W. H. ORME-JOHNSON*, E. MÜNCK[+], R. ZIMMERMANN[+], W. J. BRILL**, V. K. SHAH**, J. RAWLINGS*, M. T. HENZL*, B. A. AVERILL*[++], AND N. R. ORME-JOHNSON*

Departments of *Biochemistry and **Bacteriology, and the Center for Studies of Nitrogen Fixation, College of Agricultural and Life Sciences, The University of Wisconsin, Madison, WI 53706, and [+]Freshwater Biological Institute, Department of Biochemistry, University of Minnesota, Navarre, MN 55455. [++] Present address: Department of Chemistry, Michigan State University, East Lansing, MI 48906

ABSTRACT

This is a review of the spectroscopic and chemical evidence on the nature of the metal centers in nitrogenase. Reduction of nitrogen requires the presence of two proteins; of these the Fe protein appears to contain a single 4Fe-4S center, while about half of the ca. 30 Fe atoms in the MoFe protein seem to be present as 4Fe-4S centers. Most of the balance of the Fe can be extracted from the protein along with the Mo in the form of the "FeMo cofactor", which will reactivate nitrogenase in extracts of the nif⁻ mutant of A. vinelandii, UW-45. The Mössbauer and EPR spectra of "FeMo cofactor" preparations suggest that this cluster is the origin of the $S = 3/2$ EPR resonance detected in the enzyme, and that one such center is present per Mo atom. Although the composition of the cofactor is 8 Fe:6 labile S:Mo, attempts to dissociate or otherwise detect known 4Fe-4S or 2Fe-2S clusters in this substance failed. We conclude that a novel form of metal cluster, probably composed of Fe, Mo, and S⁻, is present in the enzyme. While we cannot now exclude the presence of, e.g., a single 2Fe-2S center or other minor species of Fe in the MoFe protein, our studies suggest that at a minimum, three 4Fe-4S centers and two 8Fe-Mo-6S cofactor clusters are present per molecule of MoFe protein. The oxidation-reduction properties of the enzyme suggest that the cofactor cluster and the 4Fe-4S clusters are not in redox equilibrium. Inhibition data with CO supports the idea that the two types of centers may be on parallel electron transfer routes within the MoFe protein, and that if the "FeMo cofactor" is present in the active site, it is involved in the part of the mechanism which evolves H_2. Electron flow through both kinds of centers appears to be essential for N_2 or C_2H_2 reduction as contrasted with H_2 evolution.

INTRODUCTION

The study of the enzyme nitrogenase presently includes attempt to identify the prothetic groups. In this regard, knowledge of this enzyme is far more primative than is the case with many other oxidative enzymes, where the presence of flavin and hemes groups often affords a reasonable basis for the design of

165

chemical and spectroscopic experiments with which to probe mechanism. It is also far more difficult to apply the powerful tools of steady-state kinetics to practically irreversable reactions, of which nitrogen reduction coupled to ATP hydrolysis is one. At the moment, then, we must rely on chemical and spectral procedures that in several instances are still under development. Here we outline the evidence for a varied set of complex prosthetic groups in the enzyme, as a prelude to future studies of electron transfer and active site chemistry. The interested reader may find a wider view of the biochemistry of this system in a number of recent reviews.[1-5]

Nitrogenase activity requires the presence of two proteins, a source of electrons (E^{ol} < -400 mv), MgATP, protons, and N_2. The minimum stoichiometry seems to be about 2MgATP hydrolyzed to MgADP and Pi per electron passed to either N_2 or H^+. One protein, the Fe protein (or "component II") has a single 4Fe-4S center per dimer of peptide chains (MW 2 x 26,000). The Fe protein binds MgATP and appears to accept electrons from ferredoxins , etc., and to supply electrons to the MoFe protein (or "component I"). Whether the complex (K_D = $10^{-7}\underline{M}$) between the MoFe and Fe proteins is the reductant of N_2 or whether the MoFe protein alone can reduce substrates is not known, but a popular working hypothesis is that the role of the Fe protein is to transduce hydrolysis energy into electron potential, and to thereby activate one or more metal cluster containing active sites on the MoFe protein. The present review addresses current evidence bearing on the nature of these metal sites.

COMPOSITION OF THE MoFe PROTEIN

It appears that MoFe protein from several sources are $\alpha_2\beta_2$ tetramers[6], leading one to suppose that there are likely to be even numbers of metal sites in the protein (this is not strictly necessary of course). The metal composition of this oxygen labile protein has been difficult to establish in all cases, with reported values ranging from 1-2 Mo and 15-40 Fe per molecule[2], but a recent study of recrystallized A. vinelandii MoFe protein[7] suggests that 2 Mo, 33 ± 1 Fe, and 27 ± 1 labile sulfur atoms are present per molecular weight of 250,000. The reported molecular weights for this MoFe protein range from 220,000 to 270,000[2]. Per atom of Mo, then, one may have to account for 16-17 Fe and 13-14 labile sulfur, presumably in one or more metal clusters, by analogy to simpler systems[8].

EPR EVIDENCE

The MoFe protein exhibits a signal arising from an S = 3/2 center with g values near 4.2, 3.6, and 2.0. This appears to amount to one spin system per Mo[9], i.e., two per molecule of A. vinelandii MoFe protein. Substitution of ^{57}Fe in the growth medium of the organism led to a protein with reproducible broadened EPR[10], showing that the EPR arises from an Fe complex. Enrichment of proteins in ^{95}Mo (I = 5/2) had no effect on the EPR; no conclusion about the participation of Mo can be reached from these data.

166

Carbon monoxide is a potent inhibitor of reduction of N_2 or C_2H_2, by nitro-genase, but has no effect on H_2 evolution[11]. It also does not affect the reduction of the center yielding EPR in the MoFe protein[12], suggesting that this center is involved in the production of H_2 by the enzyme. An examination of the EPR of nitrogenase during ATP-dependant hydrogen evolution revealed two curious signals: one, with g-values near 2.05, 1.97, and 1.93, arises when CO and nitrogenase are present in approximately 1:1 ratio. A second signal with g-values near 2.17 and 2.05 replaces the previous signal when excess CO is present. The form of these signals, observed only in the steady state of turning-over enzyme, suggests the presence of 1 or more centers able to utilize the -3 (reduced ferredoxin-like) and -1 (oxidized hipip-like) net charge states. Neither signal has been induced to quantitate above 0.5 spin per molecule of enzyme yet, as might be expected in a signal originating from a transient species, but the use of ^{57}Fe, and ^{95}Mo labelled enzyme components (and ^{13}CO) allowed us to definitely locate the origin of the spin on an Fe compound in the MoFe protein.

Finally, treatment of the MoFe protein with 80% DMSO and similar solvents in the presence of a strong reductant ($Na_2S_2O_4$) led to the development of an EPR signal near $g=2$[13], characteristic of unfolded reduced 4Fe-4S proteins.[14] Depending on the solvent, 1-3 clusters reduced per molecule were found.

MÖSSBAUER SPECTRAL EVIDENCE

Mössbauer spectroscopy should in principle afford a method to sort out the types of iron complexes in MoFe proteins, and two studies on purified proteins from Klebsiella pneumoniae[15] and A. vinelandii[9] have appeared. Despite differences in reported metal compositions, these studies as well as our unpublished work on Clostridium pasteuranum show an astonishing similarity in the overall disposition of Fe in various categories in the MoFe proteins from these sources. We estimate that fewer than 5% (i.e., 1-2 Fe at most) of the iron environments differ in any appreciable way. This suggests that MoFe protein, like the 2(4Fe-4S) ferredoxin from C. acidi-urici[16], has an all-or-none composition; i.e. that preparations are made of molecules with a full complement of metal sites, diluted perhaps by rela-tively metal-free apoprotein.

Initially, Mössbauer spectral studies were carried out on MoFe proteins as prepared, as reduced by the Fe protein[9,15], and as (reversibly) oxidized by dyes such as thionin[15]. If one assumes that all of Fe atoms have the same recoilless fraction or Mössbauer extinction coefficient, then one concludes that about 40% of the iron in MoFe proteins belongs to the class which are paramagnetic in the pro-tein as isolated, and which become EPR silent during the steady state; that is, which belong to the cluster(s) that give rise to the EPR signal. The balance of the Fe is seen to become paramagnetic during oxidation of the protein. A combined EPR and Mössbauer study (reference 13, and Münck et. al., in prepara-tion) showed that the oxidation of the protein by thionin proceeds in two

stages; that is, a stage where no change in the EPR-active iron is seen while the
EPR-silent iron becomes paramagnetic, followed by a stage where the EPR-active
center dissappears. Overall, around six electrons are removed, and the two stages
overlap; but it is apparent that there are two classes of iron present with distinct
redox potentials. Our estimate of the fractional population of Fe seen in these
experiments is given in Table I. Keep in mind that these fractions depend on the
Fe atoms having the same recoilless fraction. That this is a good assumption is
suggested by the finding by Debrunner (personal communication) that the recoilless
fractions for myoglobin and rubredoxin, two remarkably different protein environments of Fe, are the same within a few percent.

TABLE 1

MÖSSBAUER SPECIES* IN MoFe PROTEINS

Species	Mössbauer parameters		Fe*/molecule
	ΔE (mm/sec)	I.S.$^{+}$(mm/sec)	
M_{EPR} (EPR active center)	0.76	0.40	12
D	0.81	0.64	14
Fe^{2+}	3.02	0.69	4
S	1.4	0.6	2
W (oxidized D,Fe^{2+},S)	Ca. four magnetically split species		18

*Data from ref. 9,13, and Münck et. al., in preparation, for A. vinelandii, assum-
ing 32Fe, 2Mo per molecule.

+Relative to metallic iron at 300°K.

    In addition to providing evidence on the multiplicity of Fe sites in the
protein, these experiments have recently had another surprising dividend. We find
(Münck et. al., in preparation) that while both general types of Fe (i.e., M_{EPR} and
[D + Fe^{2+} + S]) are oxidized by thionin treatment of the MoFe protein (in the
absence of Fe protein)   , concentrated $Na_2S_2O_4$ (an anionic reductive substrate for
the protein) reduces only M_{EPR} at room temperature, at times up to a half hour. On
the other hand, if the cationic substrate reduced methyl viologen is supplied, both
types of iron are speedily reduced.   This preliminary experiment suggests that
for the redox states dealt with during oxidation of the native protein, the two
types of Fe centers do not exchange electrons. Combined with our previous finding
that in CO-inhibited enzyme the reduction of M_{EPR} is not impeded while N_2 reduction
is shut down completely, this suggests to us that two parallel electron flow

paths, not interconnected within the enzyme, are needed for the reduction of N_2 to ammonia.

CHEMICAL EVIDENCE

There exists a class of mutants of nitrogen fixing organisms, cell-free extracts of which will exhibit nitrogen fixation when supplemented by acid extracts of purified MoFe proteins[17]. Shah and Brill have recently shown[7] that from acid treated, dehydrated suspensions of MoFe proteins, N-methylformamide will remove a substance which will reactivate extracts from one such A. vinelandii mutant strain, UW-45. This preparation they call the "FeMo cofactor" or FeMo-co. Pienkos et. al.[18] showed that this substance is distinct from the "Mo cofactor" of Nason et. al.[19]. The FeMo-co preparation, when chromatographed on G-100 in N-methylformamide, showed coincident Mo and Fe peaks with a concentration ratio 1:8, and all of the activity expected (on a Mo basis) could be recovered from extracts of pure MoFe proteins. We have obtained the Mössbauer and EPR spectra of FeMo-co[20] shown in figures 1 and 2. In figure 1 it can be seen that the principal Mössbauer features that we ascribe to the M_{EPR} fraction of the Fe in MoFe protein are found in Mössbauer spectra of the FeMo cofactor. About 75% of the Fe in the cofactor belongs to M_{EPR}. Addition of thiophenol to solutions of simple FeS proteins in NMF rapidly produces the thiophenol-substituted FeS core compound[21], but this treatment has no effect on the cofactor by this criterion (see also figure 2). The EPR evidence (figure 2) similarly suggests that the FeMo-co is an S = 3/2 system, derived directly from the EPR center in the protein. Quantitation of the EPR spectra reveals the same number of spins per Mo in both protein and cofactor. The cofactor could be reduced and complexed with CO, based on the reversible dissappearance of the EPR signal, and the cofactor was easily and irreversibly destroyed by either O_2 or mercurials, a property consonant with the presence of 6 labile S per Mo as found by Shah and Brill.[7] Treatment of the MoFe protein with thiophenol in NMF produces a mixture whose EPR is that of a combination of the FeMo-co and 4Fe-4S centers (figure 2). Of the latter, we found approximately three per molecule of MoFe protein, by comparison with standards prepared with known 4Fe-4S ferredoxins. These experiments do not exclude the presence of 2Fe-2S centers, since under these conditions such centers are very labile and do not give appreciable EPR, due to their extremely negative redox potential (see reference 22).

EXAFS EVIDENCE

Extended x-ray fine structure spectroscopy affords a method of directly examining the environment of the Mo in nitrogenase[23]. Considerable experimentation covering several oxidation states of the enzyme and the FeMo cofactor has been carried out and is in the process of publication, but the data suggest that the Mo in the enzyme is surrounded by S atoms as nearest neighbors, and that this environment is unchanged in the cofactor (S. P. Cramer, personal communication).

169

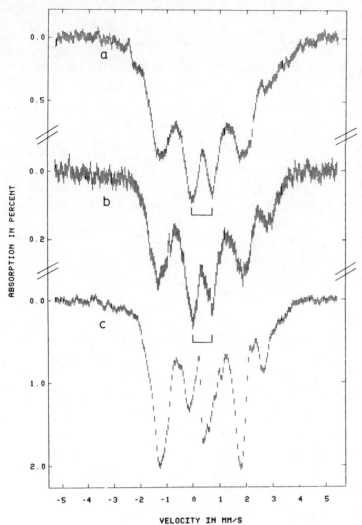

Fig. 1. Mössbauer spectra taken at 4.2 K in a field of 600 gauss applied parallel to the observed Mössbauer radiation. a) FeMo-co in NMF, b) sample of (a) after addition of 10mM thiophenol and c) spectrum of component M_{EPR} of native MoFe protein from C. pasteurianum. In spectrum c) the spectral compounds Fe^{2+}, D, and S were removed using the same procedure as described for the MoFe protein of A. vineland (ref. 9). From reference 20.

CONCLUSIONS

If we assume that the correct composition of fully active MoFe proteins is 32Fe and 2Mo, then the experiments reviewed here suggest that the metals are apportioned as indicated in Table II.

There appear to be two distinct general classes of Fe atoms, roughly the "cofactor" and "non-cofactor" kinds. The former are not previously recognized FeS centers, probably are Mo clusters, and undergo oxidation-reduction during turnover. The latter are mostly 4Fe-4S centers, but only undergo detectable oxidation reduction under special circumstances; i.e., inhibition of substrate reduction by carbon monoxide. The classes appear to be functionally quite separate in that they do not appear to exchange electrons on the time scale of catalysis. We are thus presented with at least two kinds of cluster, at least five of them definitely identified so far, per protein molecule. Further study of this system will no doubt use time-resolved spectral methods to attempt to understand the redox sequence in this enzyme system.

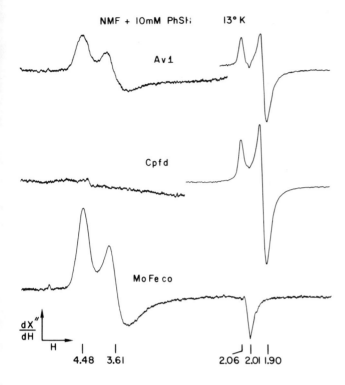

NMF + 10mM PhSH   13°K

Av1

Cpfd

MoFeco

$\frac{dX''}{dH}$

H | | | | |
   4.48 3.61   2.06 2.01 1.90

Fig. 2. Thiol effect on the EPR spectra of the MoFe protein, ferredoxin, and FeMo-co a) A. vinelandii MoFe protein in 80% NMF-20% buffer, b) C. pasteurianum ferredoxin in 80% NMF-20% buffer, c) A. vinelandii FeMo-co in NMF. All solutions contained 10 mM thiophenol plus 1mM sodium dithionite. The buffer used was .1M Tris-Cl, pH 8.0. The spectra were recorded at 13°K with a microwave power of 1mw, modulation 10 gauss at 100 KHz, and a microwave frequency of 9.2 GHz. From reference 20.

TABLE II

DISTRIBUTION OF METALS IN MoFe PROTEINS

Class	Comments	Fe/Molecule
M_{EPR}	The FeMo cofactor; undergoes oxidation-reduction during turnover; one S = 3/2 system per Mo;	12
"Other cofactor iron"	Central doublet in Mössbauer of cofactor preparation	4
$[D + Fe^{2+}+S]$	Oxidation reductions properties distinct from cofactor, composed of at least 3 4Fe-4S clusters, and up to 4 other Fe atoms not yet accounted for.	12 4

ACKNOWLEDGEMENTS

    The authors appreciate the support of the NIH through grants GM17170 and GM 22130; the NSF through grants PCM 76-24271 and PCM 77-08522; and by the Graduate Research Committee of the University of Wisconsin. E. M. was the recipient of a PHS RCDA, GM 70638.

171

REFERENCES

1. Newton, W. E., and Nyman, E. J., ed. (1976) <u>Proceedings of the 1st International Symposium on Nitorgen Fixation</u>, Washington State Press, Pullman, Washington.

2. Orme-Johnson, W. H., and Davis, L. C. (1977) In <u>Iron Sulfur Proteins</u> (W. Lovenberg, ed.), Vol. 3, Ch. 2, Academic Press, New York.

3. Barrueco, C., and Newton, W. E., ed. (1977) <u>Proceedings of the 2nd Int. Conf. N$_2$Fixation</u>, Academic Press, in press.

4. Winter, H. C., and Burris, R. H. (1976) Ann. Rev. Biochem. <u>45</u>, 409.

5. Orme-Johnson, W. H. (1977) in A. Hollaender, ed., <u>Genetic Engineering for Nitrogen Fixation</u>, Plenum Pub. Corp. New York, p. <u>317</u>.

6. Kennedy, C., Eady, R. R., and Kondorosi, E. (1976). Biochem. J. <u>155</u>, 383.

7. Shah, V. K., and Brill, W. J. (1977) Proc. Natl. Acad. Sci. US <u>74</u> 3249.

8. Orme-Johnson, W. H. (1973) Annu. Rev. Biochem. <u>42</u> 159.

9. Münck, E., Rhodes, H., Orme-Johnson, W. H., Davis, L. C., Brill, W. J., and Shah, V. K. (1975). Biochim. Biophys. Acta <u>400</u>, 32.

10. Burris, R. H., and Orme-Johnson, W. H. (1974) In <u>Microbial Iron Metabolism</u> (J. Neilands, ed.), p. 187. Academic Press, New York.

11. Rivera-Ortiz, J. M. and Burris, R. H. (1975), J. Bacteriol. <u>123</u>, 537.

12. Davis, L. C., Henzl, M. T., Burris, R. H., and Orme-Johnson, W. H. (1977) (in the press).

13. Orme-Johnson et. al., in reference 3.

14. Cammack, R., and Evans, M. C. W. (1975). Biochem. Biophys. Res. Comm. <u>67</u>, 544.

15. Smith, B. E., and Lang, G. Biochem. J., 137 (1974) 169.

16. Hong, J.-S., and Rabinowitz, J. C. (1970) J. Biol. Chem. <u>245</u> 6574.

17. Nagatani, H. H., Shah, V. K., and Brill, W. J., (1974) J. Bacteriol. <u>120</u> 697.

18. Pienkos, P. T., Shah, V. K., and Brill, W. J., Proc. Natl. Acad. Sci. US, in press (1977).

19. Nason, A., Lee, K. Y., Pan, S. S., Ketchum, P. A., Lamberti, A., and DeVries, J., (1971) Proc. Natl. Acad. Sci. USA <u>68</u> 3242.

20. Rawlings, J., Shah, V. K., Brill, W. J., Münck, E., Zimmermann, R., and Orme-Johnson, W. H., J. Biol. Chem. in press (1977).

21. Averill, B. A., Bale, R., and Orme-Johnson, W. H., J. Am. Chem. Soc. in press (1977).

22. R. H. Holm, and J. A. Ibess (1977) in <u>Iron Sulfur Proteins</u> (W. Lovenberg, ed), Vol. 3, ch. 7, Academic Press, New York.

23. S. P. Cramer, T. K. Eccles, F. Kutzler, K. O. Hodgson, and L. E. Mortenson, J. Am. Chem. Soc. <u>98</u> 1287 (1976).

ON THE MODE OF ACTION OF NITROGENASE

J.R. POSTGATE, R.R. EADY, D.J. LOWE, B.E. SMITH, R.N.F. THORNELEY, and M.G. YATES

A.R.C. Unit of Nitrogen Fixation, University of Sussex, Brighton, BN1 9QJ, U.K.

ABSTRACT

Nitrogenase, the enzyme responsible for biological nitrogen fixation, consists of two oxygen-sensitive metallo-proteins, both containing Fe and S; one containing Mo. For activity it requires Mg^{2+}, ATP and an electron donor; reduction of H^+ to H_2 accompanies N_2 reduction. The mode of action of nitrogenase is still not clear despite intensive study, but it can be arbitrarily divided into the following steps:

1. Reduction of Fe-protein.
2. Activation of Fe-protein by MgATP.
3. Binding of reducible substrate to Mo-Fe protein.
4. Association of Fe- and Mo-Fe-protein to form an active complex.
5. Transfer of electrons from Fe atoms in Fe-protein to Fe atoms in Mo-Fe protein.
6. Transfer of electrons to reducible substrate.
7. Release of reduced product, ADP, P_i and dissociation of complex.

Evidence concerning the reality, sequence and stoichiometry of these steps in nitrogenase from *Klebsiella pneumoniae* and *Azotobacter chroococcum* are surveyed and new data bearing on some are presented.

INTRODUCTION

Nitrogenase is an enzyme of unusual complexity. Structurally, it consists of not one but two distinct metallo-proteins: the Mo-Fe protein which is a heteromeric tetramer[1] and the Fe protein which is a homomeric dimer (Mortenson, personal communication). Functionally, it requires ATP, Mg^{2+} ions and a specific reductant for activity *in vivo*; the natural reductant can be replaced *in vitro* by sodium dithionite, but the functional complexity is augmented by the facts (1) that ADP, the product of ATP utilization, is inhibitory, so the enzyme shows a form of end-product inhibition; (2) that it interacts with the hydrogen ion, evolving H_2 therefrom, a reaction which accompanies nitrogen fixation to about 30% and which may become the sole reaction in the absence of other reducible substrates (eg: under Ar). An operational complexity is the fact that both the Mo-Fe protein and the Fe protein are oxygen labile, so all biochemical manipulations must be conducted in an O_2-free environment.

Despite these problems, highly purified and, in one instance, crystalline nitrogenase proteins have been obtained and a certain amount of structural information is available. Table 1 records data for nitrogenase from *Klebsiella pneumoniae* and *A. chroococcum*, the principal organisms studied in our research group.

TABLE 1

SOME PROPERTIES OF NITROGENASE PROTEINS FROM *KLEBSIELLA PNEUMONIAE* (Kp) and
AZOTOBACTER CHROOCOCCUM (Ac)

	Kp1	Ac1	Kp2	Ac2
Mol. wt.	218000	226000	66800	64000
Subunits	α 2 x 51000	(2 types)[‡]	2 x 34000	2 x 32000
	β 2 x 59000			
Mo/mole	2.0 ± 0.1	1.9 ± 0.3	0	0
Fe/mole	33 ± 3	23 ± 2	4	4
S^{2-}/mole	N.D.	20 ± 2	3.85	3.9
Specific activity[*]	2300	2000	1500	2000
epr of active species[+]	g_1 4.32	4.29	2.053	2.05
	g_2 3.63	3.65	1.942	1.94
	g_3 2.01	2.01	1.865	1.87

[‡] Judged from peptide mapping; α and β subunits not isolated.

[*] Maximum values at 30°C in nmole C_2H_4 formed from C_2H_2/mg protein/min with excess of complementary protein.

[+] $Na_2S_2O_4$ present; pH 7.0; below 18 K for proteins 1, 30 K for proteins 2; g values of proteins 1 are pH-dependent.

A working scheme for the mode of action of nitrogenase is given in Table 2 for the case of *K. pneumoniae* nitrogenase. It has been arrived at principally from Mössbauer spectroscopy[2] and e.p.r. studies[3,4] supplemented by rapid reaction kinetics[5,15,27] and is broadly consistent with the working schemes of the Wisconsin[6,7] and Purdue[8] research groups.

The scheme is divided into seven steps for present purposes:

1. Reduction of Fe-protein.
2. Binding of reducible substrate to the Mo-Fe protein.
3. Activation of reduced Fe-protein with MgATP.
4. Association of Mo-Fe protein plus bound reducible substrate with ATP-activated Fe-protein to form a complex.
5. Electron transfer within the complex from Fe atoms in Fe-protein to Fe atoms in Mo-Fe protein.
6. Electron transfer to reducible substrate.
7. Release of ADP, P_i, reduced substrate; possible dissociation of nitrogenase complex.

Among many outstanding questions are the sequential order of these processes; their stoichiometry; the rate-determining step; the factors affecting electron dis-

TABLE 2

A WORKING SCHEME FOR ONE TURNOVER OF NITROGENASE

---

1. Reduction of Kp2: $Fld^- + Kp2 \longrightarrow Fld + Kp2^-$

2. Activation of reduced Kp2: $Kp2^- + MgATP \longrightarrow (Kp2^- - MgATP)$

3. Binding of substrate by Kp1: $Kp1 + N_2 \longrightarrow (Kp1-N_2)$

4. Complex formation: $(Kp\overline{2}-MgATP) + (Kp1-N_2) \longrightarrow (Kp\overline{2}-MgATP-Kp1-N_2)$

$\downarrow$

5. Electron transfer within complex: $(Kp2^{\cdot -} -MgATP-Kp\overline{1}-N_2)$

$\downarrow$

6. Electron transfer to substrate: $(Kp2^{\cdot -} -MgATP-Kp1-\overline{N}_2)$

$\downarrow$

7. Release of products: $Kp1 + Kp2 + Mg^{2+} + ADP + P_i + NH_3$

<u>Commentary</u>: This is a hybrid scheme based on work with both Ac and Kp (Table 1);
Kp nitrogenase is chosen as an example. Superscript$^-$ denotes a reduced species.
No stoichiometries or orders of addition are implied: the complex is not necessarily
of 1:1 stoichiometry (see text). No N_2H_2 or N_2H_4 have been detected during reduction of N_2 to NH_3.

tribution between hydrogen ion and other reducible substrates; the nature and number
of protein-protein binding sites as well as sites for binding reducible substrates
and MgATP. In this contribution we shall bring together recent work bearing on
some of the steps listed.

STEP 1: REDUCTION OF Fe PROTEIN

The natural electron donor to Ac2 is a flavodoxin[12,30] and Yoch[13] was able to
isolate a similar protein from *K. pneumoniae*. Data for this step have been
obtained using Ac2 and have been reviewed recently by Thorneley *et al.*[5] Ac2
can be oxidized anaerobically by phenazine methosulphate to an epr-silent form
without serious loss of enzymic activity. It then accepts a single electron/mole
very rapidly from $SO_2^{\cdot -}$ ($k > 10^8$ M^{-1} s^{-1}) with a concomitant appearance of its epr
signal. Subsequent slower reduction steps can be observed which are not paralleled
by epr changes; in all 1.9 electrons are accepted/molecule. The extreme rapidity
of the epr-associated reduction is difficult to reconcile with the well established
fact that, in functioning nitrogenase, the bulk of the Fe protein is epr-silent[9,4].
Tso and Burris[10] had shown that *C. pasteurianum* Fe protein (Cp2) binds one molecule
of MgADP; our studies with Ac2 indicate that binding of MgADP decreases the rate
of reduction of Ac2 by $SO_2^{\cdot -}$ to the epr-active species[11].
The Fe proteins contain an Fe_4S_4 cluster[2,25]. There are several unexplained

175

features of the behavior of the Fe proteins including (1) their consistently low value for integration of epr intensity/mole (0.2 electrons for Ac2, 0.45 electrons for Kp2); (2) the ability of Ac2 to accept 2 electrons/mole though only the first has a concomitant epr change; (3) the unusual lineshape of the epr signal of Fe proteins, uncharacteristic of single epr centres. One of us (DJL) has computed spectra closely resembling those determined experimentally by assuming the presence of a rapidly relaxing paramagnetic centre separated from the Fe_4S_4 cluster by about 10 $\overset{\circ}{\text{A}}$. If the reality of a second centre is confirmed then the stoichiometry of electron transfer from Fe proteins to MoFe-proteins needs reconsideration.

STEP 2: ACTIVATION OF Fe PROTEIN BY ATP

Evidence for this step is the effect of ATP on reduced Fe protein; the epr signal changes from a rhombic to an axial form; the E_m decreases from -290 mV to -410 mV; the oxygen sensitivity increases; Fe becomes more readily extractable with $\alpha\alpha$ bipyridyl[28] and -SH groups more accessible to sulphydryl reagents[29]. Steady state studies of acetylene reduction[14,15] showed that $MgATP^{2-}$, specifically, is the activating species, that Mg_2ATP^{2-} is inactive and that free ATP^{4-} inhibits binding of $MgATP^{2-}$ competitively.

The number of binding sites for $MgATP^{2-}$ is not entirely clear; Tso and Burris[10] obtained a value of 1.7 (interpreted as 2) per mole from gel equilibration studies with Cp2. Only one of these sites bound MgADP. Thorneley and Cornish-Bowden[16] developed symmetrical and sequential models for binding of 2 moles of MgATP and competitive binding of 2 moles of MgADP per mole of Kp2 dimer from steady state kinetic data for hydrogen evolution.

STEP 3: BINDING OF REDUCIBLE SUBSTRATE

In Table 2, MoFe-protein is indicated as binding the reducible substrate before complex formation. The evidence for this is that the pK of its pH-dependent epr signals is shifted down by 0.5 pH in the presence of the reducible substrate acetylene as compared with Ar[4]. That product (C_2H_4) binding occurs is indicated by the epr studies mentioned in the context of step 6 (below). Some evidence that ATP may be involved in substrate reduction, in addition to its effect in activating Fe protein, arises from studies with analogues of ATP such as "ATP-γS"[17]. These substrates inhibit the reduction of acetylene more than the reduction of hydrogen ions.

STEP 4: FORMATION OF A COMPLEX

Our ultra centrifuge studies[18,19] showed that Kp1 and Kp2, or heterologous mixtures of Kp and Ac proteins, formed 1:1 complexes in non-functioning conditions. However, Reithel and Eady[20] showed that contact of the proteins with stainless steel could affect their sedimentation properties and that Kp1, in the absence of dithionite, dissociated from its tetrameric state. Dithionite inhibited dissociation and so did Kp2 in 1:1 molar ratio; the latter formed a complex of increased S value. However, saturating Kp2 levels for the species of higher S value were not

TABLE 3

EFFECT OF Kp2 CONCENTRATION ON SEDIMENTATION COEFFICIENT OF Kp1

Sedimentation at 20°C in 25 mM Tris-HCl, pH 7.4, + 10 mM $MgCl_2$ + 1 mM EDTA under N_2 in the M.S.E. 'Centriscan' ultra-centrifuge. (For details and rationale, see Reithel & Eady[20]).

Molar ratio Kp2/Kp1	S value of leading boundary
no Kp2	ca. 10.5
1	11.1
3	11.9
5	12.25
10	12.0

reached until a molar ratio of 5Kp2:1Kp1 (Table 3) in contrast to the earlier study. Our evidence[19] that a tight Kp1-Kp2 complex is rapidly formed is still valid, but the conclusion that only a 1:1 complex is active is less certain. It is possible that uncomplexed Kp2 can reduce a 1:1 complex, indeed, from stopped flow studies[19], the saturation ratio of Kp2 to Kp1 for protein-protein electron transfer was 3Kp2:1Kp1. Emerich and Burris[21] reached a value of 2Av2:1Av1 for the nitrogenase complex of *Azotobacter vinelandii* from studies of interference by Cp2. Evidence for the existence of a reasonably long-lived nitrogenase complex is good,[26,27] but its stoichiometry will remain ambiguous until more direct methods (eg. light or low angle x-ray scattering) can be applied to the functioning enzyme.

STEP 5: ELECTRON TRANSFER FROM Fe PROTEIN TO Mo-Fe PROTEIN

In this step Kp1 becomes super-reduced, epr silent and has a distinctive Mössbauer spectrum. The electron transfer step is very rapid ($k = 2 \times 10^2 s^{-1}$) and cannot be rate-limiting in nitrogenase function; its rate is decreased by MgADP[16]. The individual epr signals of Kp1 and Kp2 become much diminished in intensity – indicating that the oxidized form of Kp2 and the super-reduced form of Kp1 are the dominant species during turnover. When nitrogenase is functioning under Ar, a new epr signal appears[22] with g values of 2.14, 2.001 and 1.976. This signal is assignable to Fe in Mo-Fe protein and is abolished by C_2H_2; it is enchanced about 2-fold by nitrogen, which allows the calculation of an apparent binding constant of *ca* 0.1 atm, close to the apparent Km for N_2 (Eady, Thorneley and Lowe, unpublished). It integrates to only 0.02 electrons/mole and indicates the existence of a distinct transient species in functioning nitrogenase which may be involved in the H_2 evolution reaction.

STEP 6: ELECTRON TRANSFER TO REDUCIBLE SUBSTRATE

In collaboration with Professor Mortenson[23,26] we observed that hybrid nitrogenase composed of Cp2 + Kp1 showed different kinetics according to reducible substrate: H^+

TABLE 4

ELECTRON DISTRIBUTION BETWEEN HYDROGEN AND NITROGEN AS A FUNCTION OF TEMPERATURE
(Assays run under nitrogen, data from a more elaborate table of Thorneley & Eady[24]).

Temperature (^{O}C)	10	30
Kp1 (μM)	0.9	2
Kp2 (μM)	0.9	2.2
Ratio Kp1/Kp2	1	0.95
NH_3 production (nmole/mg/min)	<5	371
H_2 production (nmole/mg/min)	29	684
% electron flow to H_2	>86	55

was reduced at once, with linear kinetics; C_2H_2 was reduced with a lag of some
50 x turnover time and N_2 was reduced with a much longer lag (200 x turnover time).
More recently we have observed similar effects with homologous Kp nitrogenase[24]:
a pronounced imbalance between Kp1 and Kp2, or a low temperature at a normal ratio
of the two proteins, caused a burst of H_2 evolution and a lag in reduction of C_2H_2.
At low temperatures electron flow was not necessarily constant: acetylene stimulated electron flow to hydrogen and, after a lag, to itself; at this temperature
little N_2 was reduced and most of the electrons were directed to hydrogen evolution
(Table 4). These findings regarding the apportioning of electrons between substrates
suggest that different activation states of the complex are required for reduction
of acetylene and nitrogen compared with hydrogen ions. They are consistent with the
view that the loci of reduction of these three substrates are different, but they
do not compel that view.

If binding of C_2H_2 to the complex alters its activation state and permits increased electron flow, this might be reflected in the epr spectra of the functioning system. We find that, in an active fixing system under argon, Kp1 binds C_2H_4
(K_D = 1.3 mM), the product of acetylene reduction, to give a sharp axial epr signal
with g_{11}2.125 and g_1 = 2.000, assignable to Fe according to ^{57}Fe substitution. The
competitive interaction of these substances permits binding constants to be assessed;
the kinetics imply multiple binding sites with one approximating to the apparent
Km for acetylene reduction.

STEP 7: DISSOCIATION OF COMPLEX AND RELEASE OF PRODUCTS

The inhibitory effect of MgADP on electron input into nitrogenase (step 1) slows
down the reaction by more than 10-fold (τ<10ms to $\tau \sim$ 90ms) but the reaction remains
fast compared with the turnover time of the enzyme (τ =500ms). The most plausible
explanation of the fact that the Fe protein is predominantly oxidized in the functioning enzyme is therefore that yet another step is rate-determining, and a tempting
view is that the appropriate step is the dissociation of oxidized Fe-protein from
the complex, a process which would limit the availability of free oxidized Fe

protein to initiate step 1. MgADP does in fact influence dissociation of the complex since it augments the familiar dilution effect.[16]

CONCLUSIONS

Our opening sentence might be held by the cynical to be a sufficient conclusion. However, awareness of the complexity of a system is not incompatible with increased understanding of its nature and our findings, together with those of other research groups, are converging towards a detailed account of the workings of this unusual enzyme at a molecular level. None of the questions raised in our introduction has received a definitive answer, but we have become more confident that the working scheme represents some sort of reality: that a nitrogenase complex not only really exists but that it assumes various activation states; that these activation states influence the distribution of electrons to various substrates as well as total electron flow; that MgATP and MgADP are not only source and produce of energy input respectively, but that they stabilize active and inactive states of the complex - perhaps *via* their effect on the Fe protein - thereby exerting their widely assumed regulatory effect.

ACKNOWLEDGEMENTS

We thank Mrs. Flora Ivers for typing this manuscript. One of us (JRP) thanks the Rockefeller Foundation for financial support and Dr. H.J. Evans for hospitality at Oregon State University, where the contribution was prepared.

REFERENCES

1. Kennedy, C., Eady, R.R., Kondorosi, E. and Rekosh, D. (1976) Biochem. J. 155, 383-389.

2. Smith, B.E. and Lang, G. (1974) Biochem. J. 137, 169-180.

3. Smith, B.E., Lowe, D.J. and Bray, R.C. (1972) Biochem. J. 130, 641-643.

4. Smith, B.E., Lowe, D.J. and Bray, R.C. (1973) Biochem. J. 135, 331-341.

5. Thorneley, R.N.F., Eady, R.R., Smith, B.E., Lowe, D.J., Yates, M.G., O'Donnell, M.J. and Postgate, J.R. (1978) In: Nitrogen Fixation in Plants, 6th Long Ashton Symposium, Bristol, U.K.

6. Orme-Johnson, W.H., Hamilton, W.D., Ljones, T., Tso, M.W., Burris, R.H., Shah, V.K. and Brill, W.J. (1972) Proc. Natl. Acad. Sci., USA 69, 3142-3145.

7. Munck, E., Rhodes, H., Orme-Johnson, W.H., Davis, L.C., Brill, W.J. and Shah, V.K. (1975) Biochim. Biophys. Acta. 400, 32-45.

8. Mortenson, L.E., Walker, M.N. and Walker, G.A. (1976) Proc. 1st Intern. Symp. on Nitrogen Fixation, Pullman, Washington, USA 1, 117-149.

9. Thorneley, R.N.F., Yates, M.G. and Lowe, D.J. (1976) Biochem. J. 155, 137-144.

10. Tso, M.W. and Burris, R.H. (1973) Biochim. Biophys. Acta. 309, 263-270.

11. Yates, M.G., Thorneley, R.N.F. and Lowe, D.J. (1975) FEBS Lett. 60, 89-92.

12. Yates, M.G. (1972) FEBS Lett. 27, 63-67.

13. Yoch, D.C. (1974) J. Gen. Microbiol. 83, 153-164.

14. Thorneley, R.N.F. (1974) Biochim. Biophys. Acta 358, 247-250.

15. Thorneley, R.N.F. and Willison, K.R. (1974) Biochem. J. 139, 211-214.

16. Thorneley, R.N.F. and Cornish-Bowden, A. (1977) Biochem. J. 165, 255-262.

17. Eady, R.R., Kennedy, C., Smith, B.E., Thorneley, R.N.F., Yates, M.G. and Postgate, J.R. (1975) Biochem. Soc. Trans. 3, 488-492.

18. Eady, R.R. (1973) Biochem. J. 135, 531-535.

19. Thorneley, R.N.F., Eady, R.R. and Yates, M.G. (1975) Biochim. Biophys. Acta 403, 269-284.

20. Reithel, F.J. and Eady, R.R. (1977) Biochem. J. (in press).

21. Emerich, D.W. and Burris, R.H. (1976) Proc. Natl. Acad. Sci., USA 73, 4369-4373.

22. Yates, M.G. and Lowe, D.J. (1976) FEBS Lett. 72, 121-126.

23. Smith, B.E., Thorneley, R.N.F., Eady, R.R. and Mortenson, L.E. (1976) Biochem. J. 157, 439-447.

24. Thorneley, R.N.F. and Eady, R.R. (1977) Biochem. J. (in press).

25. Gillum, W.O., Mortenson, L.E., Chen, J.-S. and Holm, R.H. (1977) J. Amer. Chem. Soc. 99, 584-595.

26. Smith, B.E., Eady, R.R., Thorneley, R.N.F., Yates, M.G. and Postgate, J.R. (1977) Proc. 2nd Intern. Symp. on Nitrogen Fixation, Salamanca, Spain (in press).

27. Thorneley, R.N.F. (1975) Biochem. J. 145, 391-396.

28. Walker, G.A. and Mortenson, L.E. (1973) Biochem. Biophys. Res. Comm. 53, 904-909.

29. Thorneley, R.N.F. and Eady, R.R. (1973) Biochem. J. 133, 405-408.

30. Haaker, H. and Veeger, C. (1975) Eur. J. Biochem. 77, 1-10.

TWO IRON-SULPHUR CENTERS IN QH_2:CYTOCHROME c OXIDOREDUCTASE

SIMON DE VRIES, SIMON P.J. ALBRACHT AND FRANS J. LEEUWERIK

Laboratory of Biochemistry, B.C.P. Jansen Institute, University of Amsterdam, Plantage Muidergracht 12, Amsterdam, The Netherlands

ABSTRACT

1. The EPR signal at 36 K of QH_2:cytochrome c oxidoreductase reduced with ascorbate is an overlap of two signals in a 1:1 weighted ratio. Both signals are due to $[2Fe-2S]^{3-}$ centers.

2. From the signal intensity it is computed that the total concentration of Fe-S centers is equal to that of cytochrome c_1. The consequences for the basic enzymatically active unit are discussed.

3. The line shape of one of the Fe-S centers, defined as center 1, changes on further reduction of the b-c_1 complex. This change cannot be correlated with changes of the redox state of any of the cytochromes and is reversible. It is completed at a redox potential of about 50 mV at pH 7.2.

INTRODUCTION

In 1964 Rieske et al.[1-3] discovered that reduced QH_2:cytochrome c oxidoreductase (Complex III) exhibits at 100 K an EPR signal characteristic of non-heme iron. Since the isolated enzyme contains 2-3 atoms non-heme iron per molecule of cytochrome c_1[2,3] and the signal has g-values typical of a 2Fe-2S protein[4] this new redox carrier, often referred to as the Rieske Fe-S center, was generally assumed to be a [2Fe-2S] protein. This has recently been confirmed[5] with EPR studies of ^{57}Fe-containing submitochondrial particles (SMP).

An analysis of the line shape of the EPR signal shows that it is a superposition of two signals with different properties. This will be elucidated in this paper.

MATERIALS AND METHODS

Complex III was prepared from Complex II-III by the method of Hatefi et al.[6]. The protein was dissolved in 0.66 M sucrose - 50 mM Tris-HCl buffer (pH 8.0). When other pH values were required this solution was diluted 100 times in 0.66 M sucrose - 50 mM potassium phosphate buffer (pH 6.5) or in 0.66 M sucrose - 50 mM Tris-HCl buffer (pH 7.2 or 9). The protein was then precipitated with 55% saturated ammonium-sulphate and resuspended in the same medium.

Mg-ATP particles were prepared by the method of Löw and Vallin[7]. Some preparations were made from beef-heart mitochondria, that had been freed from cytochrome c as described by Tsou[8]. The succinate dehydrogenase in the particles was activated by an incubation for 20 min at $30^{\circ}C$ with 10 mM fumarate. After washing, the SMP were suspended in 0.25 M sucrose - 50 mM potassium phosphate buffer (pH 7.2) to a concentration of 100 mg of protein per ml. The cytochrome c_1 content, determined optically using a $A_{553-539}$ (red-ox) = 20.1 $mM^{-1} \cdot cm^{-1}$ (ref. 9) was about 0.35 nmol/mg protein.

EPR spectra were recorded with a Varian E-3 or E-9 spectrometer. Spectra were digitized and simulated as earlier described[10]. Low-temperature diffuse reflectance spectra were recorded with a home-built integrating sphere attached to an Aminco DW-2 spectrophotometer. Low temperatures were obtained with a gas-flow essentially as described by Lundin and Aasa[11], except that the integrating sphere instead of an EPR cavity was used to monitor the EPR tube. The spectrophotometer was connected to a microcomputer that enabled automatic baseline correction and further digital manipulations of the spectral traces.

RESULTS

The most suitable temperature for studying the EPR signal of reduced Complex III is 35-40 K. As clear from Fig. 1 the spectrum at 36 K is different from that at 85 K in relative line amplitudes and line widths. Note that the low-field shoulder at the g_z peak in the 36 K spectrum is absent at 85 K. Fig. 2 shows that the signal obtained either with isolated Complex III or with SMP reduced with ascorbate cannot be simulated with a single S = ½ signal. There are several differences: a) There is no low-field shoulder at the g_z in the simulation; b) The line shape of the g_y differs especially in the positive part; c) The broad shoulder at the high field side of the g_x is missing in the simulation; d) The relative amplitudes of the lines do not fit, although the line widths do. All these differences can be eliminated by adding a second S = ½ signal. The result is shown in Fig. 3. The extra signal is different in g-values and line widths. We will define the signal with the narrow line widths as center 1, the other as center 2. The relative intensities used for Fig. 3 were 1 for center 1 and 1.22 for center 2. The ratio varied with the preparation. The ratio needed for simulation of EPR spectra of SMP reduced with ascorbate always equalled 1:1 (not shown).

Results of absolute double integration with copper perchlorate as a standard[12] are displayed in Table 1. Complex III prepared from Com-

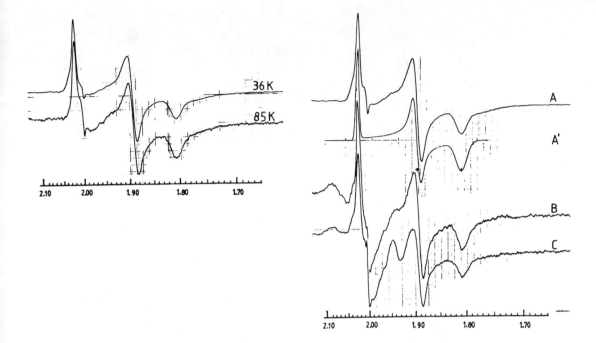

Fig. 1. Effect of the temperature on the EPR spectrum of Complex III reduced with ascorbate. Complex III (pH 8.0) was mixed with 5 mM ascorbate and frozen after 30 s at 0^OC. EPR conditions: microwave frequency (F), 9318 MHz; temperature (T), 36 K for the upper trace and 85 K for the lower one; microwave power (P), 2 mW; modulation amplitude (MA), 0.63 mT; scanning rate (SR), 25 mT/min; modulation frequency, 100 kHz for these and all other EPR spectra. The gain for the lower trace was 3.1 times that of the upper.

Fig. 2. Comparison of the EPR signal of the "Rieske Fe-S" center in Complex III and submitochondrial particles (SMP) with a $S = \frac{1}{2}$ simulation. A: same spectrum as upper trace in Fig. 1. A': computed line shape of a single $S = \frac{1}{2}$ signal with best fitting parameters: $g_{z,y,x}$ = 2.0188, 1.8914, 1.8046 and widths (z,y,x) = 1.25 mT, 2.4 mT, 4.2 mT. B: Mg-ATP SMP, low in cytochrome c content (pH 7.2) + 2.5 μM PMS (phenazine methosulphate) + 2.5 mM KCN mixed with 90 mM fumarate + 0.9 mM succinate for 15 sec at 20^OC and then frozen in liquid nitrogen. C: Mg-ATP SMP as in B, with added cytochrome c (20 μM), (pH 7.2) mixed with 12 mM ascorbate plus 3 μM TMPD for 1 min at 0^OC before freezing. EPR conditions: For A: same as in Fig. 1, upper trace. For B and C: F, 9316 MHz; T, 36 K; P, 5 mW; MA, 0.63 mT; SR, 12.5 mT/min.

plex II-III usually contained 10% less Fe-S centers than SMP. When the complex was prepared from Complex I-III, only 50-60% of the total Fe-S was recovered relative to cytochrome c_1.

Rieske et al.[1,2] already observed that the line shape of the EPR signal is dependent on the reductant used. This was reported in somewhat more detail by Orme-Johnson et al.[13]. The isolated Fe-S protein does not show this effect[3].

We have systematically studied this phenomenon to find the reason for the line-shape change. A typical example of a reducing titration is shown in Fig. 4. As can be seen the most prominent change is that

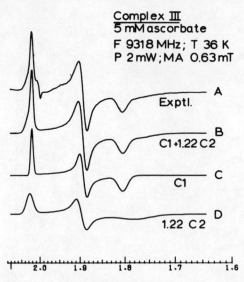

Complex III
5 mM ascorbate
F 9318 MHz; T 36 K
P 2 mW; MA 0.63 mT

A   Exptl.

B   C1+1.22 C2

C   C1

D   1.22 C2

Fig. 3. Simulation of the EPR signal of reduced Complex III. A: Experimental
spectrum of Complex III as in Fig. 1 upper trace. B: simulation as a superposition
of two $S = \frac{1}{2}$ signals, plotted separately as C and D. The relative weight for C and
D are 1 and 1.22, respectively. EPR conditions for A: as in Fig. 1 upper trace.
Simulation parameters for C: $g_{z,y,x}$ = 2.0188, 1.8914, 1.8046; width (z,y,x) = 1.25 mT,
2.4 mT, 4.2 mT. Parameters for D: $g_{z,y,x}$ = 2.0242, 1.8946, 1.7746; width (z,y,x) =
3.1 mT, 3.4 mT, 13 mT.

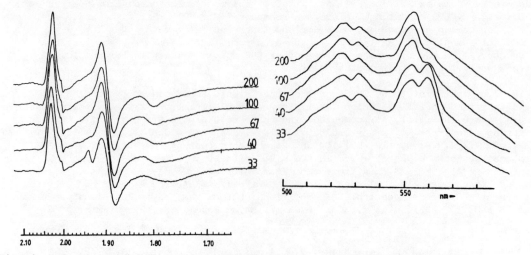

Fig. 4. Reductive titration of Complex III with the succinate-fumarate couple.
Complex III (pH 7.2) was mixed with 4 μM PMS, 90 mM fumarate and 0.9 mM succinate
for 5 min at 20°C and then frozen. After recording the EPR and optical spectra,
the EPR tube was thawed and succinate was added to lower the fumarate/succinate
ratio. After 5 min at 20°C the contents were again frozen. This was repeated several times. The fumarate/succinate ratio is indicated in the figure. A: EPR spectra
at F, 9.3 GHz; T, 36 K; P, 2 mW; MA, 0.63 mT; SR, 25 mT/min. The gain is the same
for all spectra. B: Diffuse reflectance spectra at T, 110 K; SR, 0.2 nm/sec; slit
width (SW), 1.0 nm.

of the g_x of center 1. This change seems to accompany the reduction
of cytochrome b_{562}, as earlier suggested by Orme-Johnson et al.[13].

TABLE 1

CONCENTRATION OF THE Fe-S CENTERS IN QH_2:CYTOCHROME c OXIDOREDUCTASE

Preparation	Reductant	Total Fe-S signal µM (S = ½)	Cytochrome c_1 concentration µM	Ratio Fe-S/c_1
Complex III	Ascorbate	229	257	0.89
Complex III	Succinate	231	257	0.90
Complex III	Dithionite	261	304	0.86
Mg-ATP SMP	Ascorbate	34.6	33.5	1.03
Mg-ATP SMP	Succinate	33.7	33.5	1.01

The same effect was seen in SMP. That this effect is reversible is shown in Fig. 5. Here we show the results with SMP, but Complex III behaves in exactly the same way. Again it can be seen that it is the

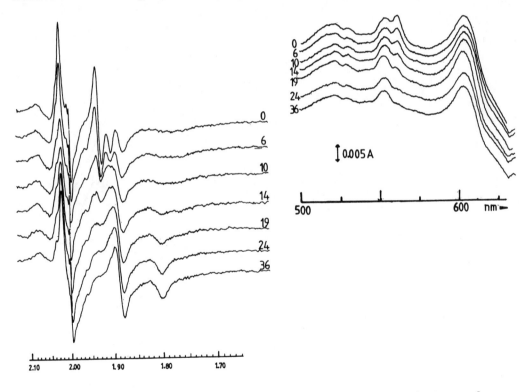

Fig. 5. Oxidative titration of Mg-ATP SMP with the fumarate-succinate couple. Mg-ATP SMP (pH 7.2) in the presence of 2.5 mM KCN were mixed with 2.5 mM succinate for 5 min at 20°C before freezing. The further procedure is as in Fig. 4, but now fumarate was added. The fumarate/succinate ratio is indicated in the figure. A: EPR spectra at F, 9.3 GHz; T, 36 K; P, 5 mW; MA, 0.63 mT; SR, 12.5 mT/min. The gain is the same for all spectra. B: Diffuse reflectance spectra at T, 110 K; SR, 1 nm/sec; SW = 2.0 nm.

line shape of center 1 that is affected. Owing its large line width, it is not possible to detect in Fig. 5A whether the line shape of center 2 is affected. However, since the line shape obtained with succinate (or dithionite) cannot be simulated satisfactorily by one $S = \frac{1}{2}$ signal, it follows that the broadened line shape of center 1 is different from the shape of center 2 and center 1 and 2 are thus different in shape under all conditions used. The total signal intensity does not change when the line shape change of center 1 occurs.

When the pH was varied, no correlation of the line shape change of center 1 and the degree of reduction of cytochrome b_{562} could be detected. An example is shown in Fig. 6. At pH 6.5 the change of line shape occurs without a redox change of the (oxidized) cytochrome b_{562}. This is also the case at pH 8.0 and 9.0, when cytochrome b_{562} remains reduced. Under all these conditions the cytochromes b_{566} and b_{558} were not reduced as judged from the EPR absorption around $g = 3.8$ at 15 K[13].

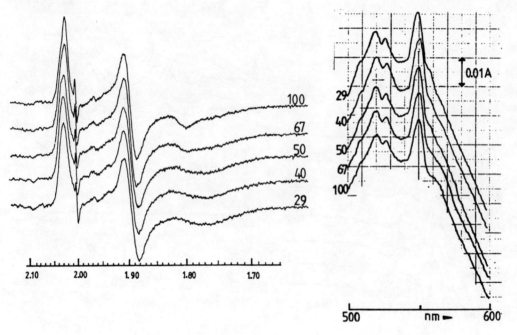

Fig. 6. Reductive titration of Complex III with the succinate-fumarate couple. Condition as in Fig. 4, except for the pH which was 6.5.

DISCUSSION

The findings described above create a problem concerning the basic enzymatic unit of QH_2:cytochrome c oxidoreductase, since the ratio of the intensities of the two Fe-S signals is precisely 1:1 in SMP, whereas their total concentration is equal to that of cytochrome c_1.

Important thereby is the molecular weight of the isolated Complex III. Rieske[14] reported a sedimentation coefficient ($s_{20, w}^o$) of 10.2 S. Using the average partial specific volume of globular proteins (0.75 cm^3 per g) and a diameter of 8.5 nm, Rieske estimated the molecular weight to be 250 000. Preliminary experiments with our Complex III preparations at 8 and 25 mg of protein per ml gave similar $s_{20, w}^o$ values as those found by Rieske[14], but since isolated Complex III will undoubtedly have detergent molecules bound, the partial specific volume of this detergent-lipoprotein complex can differ substantially from the average value of 0.734 cm^3 per g. This was recently demonstrated for isolated Complex I, which was found to contain 2 FMN molecules per molecule[15]. Further work is required to establish the molecular weight.

If the molecular weigth is insufficient to permit 2 molecules of cytochrome c_1 per protein molecule, then the alternative possibility, namely the existence of two different types of Complex III both with a MW of 250 000, one of them containing center 1 and the other center 2 must be considered. Since 3-4 cytochrome b types exist[16,17] one cannot predict how these would be divided over the two different complexes. From EPR studies we have strong indications that the stoichiometry of the 3-4 b-type cytochromes would allow such a division (S. de Vries and S.P.J. Albracht, unpublished). In any case, the stoichiometry within the total complex or complexes is $b:c_1$: center 1: center 2 = 4:2:1:1 which makes the existence of 3-4 different b-type cytochromes more plausible.

We would like to point out that existence of two different EPR signals, very close in g-values, does not necessarily mean that the two different clusters are embedded in different polypeptide chains. It simply indicates that the immediate environment of both clusters is slightly different.

We do not know what triggers the change of line shape of center 1. It is not related to redox changes of the cytochromes. At all pH values used (6.5, 7.2, 8.0 and 9.0) the change occurred at roughly the same fumarate/succinate ratio. This indicates that the broadening of center 1, that is completed at an estimated redox potential of 50 mV at pH 7.2, might parallel the redox change of a component having the same pH dependency of the midpoint potential as the fumarate/succinate couple. We will investigate whether Q-10 could be this redox component.

Since the g_z lines of center 1 and 2 nearly coincide, the analysis of the overall g_z line shape observed in ^{57}Fe-containing SMP of

Candida utilis[5] indicates that both centers are [2Fe-2S] types.

We like to point out that centers 1 and 2 have no relation to the signal described by Lee and Slater[18] in pigeon heart mitochondria. Complex III either reduced by ascorbate or by dithionite did not show extra EPR signals on lowering the temperature down to 7 K.

ACKNOWLEDGEMENTS

We thank Prof. E.C. Slater for his valuable criticism and for reading the manuscript. We also like to thank Mr. H.P.M. Plat for carrying out the analytical ultracentrifuge experiments and Mr. F.P. A. Mol and Mr. H. de Haan for their help with the construction of the optical reflectance unit. Part of this work has been supported by grants from the Netherlands Organization for the Advancement of Pure Research (Z.W.O.) under the auspices of the Netherlands Foundation for Chemical Research (S.O.N.).

REFERENCES

1. Rieske, J.S., Hansen, R.E. and Zaugg, W.S. (1964) J. Biol. Chem. 239, 3017-3022

2. Rieske, J.S., Zaugg, W.S. and Hansen, R.E. (1964) J. Biol. Chem. 239, 3023-3030

3. Rieske, J.S., MacLennan, D.H. and Coleman, R. (1964) Biochem. Biophys. Res. Commun. 15, 338-344

4. Gibson, J.F., Hall, D.O., Thornley, J.H.M. and Whatley, F.R. (1966) Proc. Natl. Acad. Sci. U.S. 56, 987-990

5. Albracht, S.P.J. and Subramanian, J. (1977) Biochim. Biophys. Acta 462, 36-48

6. Hatefi, Y., Haavik, A.G. and Jurtshuk, P. (1961) Biochim. Biophys. Acta 52, 106-118

7. Löw, H. and Vallin, I. (1963) Biochim. Biophys. Acta 69, 361-374

8. Tsou, C.L. (1952) Biochem. J. 50, 493-499

9. Berden, J.A. and Slater, E.C. (1970) Biochim. Biophys. Acta 216, 237-249

10. Albracht, S.P.J., Dooijewaard, G., Leeuwerik, F.J. and Van Swol, B. (1977) Biochim. Biophys. Acta 459, 300-317

11. Lundin, A. and Aasa, R. (1972) J. Magn. Res. 8, 70-73

12. Aasa, R. and Vänngård, T. (1975) J. Magn. Res. 19, 308-315

13. Orme-Johnson, N.R., Hansen, R.E. and Beinert, H. (1974) J. Biol. Chem. 249, 1928-1939

14. Rieske, J.S. (1976) Biochim. Biophys. Acta 456, 195-247

15. Dooijewaard, G., De Bruin, G.J.M., Van Dijk, P.J. and Slater, E.C. (1978) Biochim. Biophys. Acta, in press

16. Berden, J.A., Opperdoes, F.R. and Slater, E.C. (1972) Biochim. Biophys. Acta 256, 594-599

17. Davis, K.A., Hatefi, Y., Poff, K.L. and Butler, W.L. (1973) Biochim. Biophys. Acta 325, 341-356

18. Lee, I.-Y. and Slater, E.C. (1974) Biochim. Biophys. Acta 347, 14-21

ON THE MULTIPLICITY OF MICROSOMAL N-OXIDASE SYSTEMS

JOHN W. GORROD

Department of Pharmacy, Chelsea College (University of London), Manresa Road, London SW3 6 LX, U.K.

ABSTRACT

Work leading to the recognition of multiple enzyme systems able to carry out the biological oxidation of nitrogen in organic molecules is reviewed. Attempts at the differentiation of these processes and correlation with the pKa of the susceptible nitrogen are described, together with recent work on the enzymology of two distinct tertiary amine oxidase systems.

INTRODUCTION

Metabolic attack on nitrogen in organic molecules of diverse chemical structure occurs in a wide variety of biological systems[1,2,3]. Many of these reactions can be carried out in vitro by microsomal preparations fortified with NADPH and molecular oxygen[4,5,6]. Early work on the mechanism of the oxidation of nitrogen in organic molecules used trimethylamine[7] or dimethyl-aniline[8] as substrate. The results from these systems were compared with data obtained using aromatic primary amines[9], amides[10], carbamates[11] and heterocyclic compounds[12]. More recently it has been recognised that aliphatic primary and secondary amines are also substrates for metabolic oxidation of the constituent nitrogen. Other groups which are biologically N-oxidised are imines[14], hydrazines[15], barbiturates[16] and benzodiazepines[17]. From the foregoing it can be seen that the substrates used possessed very different physico-chemical properties. The results which were obtained precluded a single enzyme system being involved in all types of "nitrogen" oxidation.

DISCUSSION

It was in an effort to understand the factors which governed the oxidation of particular types of nitrogen compounds that the published data on this process was examined. Using a number of parameters an attempt was made to differentiate between these processes and relate them to the structure of the nitrogenous compound oxidised. In order to characterise the various nitrogen types I used a rather simplistic scheme based on the pKa of the metabolically vulnerable nitrogen[18] (Fig. I). Using the parameters referred to in Table I it appeared that compounds in group I and IIc were predominantly oxidised by a flavoprotein system, whereas aromatic amines (IIa),

	PRIMARY (a)	SECONDARY (b)	TERTIARY (c)
GROUP I (pK_a 8–11)	RCH_2NH_2	RCH_2NHR'	$RCH_2NR'R''$
PRODUCT	RCH_2NHOH	$RCH_2N(OH)R'$	$RCH_2N(\rightarrow O)R'R''$
GROUP II (pK_a 1–7)	$ArNH_2$	$ArNHR$	$ArNRR'$
PRODUCT	$ArNHOH$	$[ArN(OH)R]$	$ArN(\rightarrow O)RR'$
GROUP III (pK_a below 1)	$R\overset{\|}{\underset{O}{C}}NH_2$	$R\overset{\|}{\underset{O}{C}}NHR'$	$R\overset{\|}{\underset{O}{C}}NR'R''$
PRODUCT	$R\overset{\|}{\underset{O}{C}}NHOH$	$R\overset{\|}{\underset{O}{C}}N(OH)R'$	[ ]

Fig. 1. Differentiation of the various types of nitrogen and their N-oxidation products.

carbamates (IIIa) and amides (IIIb) were oxidised via a cytochrome P-450 mediated system. Since this concept was proposed new data has become available[15,16,17] which indicates that the original classification does not provide a framework in which all the structures which can be N-oxidised can be meaningfully considered.

TABLE I SOME PARAMETERS USED IN THE DIFFERENTIATION OF VARIOUS TYPES OF N-OXIDATION

PARAMETERS	GROUP I and IIc	GROUP III and IIa
Age of animal	Present at birth and in foetal tissue	Low at birth Absent in foetal tissue
pH Optima	8.3	7.4
Inducing agents	No effect	Increased by pretreatment with PB or MC
Inhibitory agents	Not inhibited by SKF 525A or CO	Inhibited by SKF 525A and CO
Treatment of microsomes	Not affected by mild treatment i.e. ageing detergents UV light ultrasonics	Activity decreased by mild treatment i.e. ageing detergents UV light ultrasonics
Purified enzyme	FAD containing flavoprotein (Ziegler)	Cytochrome P-450 (Lotlikar)

This classification, which was based on the pKa of the molecule oxidised, was originally used as it was thought that the pKa would give a useful indication as to the availability of the nitrogen lone pair electrons. It appears from the structures cited above that the hybridisation of the nitrogen atom has also to be considered as recent work[20,21] has shown that even when compounds have very similar pKa values the enzymology may be different. This is exemplified by results obtained on the N-oxidation of N-ethyl-N-methylaniline (pKa 5.95) and 3-methyl-pyridine (pKa 5.69).

Many oxidative reactions involving drugs and foreign compounds can be inhibited by SKF525A (β-diethylaminoethyl diphenylpropylacetate). Therefore it was of interest to compare the effect of this compound on various types of N-oxidation. It had earlier been shown to be without effect on the N-oxidation of tertiary aliphatic and aromatic amines[18], and certain aliphatic secondary and primary amines[13], although SKF525A inhibits the N-oxidation of phentermine, a primary aliphatic amine, and an imine[14]. SKF525A also inhibits the N-hydroxylation of primary, but not secondary or tertiary aromatic amines[23]. The failure of SKF525A to inhibit aromatic tertiary amine-N-oxidation was confirmed[21] whereas the N-oxidation of heterocyclic aromatic amines was found to be inhibited[20]. Another compound, widely used as a metabolic inhibitor is DPEA (2:4-dichloro-6-phenylphenoxyethylamine). When the N-oxidation of the two types of tertiary amine was examined in the presence of DPEA, the tertiary aniline N-oxidase was activated whereas the heterocyclic-N-oxidase was inhibited. These effects were accentuated as the concentration of DPEA in the incubate was increased. (Table 2). The oxidation of carbon in both molecules, i.e. N-demethylation or carbinol formation, was inhibited at all concentrations of DPEA used. The activation observed using N-methyl-N-ethylaniline as substrate is reminiscent

TABLE 2   THE EFFECT OF INCREASING CONCENTRATIONS OF DPEA ON THE C- AND N-OXIDATION OF N-ETHYL-N-METHYLANILINE AND 3-METHYLPYRIDINE

DPEA	N-Ethyl-N-methylaniline		3-Methylpyridine	
Concn.	N-oxidation	N-demethylation	N-oxidation	C-oxidation
Nil	100	100	100	100
10^{-5}M	126	87	94	76
10^{-4}M	187	65	83	56
10^{-3}M	222	35	22	18

DPEA = 2:4-dichloro-6-phenylphenoxyethylamine

of the activation of purified porcine amine oxidase which occurs with a number of primary aliphatic amines[24], and is maximal with n-octylamine. A similar activation of N-oxidation of the tertiary aniline was shown with octylamine (Table 3), a compound which inhibited pyridine

N-oxidation and the oxidation of the carbon centres. Naphthylthiourea, a substance reported to be a good inhibitor of the purified amine oxidase[25] did inhibit the N-oxidation of N-ethyl-N--methylaniline to a greater extent than the other oxidations studied. Incorporation of cyanide into the incubate failed to significantly effect either N-oxidation reactions, but seemed to stimulate the oxidation of 3-methylpyridine to the carbinol. Quinuclidine, a compound in which the nitrogen lone pair should be minimally hindered also failed to differentiate between the processes involved in the oxidation of the two types of substrate. Surprisingly, Bromazepam, a cyclic imine, caused a stimulation of the N-oxidation of 3-methylpyridine and an inhibition of the N-oxidation of the tertiary aniline. Bromazepam also caused an activation of the N-hydroxylation of some aromatic amines[26].

TABLE 3   THE EFFECT OF CERTAIN COMPOUNDS ON THE MICROSOMAL C- AND
N-OXIDATION OF N-ETHYL-N-METHYLANILINE AND 3-METHYLPYRIDINE.
(RESULTS ARE EXPRESSED AS PERCENTAGE ACTIVITY OF THE PREPARATION
WITHOUT ANY ADDITION).

Substrate	N-Ethyl-N-methylaniline		3-Methylpyridine	
Addition (10^{-4}M)	N-oxidation	N-demethylation	N-oxidation	C-oxidation
Octylamine	225	30	57	40
Naphthylthiourea*	58	95	61	55
Bromazepam	33	100	155	65
Cyanide	103	102	114	140
Amylobarbitone	76	65	76	50
Quinuclidine	110	98	92	88

* 10^{-5}M

The results obtained above, suggested that the pyridines were oxidised via a cytochrome P-450 system, therefore the metabolism of the two substrates were studied in a system in which carbon monoxide was introduced into the atmosphere. The results (Table 4) show that under these conditions pyridine N- and C-oxidation and the N-demethylation of N-ethyl-N-methylaniline were inhibited whereas the N-oxidation of the latter substrate was not. This clearly implicates a form of cytochrome P-450 in the N-oxidation of aromatic heterocyclic amines. In order to gain further evidence as to whether cytochrome P-450 or P-448 was involved in this reaction, experiments were carried out using hepatic microsomal preparations derived from rabbits which had been pretreated with either phenobarbitone or 3-methylcholanthrene. These results (Table 5) clearly show that pretreatment with phenobarbitone but not 3-methylcholanthrene induced pyridine-N-oxidase indicating that cytochrome P-450 was involved rather than cytochrome P-448. Similar results are shown for the N-oxidation of 4-substituted anilines[26], however the N-oxidation of tertiary anilines was only slightly increased by phenobarbitone pretreatment. No correlation could be

TABLE 4   THE EFFECT OF CARBON MONOXIDE ON THE C- AND N-OXIDATION OF
N-ETHYL-N-METHYLANILINE AND 3-METHYLPYRIDINE

Incubation atmosphere Incubate	N-Ethyl-N-methylaniline		3-Methylpyridine	
CO : O_2	N-oxidation	N-demethylation	N-oxidation	C-oxidation
5 : 1	116	56	44	35
10 : 1	100	40	36	25

found between either the pKa or the partition coefficient of the substrate and the effect of
inducing agents on the N-oxidations studied.

TABLE 5   THE EFFECT OF PRETREATING RABBITS WITH EITHER 3-METHYLCHOLANTHRENE
OR PHENOBARBITONE ON THE N-OXIDATION OF SUBSTITUTED PRIMARY AND
TERTIARY ANILINES AND PYRIDINES

Substrate	N-Ethyl-N-methylanilines			Pyridines			Anilines		
Substituent	4-CH$_3$	4-H	4-Cl	3-CH$_3$	3-H	3-Cl	4-CH$_3$	4-H	4-Cl
pKa	6.39	5.95	5.15	5.65	5.20	2.84	5.12	4.68	4.05
Log P	2.86	2.74	3.60	1.25	0.71	1.39	1.46	0.99	1.86
Methylcholan-threne[1]	81	81	87	98	106	94	200	114	148
Phenobarbitone[2]	112	150	188	700	1300	340	1960	1240	1430

Results are expressed as percentage change compared with control animal and are related to
activity per gram of liver incubated at 37° for 10 mins.   1. 40 mg.Kg.i.p. for 1 day,   2. 80 mg.
Kg. i.p. for 3 days.

It has been suggested that the stimulatory effect on certain oxidations sometimes observed when
NADH is added to microsomal preparations already fully saturated with NADPH is due to the involvement of cytochrome b$_5$[27] , such an involvement has never been proposed for the flavoprotein
system. The results obtained with N-ethyl-N-methylaniline and 3-methylpyridine are shown in
Figure 2. Maximal metabolism was determined to occur with about 5µmols NADPH present in the
incubate. In order to ensure saturation of this pathway the coenzyme was increased to 10 µmols
and when 2 µmols NADH were added a stimulation of N-demethylation (28%) and N-oxidation of
the pyridine (13%) were observed. No stimulation of the N-oxidation of the tertiary aniline was
observed with the addition of between 2-10µmols NADH. The C-oxidation of 3-methylpyridine
was not stimulated with 2 µmols NADH but was (12%) when 4 µmols of NADH was used in addition
to 10µmols NADPH.

These results are further evidence as to the differences in the enzymology of N-oxidation. The
present situation is summarised in Table 6 where similarities can be seen in the properties of the

enzymes involved in Group I and IIc, which are different from those involved in the N-oxidation of Group IIa, Pyridines and the Imine. The N-oxidation of Group IIIb appears to be mediated via cytochrome P-448 as the terminal oxidase. It may be that Group I substrates will have to be further differentiated as it has been claimed[28] that most primary aliphatic amines are not substrates for the purified amine oxidase and indeed evidence has been presented for the involvement of cytochrome P-450 in the N-oxidation of phentermine[22] and β,β-difluoroamphetamine[29]. Further work will have to be performed in order to see where benzodiazepines[17], barbiturates[16] and other nitrogen containing compounds fit in with the three processes so far indicated. A further factor is the possibility that certain compounds may be substrates for more than one enzyme[19,30]. In these cases the enzymology of the oxidation will depend upon the amounts of the various enzymes present as well as the affinity of the substrate for the enzyme. Such a phenomenon could explain the slight induction of the tertiary aniline N-oxidase we observe after pretreatment of rabbits with phenobarbitone (Table 5), whereas most of the other evidence points to a non-inducible flavoprotein system being involved in this N-oxidative reaction.

Irrespective of the number of N-oxidative enzymes that exist, the problem of their actual mechanisms have still to be resolved. It has been proposed that superoxide anion is the active oxygenating species in cytochrome P-450[31] mediated oxidations. However, our experiments in which superoxide dismutase either on its own or with catalase have been incorporated into the microsomal incubate failed to show an inhibition of either the flavoprotein or cytochrome P-450 N-oxidation (Table 7). It may be that the use of membrane bound N-oxidases precluded the

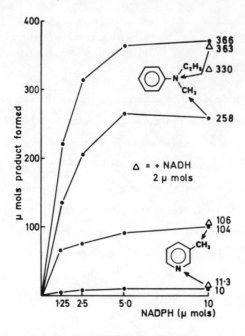

Fig. II. The influence of increasing NADPH and additional NADH on C- and N-oxidation of N-ethyl-N-methylaniline and 3-methylpyridine

TABLE 6   THE EFFECT OF PRETREATMENTS, INHIBITORS AND ACTIVATORS ON THE IN VITRO N-OXIDATION OF THE VARIOUS TYPES OF NITROGEN

Substrate Test	Pretreatment		Inhibitors and Activators					Species Guinea-pig
	Pheno-barbitone	3-methyl-cholanthrene	Carbon monoxide	SKF 525A	DPEA or Octylamine	Naphthyl-thiourea	Bromaz-epam	
Group I	Nil	Nil	Nil	Nil	Inc	Inh	?	Present
Group IIc	Nil	Nil	Nil	Nil	Inc	Inh	Nil	Present
Group IIa	Inc	Nil	Inh	Inh	Inh	Inh	Inc	Present
Pyridines	Inc	Nil	Inh	Inh	Inh	Inh	Inc	Present
Imine	Inc	Nil	Inh	Inh	Inh	?	?	?
Group IIIb	Nil	Inc	Inh	?	?	?	?	Absent

Inh = inhibited   Inc = increased

penetration of either superoxide dismutase or catalase to the site where the oxygen was activated. Superoxide anion is also generated during the breakdown of flavin peroxide[32]. Whether this is the mechanism by which nitrogen in organic compounds is oxidised or whether direct oxidation via

TABLE 7   THE INFLUENCE OF SUPER OXIDE DISMUTASE (SOD) AND CATALASE (CAT) ON THE C- AND N-OXIDATION OF N-ETHYL-N-METHYLANILINE AND 3-METHYL-PYRIDINE

Addition Route	N-Ethyl-N-Methylaniline		3-Methylpyridine	
	N-Oxidation	N-Demethylation	N-Oxidation	C-Oxidation
SOD 200 U	104	122	100	85
CAT 2000 U	101	101	98	102
SOD & CAT	103	117	92	87

flavinperoxide occurs is not at present known.   Another proposal[33] is that aliphatic amines form a tertiary complex with flavoprotein and molecular oxygen by the donation of a lone pair electron. Such a reaction may not be possible with weakly basic or acidic nitrogenous compounds where delocalisation of the electrons occurs.   Further work will be required before all the questions involved in the oxidation of nitrogen in organic molecules can be answered.

ACKNOWLEDGEMENTS

    The work of my unit is supported by the Science Research Council, the Cancer Research Campaign and the Council for Tobacco Research U.S.A. inc.   Unpublished experimental data cited was obtained by L.A.Damani, L.H.Patterson and M.R.Smith and will be published in full elsewhere.

REFERENCES

1.   Polonovski, M. (1930) Bull. Soc. chim. Belges 39, 1.

2.   Culvenor, C.C.J. (1953) Revs. Pure & Appl. Chem. 3, 83.

3.   Bickel, M.H. (1969) Pharmacol. Revs. 21 325.

4.   Weisburger, J.H. & Weisburger, E.K. (1971) in Handbook of Experimental Pharmacology 28(2) 312 ed. Brodie, B.B. & Gillette, J.R.

5.   Biological Oxidation of Nitrogen in Organic Molecules (1972) ed. Bridges, J.W., Gorrod, J.W. & Parke, D.V.   publ. Taylor & Francis, London, U.K.

6.   Biological Oxidation of Nitrogen in Organic Molecules (1978) ed. Gorrod, J.W. publ. Elsevier/North Holland, Amsterdam.

7.   Baker, J.R., Struempler, A. & Chaykin, S. (1963) Biochim. Biophys. Acta 71, 58.

8.   Ziegler, D.M. & Pettit, F.H. (1964) Biochem.Biophys. Res. Commun. 15, 188.

9.   Uehleke, H. (1960) Arch. exp. Path. Pharmak. 241, 150.

10. Miller, E.C. & Miller, J.A. (1966) Pharmacol. Revs.

11. Boyland, E. & Nery, R. (1965) Biochem. J.,94, 198.

12. Gorrod, J.W. (1971) Xenobiotica, 1, 349.

13. Coutts, R.T. & Beckett, A.H. (1977) Drug. Metab. Revs., 6, 51.

14. Parli, C.J., Wang, N. & McMahon, R.E. (1971) J.Biol.Chem., 246, 6953.

15. Prough, R.A. (1973) Arch. Biochem. & Biophys., 158, 442.

16. Tang, B.K., Inaba, T. & Kalow, W. (1975) Drug Met. & Disp., 3, 479.

17. Schwartz, M.A., Postma, E., Kolis, S.J. & Leon, A.S. (1973) J.Pharm.Sci., 62, 1176.

18. Gorrod, J.W. (1973) Chem.-biol.Interact. 7, 289.

19. Gorrod, J.W. (1974) Pharmaz. Ztg., 119S, 1973.

20. Damani, L.A. (1977) Ph.D. Thesis University of London.

21. Patterson, L.H. (1978) Ph.D. Thesis University of London.

22. Cho, A.K., Lindeke, B. & Sum, C.Y. (1974) Drug. Metab. & Disp., 2, 1.

23. Kampffmeyer, H. & Kiese, M. (1964) Arch. exptl. Pathol. & Pharmakol., 246, 397.

24. Ziegler, D.M., Poulsen, L.L. & McKee, E.M. (1971) Xenobiotica, 1, 523.

25. Ziegler, D.M. & Mitchell, C.H. (1972) Arch. Biochem. Biophys., 150, 116.

26, Smith, M.R. (1978) Ph.D. Thesis University of London.

27. Cohen, B. S. & Estabrook, R.W. (1971) Arch. Biochem. Biophys., 143, 54.

28. Ziegler, D.M., McKee, E.M. & Poulsen, L.L. (1973) Drug. Metab. Disp., 1, 314.

29. Parli, C.J. & Wang, N. (1972) Abst. 5th Internat. Pharmacol. Congress,
San Francisco, p.176.

30. Hlavica, P. & Kehl, M. (1977) Biochem. J., 164, 487.

31. Stroebel, H.W. & Coon, M.J. (1971) J.Biol.Chem., 246, 7826.

32. Hemmerich, P. & Wessiak, A. (1976) in Flavins and Flavoproteins ed. Singer, T.P.
Elsevier, Amsterdam, p.9.

33. Beckett, A.H. & Belanger, P.M. (1975) J.Pharm. Pharmac., 27, 547.

HEMOPROTEINS

STRUCTURE AND FUNCTION OF SIROHEME AND THE SIROHEME ENZYMES

LEWIS M. SIEGEL

Department of Biochemistry, Duke University School of Medicine, and the Veterans Administration Hospital, Durham, North Carolina, 27705

SULFITE AND NITRITE REDUCTASES

Sulfite reductases catalyze the 6-electron reduction of SO_3^{2-} to H_2S, a key reaction in the assimilation of sulfate by plants and microorganisms[1]. The enzyme has been highly purified from a variety of sources, including both aerobic (E. coli, Salmonella typhimurium) and anaerobic (Desulfovibrio and Desulfotomaculum species) bacteria, fungi (yeast), and higher plants (spinach)[1]. Assimilatory nitrite reductases catalyze the 6-electron reduction of NO_2^- to NH_4^+, a key reaction in the assimilation of nitrate by plants and many microorganisms. The enzyme has been highly purified from plant sources (spinach, vegetable marrow, Chlorella) and partially purified from fungi (Neurospora)[2]. Despite the similarities between sulfite and nitrite reductases outlined in the following sections, it should be made clear that these two enzymes are physically distinct entities. Thus, in our laboratory, both sulfite and nitrite reductase have been isolated from the same organism, spinach. The two enzymes differ not only in primary substrate specificity, but in the molecular weights of subunits (60,000 for nitrite reductase vs. 69,000 for sulfite reductase) and the enzyme as isolated (nitrite reductase is a monomer, sulfite reductase a dimer), absorption spectra, and perhaps most striking, in immunological reactivity. Indeed, no cross reaction at all can be detected between spinach sulfite and nitrite reductases when the enzymes are reacted with rabbit antisera prepared with either enzyme serving as antigen.

SIROHEME: ITS STRUCTURE AND ROLE IN CORRIN BIOSYNTHESIS

Absorption and EPR spectra of sulfite and nitrite reductases indicate that both enzymes are hemoproteins, with the heme primarily or exclusively in the high spin ferric state in the enzymes as isolated[3,4]. The heme is not covalently bound, however, and chemical analyses of the extracted heme reveal that both sulfite and nitrite reductases contain a new type of heme prosthetic group, termed "siroheme".[5-9] The currently accepted structure for siroheme is shown in Figure 1. Some of the work on which this structure is based is outlined below.

Murphy, et al[6] reported in 1973 the results of an extensive investigation of the chemical properties of siroheme and its demetallated porphyrin derivative (termed "sirohydrochlorin") derived by extraction of E. coli sulfite reductase with acid

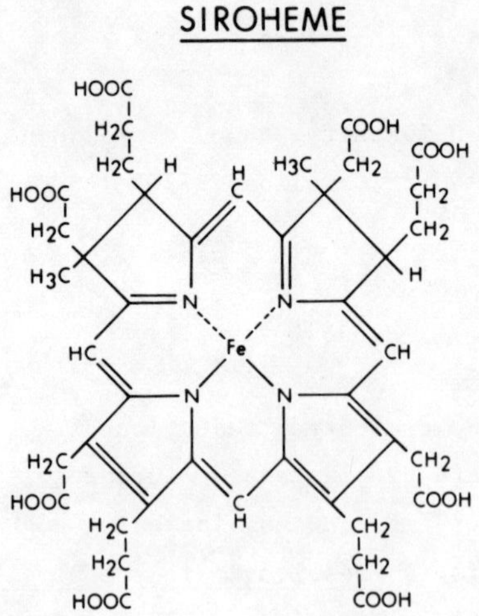

SIROHEME

Fig. 1. Structure of siroheme (based on the recent work of Scott, et al[10]).

acetone. They concluded that siroheme has the following structural features:

(a) Its molecular formula is $FeC_{42}H_{44}N_4O_{16}$, equivalent to that of a Fe-uro-porphyrin to which the elements $(CH_3 + H)_2$ have been added.

(b) It contains two saturated pyrrole rings which are adjacent in the otherwise fully cyclically conjugated porphyrin macrocycle. Thus siroheme can be formally classified as an Fe-urotetrahydroporphyrin of the "isobacteriochlorin" type.

(c) It contains eight carboxylic acid groups, associated on the basis of mass spectrometric fragmentation patterns with acetic and propionic acid side chains. This pattern of side chains is characteristic of uroporphyrins.

(d) Its metal derivatives are unusually stable to photooxidation under conditions which normally convert tetrahydroporphyrins to dihydroporphyrins. It was concluded that the two partially saturated pyrrole rings are not simply "reduced", but are in fact substituted with something more difficult to remove by photooxidation than hydrogen atoms.

Murphy, et al[6] proposed that siroheme is, in fact, an Fe-dimethyl-urotetrahydro-porphyrin, with the element $(CH_3 + H)$ added across each of two adjacent pyrrole rings. Such a structure could plausibly be derived by two successive methylations of uroporphyrinogen III, an intermediate in the biosynthesis of all known heme, chlorophyll, and corrin compounds[11], followed by oxidation (spontaneous or enzyme-catalyzed) of the substituted hexahydroporphyrin macrocycle to its most fully conjugated state, in this case at the oxidation level of a tetrahydroporphyrin. The structure proposed by Murphy, et al[6] contained an arbitrary assignment of the methyl groups to particular pyrrole rings, and was in fact an isomer of that shown in Figure 1. Subsequent work in our laboratory[12], in which radioactive siroheme was isolated from an E. coli methionine-requiring mutant grown in the presence of ^3H_3C-labelled L-methionine, has confirmed the presence of two methionine-derived methyl groups in

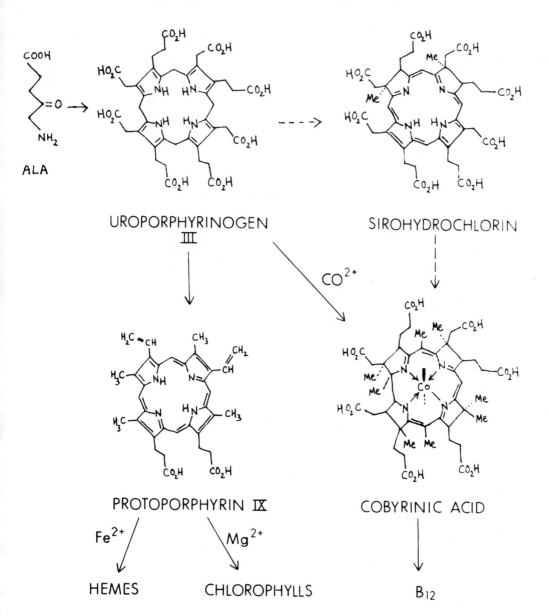

ALA

UROPORPHYRINOGEN III

SIROHYDROCHLORIN

CO^{2+}

PROTOPORPHYRIN IX

COBYRINIC ACID

Fe^{2+} Mg^{2+}

HEMES   **CHLOROPHYLLS** B_{12}

Fig. 1A. Structural relationships between the porphyrinogens, porphyrins, sirohydrochlorin (demetallated siroheme), cobyrinic acid (a corrin), and the commonly observed hemes, the chlorophylls, and vitamin B_{12}. The biosynthetic relationships known prior to this work are indicated by solid arrows. The biosynthetic relationships established by the work of Scott, Irwin, and Siegel[10] with the cell-free system from <u>Propionobacterium shermanii</u> are shown with dashed arrows. See text for details.

siroheme. The essential features of the proposed structure have also been confirmed by proton magnetic resonance spectra of sirohydrochlorin octamethyl ester.[10,13] Such spectra also narrowed the number of possible isomers for siroheme through the observation that there is no detectable coupling between either of the methine H atoms in the two reduced pyrrole rings and the readily identified methylene H's of the acetyl side chains attached to the reduced pyrrole rings. Thus, only isomers in which both acetyl and methyl groups are attached to the same carbon atom in two adjacent reduced pyrrole rings need be considered.

On the basis of the proposed structure for siroheme, Murphy et al[6] proposed that siroheme biosynthesis must represent a new branch of heme metabolism, since it is most unlikely that the siroheme structure can be derived metabolically from protoporphyrinogen IX, the common precursor of all protein associated heme (and chlorophyll) compounds other than siroheme. They also pointed out that the structure for siroheme bears some resemblance to the corrin series of tetrapyrroles, in that both types of compound are highly carboxylated with the side chain pattern of a modified uroporphyrinogen III, contain multiple reduced pyrrole rings, and are multiply methylated. It was noted that Clostridia, thought to be primitive organisms devoid of known heme compounds, are reportedly able to produce both corrins and a sulfite reductase with a heme-like absorption spectrum[8]. It was therefore suggested that: (a) siroheme may be an "ancestral" type of heme compound suited to a primitive photoreducing atmosphere, and (b) sirohydrochlorin (or, more probably, its porphyrinogen analogue) may be a "missing link" joining the metabolism of the porphyrins to that of the corrins. (See Fig. 1A for structures.)

Recent work in several laboratories[10,14,15] has led to a striking confirmation of the suggestion that sirohydrochlorin (or a closely related derivative) may serve as an intermediate in B_{12} biosynthesis. The propionic acid bacteria, as well as the Clostridia, synthesize B_{12} coenzymes in large quantities. Scott, et al[16] have developed a cell-free system from Propionobacterium shermanii which can catalyze biosynthesis of the corrin cobyrinic acid (a B_{12} precursor) when provided with uroporphyrinogen III or its precursors (e.g., 4-amino-levulinic acid (ALA)), S-adenosyl-methionine (SAM), cobalt, and reducing agents. Subsequently, Scott, et al[10] and Deeg, et al[14] found that such extracts accumulate substantial amounts of a material with a strong orange fluorescence. (Whole cells of P. shermanii have been reported by Bykhovsky, et al[17] to synthesize a related group of fluorescent materials, termed "corriphyrins", and to excrete them into the medium during active B_{12} synthesis. Addition of B_{12} to the medium inhibits production of the corriphyrins.)

In collaboration with our laboratory, Scott and Irwin[10] isolated almost one milligram of the orange fluorescent metabolite accumulated by cobyrinic acid synthesizing P. shermanii extracts. Comparison of the ultraviolet, circular dichroism, mass, and proton magnetic resonance spectral data for the methyl ester of the P. shermanii metabolite with that for authentic sirohydrochlorin octamethyl ester prepared from E. coli sulfite reductase leaves no doubt that the substances, which also show complete correspondence in R_f values on thin layer chromatography, are identical. High resolution mass spectrometry of the molecular ion of the P. shermanii metabolite methyl ester revealed a composition identical to that of sirohydrochlorin octamethyl ester[6,10]. Deeg, et al[14] have independently proved the structure of the orange fluorescent metabolite accumulated by extracts of both P. shermanii and Clostridium tetanomorphum; their structure is identical with that of sirohydrochlorin. (Battersby, et al[15] have shown that the "corriphyrins" isolated from whole cells of P. shermanii are in fact lactonized derivatives of sirohydrochlorin.)

With the isolation of sirohydrochlorin from P. shermanii extracts, it was now possible for Scott, Irwin, and Siegel[10] to prepare radioactive sirohydrochlorin by incubation of the extract with either $(4-^{14}C)$-ALA and non-radioactive SAM or with non-radioactive ALA and (^3H_3C)-SAM. The radiolabelled sirohydrochlorin was isolated and reincubated as single or double-labelled material with the cell extract, either with or without prior reduction of the tetrahydroporphyrin to the porphyrinogen state with sodium amalgam. Both reduced and non-reduced sirohydrochlorin were incorporated by the P. shermanii system in similar radiochemical yield into cobyrinic acid. Of crucial importance is the fact that the doubly-labelled sirohydrochlorin ($^3H/^{14}C$ ratio = 3.45) was incorporated into cobyrinic acid ($^3H/^{14}C$ ratio = 3.41) intact. It may therefore be concluded that sirohydrochlorin (or a compound to which it can be readily converted) is on the pathway of corrin biosynthesis from the porphyrinogen uroporphyrinogen III. (See Fig. 1A for structures.)

The finding that sirohydrochlorin is a "missing link" between porphyrin-hemes and corrins may have a profound evolutionary implication for the understanding of the development of both heme and corrin prosthetic groups. The association of all known sulfite reductases with siroheme[1], and the discovery of ancient (3×10^9 yr old) sulfur deposits with $^{34}S/^{32}S$ ratios indicating that they are products of biological rather than geological sulfate reduction, supports the primitive biochemical nature of siroheme[1].

The finding that sirohydrochlorin is incorporated intact into the corrins also indicates that the pattern of methylation of the pyrrole rings in sirohydrochlorin (and thus in siroheme) must be compatible with the final pattern of ring methylation in B_{12}. When this restriction on possible isomers is taken into account, together with the restriction that decarboxylation of an acetic acid side chain to form one of the C-12 methyl groups of vitamin B_{12} is mechanistically plausible only if the acetic acid group is attached to an unsaturated pyrrole ring, only one isomer remains possible for siroheme which is compatible with both the bio-incorporation and proton magnetic resonance data, the structure shown in Figure 1.

It must be emphasized that despite the general association of the novel siroheme group with sulfite and nitrite reductases, we have as yet no understanding of the mechanistic basis for this association. It remains conceivable that the siroheme enzymes are vestiges of a more primitive type of heme environment, such that once a protein evolved to fold about a heme with such highly polar side chains, that protein-heme association became "locked in" in further evolution. Obviously, much more work on the chemical properties of siroheme and its possible catalytic behavior in multielectron reduction reactions is needed.

ROLE OF SIROHEME IN CATALYSIS BY SULFITE AND NITRITE REDUCTASES

Enzyme-bound siroheme is reducible by agents which serve as electron donors for catalytic sulfite reduction (e.g., NADPH, reduced methyl viologen. and dithionite) with E. coli sulfite reductase[5], and stopped-flow and preliminary rapid freeze EPR experiments reported by Siegel and Kamin[18] showed that the rate of heme reduction in that enzyme upon treatment with NADPH was sufficient (although marginally so) to permit inclusion of the heme as a kinetically significant participant in the process of enzymic sulfite reduction. Rapid kinetic studies for nitrite reductases have not as yet been reported.

The E'_o values for the siroheme components of E. coli sulfite reductase, -345 mv, and spinach nitrite reductase, -50 mv,[19,20] obtained by potentiometric titration experiments, are also compatible wuth roles for siroheme in reduction of sulfite to H_2S (average E'_o = 120 mV) and of nitrite to NH_4^+ (average E'_o = +330 mV), respectively.

A number of ligands can bind to assimilatory sulfite and nitrite reductases with alterations in the optical and/or EPR spectra chacteristic of the enzyme-bound heme[2,4,5]. These ligands, which include cyanide and CO, have been shown to inhibit sulfite and nitrite reduction in parallel with formation of spectrophotometrically detectable complexes with E. coli sulfite reductase[5,22] and spinach nitrite reductase[2]. The optical absorption spectra of siroheme-CO complexes are quite characteristic, with the absorption peaks shifted to longer wavelengths and the alpha-band intensified into a hemochromogen-type spectrum[21]. Complexes of CO with sulfite[5] and nitrite[8] reductases characteristically exhibit these hemochromogen-type peaks and wavelength shifts, a result which strongly suggests that CO is binding to the heme component of these enzymes. Rueger and Siegel[3] have shown that there is only a single binding site for $^{14}CN^-$ per heme in E. coli sulfite reductase. Cyanide binding leads to a change in spin state of the heme, indicating that cyanide must bind at or near the heme iron. These results lead to the conclusion that the siroheme component must be involved in electron flow to sulfite and nitrite.

Complexes between sulfite reductase and sulfite[3] and between nitrite reductase and nitrite[2,9] have been detected by means of the spectral alterations induced in each enzyme upon incubation with substrates. Rueger and Siegel[3] have shown that only one $^{35}SO_3^=$ per heme can be bound to E. coli sulfite reductase hemoprotein. Precomplexation of the enzyme with cyanide or CO prevents $^{35}SO_3^=$ binding. The enzyme-sulfite complex is catalytically active, and the bound sulfur is reduced to H_2S in preference to exogenously added sulfite. The optical spectrum of the sulfite complex with E. coli sulfite reductase is very similar to that of the low spin ferri-heme–cyanide complex of the enzyme, although no comparable EPR signal has been detected for the enzyme–sulfite complex. Rueger and Siegel[3] have provided evidence that the sulfur of sulfite remains bound to the enzyme during the entire course of its reduction; distinct spectrophotometric species appear as transients during sulfite reduction, such species possibly representing intermediates in the 6-electron transfer process.

Vega and Kamin[2] have found changes in the heme absorption bands of spinach nitrite reductase upon reaction of the enzyme with NO_2^- or NH_2OH. Both types of spectral species were detected during a prolonged turnover reaction involving enzyme, nitrite, and dithionite (as electron donor). Vega and Kamin suggested that the enzyme proceeds through both bound nitrite and hydroxylamine complexes during the course of nitrite reduction. Upon reaction of spinach nitrite reductase with NO_2^- and dithionite, a new EPR species with principal g value at 2.07 has been seen by Aparicio, et al[4]. These authors suggested that this signal may be due to a ferroheme-NO complex of the enzyme.

These results indicate that the binding site for sulfite and nitrite to their respective reductases is either identical to or closely associated with the siroheme iron atom. They suggest furthermore that the bound substrate may maintain this close association with siroheme during the course of its reduction.

ROLE OF IRON–SULFUR CENTERS IN NITRITE AND SULFITE REDUCTASES

Although much effort has gone into studies of the role of the novel siroheme prosthetic group in catalysis by the siroheme enzymes, we have recently focussed attention on a second type of prosthetic group which has been found in all sulfite and nitrite reductases sufficiently puified to permit analysis for it: the iron-sulfur center (Fe/S).

Thus, plant nitrite reductases (which utilize reduced ferredoxin as the physiological electron donor) are monomeric polypeptides of molecular weight 60,000, containing approximately one mole of siroheme and 2-3 moles of non-heme iron/acid labile sulfur per mole enzyme[2]. The simplest sulfite reductase yet obtained in homogeneous form has been derived by dissociation (in 5M urea) of the large (670,000 molecular weight) hemoflavoprotein NADPH-sulfite reductase complex from E. coli.[23] The isolated heme-containing component is a monomeric polypeptide, molecular weight 54,000, which contains one mole of siroheme and one $Fe_4S_4^*$ center per mole of polypeptide.[3,23] Although the hemoprotein subunit, unlike its parent hemoflavoprotein complex, cannot utilize NADPH as an electron donor, it can catalyze the complete reduction of sulfite to H_2S (without detectable release of intermediates) if supplied with an artificial electron donor, such as reduced methyl viologen (MVH). The ferredoxin-linked sulfite reductase from spinach, which we have recently isolated, appears to be a dimer of the E. coli hemoprotein type of structure, since it contains two identical polypeptide subunits, 2 siroheme, and 7-8 non-heme iron/acid labile sulfur per enzyme molecule. More complicated sulfite reductases than the aforementioned are known, such as the flavin-containing enzymes from E. coli, S. typhimurium, and yeast[1], and the respiratory sulfite reductases from Desulfovibrio and Desulfotomaculum species (which enzymes are composed of non-identical subunits and contain 1-2 siroheme and 14-16 Fe/S per molecule[24]), however the additional structure seems generally to be associated with a requirement for linking the siroheme-iron/sulfur center polypeptide with a physiological source of reducing power.

It is clear then that the minimal catalytic unit for both sulfite and nitrite reduction via a 6-electron transfer reaction is remarkably simple: a single polypeptide containing one molecule of siroheme and one iron-sulfur center.

Unlike the siroheme components of sulfite and nitrite reductases, which exhibit apparent E'_o values in keeping with the natural electron donors and acceptors for the reactions catalyzed, the iron-sulfur components of these enzymes exhibit remarkably negative reduction potentials. Thus, little or no EPR signal attributable to reduced Fe/S centers can be observed when reductants are added to either E. coli sulfite reductase (NADPH, MVH, and dithionite have been used) or spinach nitrite reductase[4] in the absence of added heme-ligating compounds. The signals which are observed (equivalent to a maximum of 0.06 reduced $Fe_4S_4^*$ per heme in the presence of a 900-fold excess of NADPH per heme in E. coli sulfite reductase) are typical of reduced ferredoxins (g values at 2.04, 1.935, and 1.91 with E. coli sulfite reductase). Potentiometric titrations of both spinach nitrite reductase by Stoller, et al[20] and E. coli sulfite reductase by our laboratory have indicated E'_o values of approximately -550 mV for the respective Fe/S centers.

We have found that addition of heme ligands such as CO or cyanide to E. coli sulfite reductase renders the Fe/S centers of the enzyme more readily reducible, the midpoint poential of the Fe/S shifting approximately +200 mV (i.e., near that of the siroheme moiety) in the presence of 1 atm CO. The effect of cyanide is much smaller. Similar effects have been noted by R. Krueger in our laboratory with the sulfite reductase of spinach and by Aparicio, et al[4] with cyanide and spinach nitrite reductase.

Even more striking than the quantitative changes in the Fe/S EPR signals elicited by heme ligands is the change in shape of the reduced Fe/S signals induced by CO or cyanide interaction with E. coli sulfite reductase. Figure 2 shows a set of Fe/S center EPR spectra obtained for enzyme treated with NADPH alone or with NADPH + dithionite, in the absence (A and A') or presence of

NADPH   NADPH + S₂O₄⁼

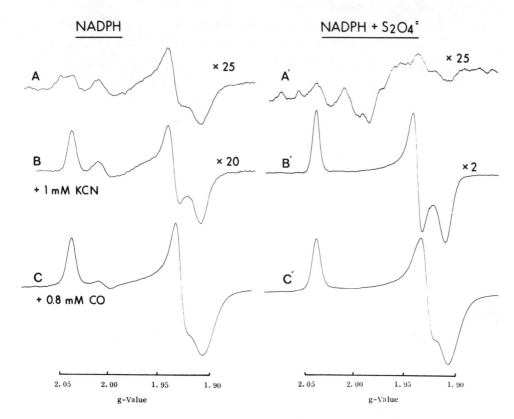

Fig. 2. Effect of heme ligands on EPR spectra of E. coli sulfite reductase
iron-sulfur centers. Spectra A, B, and C: Solutions, 0.25 ml each, containing
E. coli NADPH-sulfite reductase (0.050 mM in heme), 5 mM NADPH, and the
indicated ligands, in 0.1 M potassium phosphate buffer (pH 7.7), were
incubated anaerobically for 60 min at 23C, and the solutions then frozen in
liquid N₂. Spectra A', B', and C': Solutions identical to those above,
incubated in parallel for 60 min, were made 5 mM in sodium dithionite,
incubated at 23C for a further 5 min, and then frozen in liquid N₂. Ligands
present: A and A': none. B and B': 1 mM KCN. C and C': 0.8 mM CO.
EPR spectra were recorded with a Varian E-9 spectrometer at 20K under the
following conditions: microwave frequency, 9.2 GHz; modulation frequency,
100 KHz; modulation amplitude, 10 gauss; microwave power, 50 mW; magnetic field
sweep rate, 100 gauss per min; time constant, 0.3 sec. Relative receiver gains
were: A and A', 25; B, 20; B', 2; C and C', 1.

either cyanide (B and B') or CO (C and C'). It can be seen that cyanide renders
the reduced Fe/S EPR signal more rhombic than that of unligated enzyme, while CO
binding renders the signal more axial. It can be seen that whereas cyanide
only increases the intensity of the NADPH-induced Fe/S signal by approximately
2-fold, CO increases the intensity in the presence of NADPH by about 30-fold.
Both ligands serve to block enzyme turnover in the presence of dithionite
(which gives rise to sulfite on reduction of the enzyme), hence dithionite can

serve its usual role as a powerful reductant in the presence but not the absence of agents which inhibit sulfite reductase activity. Note that there is _less_ reduced iron-sulfur center in the presence of dithionite than in its absence if enzyme turnover is permitted (compare A and A'); this result suggests that the iron-sulfur centers may indeed undergo oxidation-reduction during sulfite turnover.

The qualitative and quantitative effects of cyanide and CO on the Fe/S center spectra are maintained even if excess ligand is removed from sulfite reductase by gel filtration prior to addition of the NADPH (or dithionite) in an experiment otherwuse identical to that shown in Figure 10. If such isolated CO- or cyanide-complexed enzyme is incubated in the presence of the other ligand (i.e., cyanide-enzyme with CO, or CO-enzyme with cyanide) in an experiment otherwise like that of Figure 10, it is found that the spectral properties (both line shape and degree of reduction) are characteristic of the original bound ligand and not of the exogenous heterologous ligand. Thus, we must conclude that there is an interaction between the heme and iron-sulfur centers of E. coli sulfite reductase which is sensitive to the nature of the heme ligand present. Since there is excellent evidence that CO and CN^- interact with the heme of sulfite reductase, and that only one cyanide binds per hemoprotein subunit in the E. coli enzyme[3], there can be no question of separate ligand association sites for heme and iron-sulfur centers. It may even be possible that ligands such as CO and cyanide (and substrates as well?) bridge the heme and iron-sulfur centers. In this context, it is interesting to note the report of Erbes, et al[25] that there is direct interaction between the Fe_4S_4* center of bacterial hydrogenase and CO, as measured by comparison of EPR spectra of enzyme treated with ^{12}CO and ^{13}CO. Such experiments are now in progress with E. coli sulfite reductase.

DETECTION OF POSSIBLE INTERMEDIATES IN SULFITE REDUCTION

Careful stoichiometry studies performed during the course of sulfite reduction catalyzed by the E. coli[22], yeast[26], and spinach[27] sulfite reductases, and during nitrite reduction catalyzed by plant nitrite reductases[2], have consistently shown the consumption of six electrons from added reductant for each molecule of substrate consumed or product (H_2S or NH_4^+) formed. Yet the simplest sulfite and nitrite reductases are molecules with a probable maximum electron storage capacity of two per catalytically active unit. Obviously, then, these 6-electron transfer reactions must proceed through enzyme-bound intermediates. As mentioned previously, Vega and Kamin[2] have suggested, on the basis of spectrophotometric evidence, that the 4-electron reduction product of nitrite reduction, NH_2OH, is in fact an enzyme-bound intermediate during the 6-electron reduction of nitrite to NH_4^+.

The studies described below were initiated following reports by Lee, et al[28-30]

210

that the sulfite reductases from three strains of <u>Desulfovibrio</u> produced trithionate ($S_3O_6^=$) as the major reaction product when incubated with sulfite and MVH (as electron donor), and of Trudinger[31] that the sulfite reductase from <u>Desulfotomaculum nigrificans</u> another sulfate reducing anaerobic bacterium, produced H_2S (the only product looked for) from sulfite and MVH, but with an MVH utilized/H_2S produced stoichiometry of 12 instead of the expected 6.

J. Saks of our laboratory found that when purified sulfite reductases from three different respiratory sulfate reducing bacteria (<u>Desulfovibrio gigas</u>, <u>Desulfovibrio desulfuricans</u> Norway strain, and <u>Desulfotomaculum nigrificans</u>) were incubated with various concentrations of MVH and $^{35}SO_3^=$ and the reaction products analyzed, two radioactive products in addition to H_2S were formed: trithionate and thiosulfate ($S_2O_3^=$). The proportion of electrons appearing in trithionate as opposed to H_2S was found to increase with increasing sulfite concentration, if the (MVH) was maintained constant, and to decrease with increasing MVH concentration, if the ($SO_3^=$) was maintained constant. Thiosulfate formation was maximal as a proportion of the total number of electrons consumed at intermediate ratios of (MVH)/($SO_3^=$). A typical result (for the thermophilic <u>D. nigrificans</u> enzyme) is shown in Figure 3.

Kinetic studies showed that each of the respiratory sulfite reductases exhibited a turnover number under the conditions of our experiments for MVH-sulfite reduction which was 1-2 orders of magnitude less than that of the <u>E. coli</u> sulfite reductase under comparable assay conditions. The respiratory sulfite reductases were readily saturated with sulfite (K_M values were on the order of 0.1 mM), but could not be saturated with MVH under the conditions of our experiments (up to 0.7 mM MVH was tested). Thus, in our experiments, reduction generally proceeded to H_2S more efficiently as the overall rate of electron consumption increased.

Control experiments showed that respiratory sulfite reductases could not reduce either trithionate or thiosulfate to H_2S with MVH as donor. Similarly, non-radioactive trithionate when added to reaction mixtures containing $^{35}SO_3^=$, MVH, and enzyme, did not dilute the incorporation of radioactivity into H_2S. (<u>E. coli</u> sulfite reductase, when incubated with MVH and various concentrations of $^{35}SO_3^=$ in parallel with the respiratory enzymes, produced only $H_2^{35}S$.) Thus, $S_3O_6^=$ and $S_2O_3^=$ are not true intermediates of sulfite reduction catalyzed by the respiratory sulfite reductases, but are rather by-products of sulfite reduction.

Kobayashi, et al[32] have suggested that trithionate (structure $^-O_3S-S-SO_3^-$) and thiosulfate (structure $^-S-SO_3^-$) may be in fact sulfite adducts of highly reactive 2-electron ($SO_2^=$) and 4-electron ($SO^=$) reduction products of sulfite. Such products should normally be enzyme-bound during sulfite reduction, but it may be possible for them to escape from the enzyme and react with the efficient nucleophile sulfite if the lifetime of such intermediates is artificially prolonged, such as with low levels of reductant and high levels of nucleophile added to an enzyme reaction with intrinsically low turnover number. Preliminary

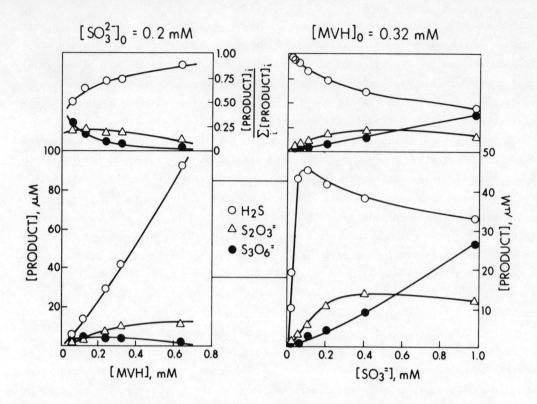

Fig. 3. Distribution of ^{35}S-labelled trithionate, thiosulfate, and H_2S during the course of sulfite reduction catalyzed by a respiratory sulfite reductase. Reaction mixtures, 1.0 ml, containing the indicated concentrations of $Na_2{}^{35}SO_3$, reduced methyl viologen (MVH), and 0.1 mg of purified <u>Desulfotomaculum nigrificans</u> sulfite reductase, in 0.1 M potassium phosphate buffer (pH 7.0), were incubated anaerobically in one cm cuvettes at 55C for 90 min. After the incubation, an aliquot of the reaction mixture was removed and the remaining solution vigorously bubbled with argon for 20 min to remove $H_2{}^{35}S$. The difference in radioactivity between these two solutions allowed calculation of the amount of $H_2{}^{35}S$ formed. The bubbled reaction mixture was then analyzed by the method of Kelley, et al[37] which defines specific conditions for the selective conversion of trithionate and thiosulfate to SCN^-. The $^{35}SCN^-$ formed was separated from other radioactive sulfur compounds by descending paper chromatography in pyridine:butanol:H_2O:NH_3 (20:40:40:5). The paper chromatogram was cut into narrow strips and the radioactivity measured.

experiments in which addition of KCN to reaction mixtures containing $^{35}SO_3{}^=$, MVH, and <u>D</u>. <u>gigas</u> sulfite reductase resulted in significant production of $^{35}SCN^-$ tend to support the possibility of trapping reaction intermediates with strong nucleophiles. Control experiments showed that under the conditions of these experiments reaction between trithionate, thiosulfate, or H_2S with cyanide to yield SCN^- was negligible.

These studies give an indication of the type of intermediates which may be bound to sulfite reductases as sulfite is being reduced to H_2S. They suggest, but hardly prove, that the sulfite reductase reaction, at least with the respiratory enzymes studied, proceeds in classical one- or two-electron stages, with intermediates normally remaining enzyme-bound. (Chambers and Trudinger[36] have shown that neither trithionate nor thiosulfate are normally formed by cells of <u>Desulfovibrio</u> catalyzing sulfite reduction under physiological conditions, presumably with a "better" electron donor than MVH.) Whether the same types of intermediates are transiently present in assimilatory sulfite reductases (and nitrite reductases) remains to be determined.

ACKNOWLEDGEMENTS

Veterans Administration Project number 7875-01. These studies were supported in part by research grant AM-13460 from the National Institutes of Health.

REFERENCES

1. Siegel, L.M. (1975) in Metabolic Pathways, 3d edition, vol. 7, Greenberg, D.M., ed., Academic Press, New York, p. 217.

2. Vega, J.M. and Kamin, H. (1977) J. Biol. Chem., 252, 896.

3. Rueger, D.C. and Siegel, L.M. (1976) in Flavins and Flavoproteins, Singer, T.P., ed., Elsevier, Amsterdam, p. 610.

4. Aparicio, P.J., Knaff, D.B., and Malkin, R. (1975) Arch. Biochem. Biophys., 169, 102.

5. Siegel, L.M., Murphy, M.J., and Kamin, H. (1973) J. Biol. Chem., 248, 251.

6. Murphy, M.J., Siegel, L.M., Kamin, H., and Rosenthal, D. (1973) J. Biol. Chem., 248, 2801.

7. Murphy, M.J. and Siegel, L.M. (1973) J. Biol. Chem., 248, 6911.

8. Murphy, M.J., Siegel, L.M., Tove, S.R., and Kamin, H. (1974) Proc. Nat. Acad. Sci., U.S.A., 71, 612.

9. Vega, J.M., Garrett, R.H., and Siegel, L.M. (1975) J. Biol. Chem., 250, 7980.

10. Scott, A.I., Irwin, A.J., and Siegel, L.M. (1977) J. Amer. Chem. Soc., in press.

11. Scott, A.I., Yagen, B., Georgopapadakou, N., Ho, K.S., Klioze, S., Lee, E., Lee, S.L., Temme, G.H., Townsend, C.A., and Armitage, I.M. (1975) J. Amer. Chem. Soc., 97, 2548.

12. Siegel, L.M., Davis, P.S., and Murphy, M.J. (1977) Biochem. J., in press.

13. Battersby, A.R., Jones, K., McDonald, E., Robinson, J.A., and Morris, H.R. (1977) Tetrahedron Letters, 2213.

14. Deeg, R., Kriemler, H.D., Bergmann, K.H., and Muller, G. (1977) Z. Physiol. Chem., 358, 339.

15. Battersby, A.R., McDonald, E., Morris, H.R., Williams, D.C., Bykhovsky, V. Ya., Zaitseva, N.I., and Bukin, V.N. (L977) Tetrahedron Letters, 2217.

16. Scott, A.I., Yagen, B., and Lee, E. (1973) J. Amer. Chem. Soc., 95, 5761.

17. Bykhovsky, V. Ya., Zaitseva, N.I., and Bukin, V.N. (1975) Doklady Acad. Sci. U.S.S.R., 224, 1431.

18. Siegel, L.M. and Kamin, H. (1968) in Flavins and Flavoproteins, Yagi, K., ed., University Park Press, Baltimore, p. 15.

19. L.M. Siegel and M.J. Barber, unpublished data.

20. Stoller, M.L., Malkin, R., and Knaff, D.B. (1977) F.E.B.S. Letters, 81, 271.

21. Murphy, M.J., Siegel, L.M., and Kamin, H. (1974) J. Biol. Chem., 249, 1610.

22. Siegel, L.M., Davis, P.S., and Kamin, H. (1974) J. Biol. Chem., 249, 1572.

23. Siegel, L.M. and Davis, P.S. (1974) J. Biol. Chem., 249, 1587.

24. Saks, J., and Siegel, L.M., unpublished data.

25. Erbes, D.L., Burris, R.H., and Orme-Johnson, W.H. (1975) Proc. Nat. Acad. Sci., U.S.A., 72, 4795.

26. Yoshimoto, A. and Sato, R. (1968) Biochim. Biophys. Acta, 153, 555.

27. Asada, K., Tamura, G., and Bandurski, R.S. (1969) J. Biol. Chem., 244, 4904.

28. Lee, J.P., LeGall, J., and Peck, H.D., jr. (1973) J. Bacteriol., 115, 529.

29. Lee, J.P., Yi, C.S., LeGall, J., and Peck, H.D., jr. (1973) J. Bacteriol., 115, 453.

30. Lee, J.P. and Peck, H.D., jr. (1971) Biochem. Biophys. Research Commun., 45, 583.

31. Trudinger, P.A. (1970) J. BActeriol., 104, 158.

32. Kobayashi, K., Seki, Y., and Ishimoto, M. (1974), J. Biochem., Tokyo, 75, 519.

33. Jones, H.E., and Skyring, G.W. (1975) Biochim. Biophys. Acta, 377, 52.

34. Akagi, J.M., Chen, M., and Adams, V.A. (1974) J. Bacteriol., 120, 240.

35. Drake, H.L., and Akagi, J.M. (1976) J. Bacteriol., 126, 733.

36. Chambers, L.A. and Trudinger, P.A. (1975) J. Bacteriol., 123, 36.

37. Kelly, D.P., Chambers, L.A., and Trudinger, P.A. (1969) Anal. Chem., 41, 898.

MECHANISM OF ACTION OF CYTOCHROME c PEROXIDASE

TAKASHI YONETANI

Department of Biochemistry and Biophysics, School of Medicine, University of Pennsylvania, Philadelphia, PA. 19104 (U.S.A.)

ABSTRACT

The protoheme prosthetic group of cytochrome c peroxidase (CCP) is located in a hydrophobic crevice and coordinated to a histidyl group as the fifth proximal ligand of the heme iron. The distal position appears to be a non-histidyl group. The sixth coordination site of the heme iron is either unoccupied or is loosely occupied by a rapidly exhange H_2O molecule.

Redox titration, optical absorption, electron paramagnetic resonance and Mossbauer spectroscopic measurements of Compound ES, the red peroxide enzyme intermediate are consistent with the previously proposed redox state of Compound ES containing a ferryl heme iron, Fe (IV), and a free radical in the apoprotein moiety, R^*.

Fluorescence and nuclear magnetic resonance studies of the interaction of cytochrome c with CCP exclude the direct contact between the heme groups of these two hemoproteins. Therefore, the redox reaction between them must be carried out by electron transfer via the protein moiety or quantum mechanical tunneling.

INTRODUCTION

Heme-containing hydroperodixases (catalases and peroxidases) are a two-equivalent oxidoreductase which uses hydrogen peroxide as a primary oxidizing substrate. They share a common feature of forming enzyme intermediate(s) containing higher oxidation state of iron upon reaction with hydrogen peroxide. The secondary reducing substrate is another hydrogen peroxide for catalase and various reducing agents for peroxidases, as shown in Figure 1.

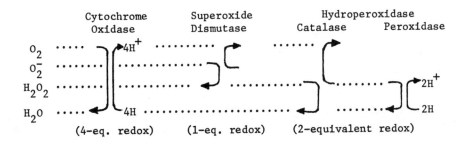

Fig. 1. Comparison of reactions catalyzed by cytochrome oxidase, superoxide dismutase, catalase, and peroxidase.

Cytochrome c peroxidase (CCP) possesses two notably unique features among hydro-peroxidases. CCP forms a relatively stable red peroxide intermediate (Compound ES), rendering it for detailed physical and chemical characterization. The secondary substrate of CCP is ferrocytochrome c. This makes it feasible to investigate the mode of interaction of two redox proteins and the mechanism of electron transfer reactions in a highly purified state. This paper describes the current state of our understanding on the structure of the heme environment in CCP, the chemical nature of Compound ES, and the mode of interaction between CCP and cytochrome c.

MATERIALS AND METHODS

CCP was purified from baker's yeast and thrice crystallized by isoelectric dia-lysis, as described previously[1]. Horseradish peroxidase (HRP, Type V) was purchased from Sigma and used without further purification.

Spectroscopic measurments were performed using Cary 118C spectrophotometer, Perkin-Elmer MPF-2A fluorescence spectrometer, Varian E109 EPR spectrometer, and Varian 220 MHz NMR spectrometer.

RESULTS AND DISCUSSION

Heme Environment. The prosthetic group of CCP is protoheme, which is non-covalent-ly bound to the apoprotein moiety[2]. The fluorescence study of the interaction of apo-CCP with anilino-naphthalene sulfonate[3] indicates that the heme-binding site is as hydrophobic as that of myoglobin and hemoglobin. The effects of chemical modification of heme side chains on the enzymic activity of CCP indicate that the enzymic activity is relatively insensitive to the modification of porphyrin side chains at positions 2 and 4 and is greatly diminished by that at position 6 and 7 (the propionic acid side chains), suggesting that these propionic acid groups may be facing the exterior of the molecule[4]. The identification of a histidyl group as the fifth heme ligand, which was made by electron paramagnetic resonance (EPR) examination of the CCP nitric oxide complex[5], has been recently confirmed by the x-ray structural study[6]. The x-ray study indicates further that the distal group near the sixth ligand-binding site is not a histidyl group and the sixth coordination position of the heme iron appears unoccupied in the resting state. The apparent absence of a distinctive acid-alkaline transition in CCP as measured by spectrophoto-metry[7] and nuclear magnetic resonance (NMR) spectroscopy[8] is consistent with the x-ray data. These observations suggest that the conventionally assumed H_2O molecule is either relatively loosely bound to the heme iron or absent at neutral pHs.

Chemical Nature of Compound ES (Peroxide Compound). Cytochrome c peroxidase reacts with a stoichiometric amount of hydroperoxide to form a red Peroxide Compound, which will be referred to hereafter as Compound ES. The formation of Compound ES from the enzyme and hydroperoxidases is very rapid ($k_1 > 10^7 \sim 10^8$ M^{-1} sec^{-1}). No intermediate, which preceeds Compound ES, has been thus far detected. In the absence of reductants, Compound ES is highly stable. The rate constant of its spontaneous decay is of the order of $10^{-5} sec^{-1}$. The primary peroxide compound (Compound I) of horseradish peroxidase decays much faster at a rate of $10^{-3} sec^{-1}$ [9]. Titrations of Compound ES with reductants such as ferrocytochrome c[10,11], and ferrocyanide[12,13] have established that Compound ES is two oxidizing equivalents above the original ferric enzyme. The absorption spectrum of Compound ES is essentially identical to that of Compound II of horseradish peroxidase which contains one oxidizing equivalent per mole in the form of Fe(IV). In addition, EPR examinations have revealed that Compound ES contains a stable free radical, the spin concentration of which is approximately one equivalent per mole. Therefore, it is reasonable to conclude that two oxidizing equivalents in Compound ES are maintained in the form of Fe(IV) and a free radical of a protein group (R*). Compound ES formed from alkyl-hydroperoxide is indistinguishable from that derived from hydrogen peroxide. Furthermore, the formation of Compound ES upon reaction with ethyl hydroperoxide is accompanied by the release of 1 mole of ethyl alcohol per mole. These observations indicate that the O----O bond in the primary substrate, hydroperoxide, has been broken upon the formation of Compound ES and that Compound ES is not a so-called reversible enzyme-substrate complex, but an enzyme intermediate carrying two oxidizing equivalents per mole in a form other than the original substrate. In other words, a two-equivalent electron transfer (or redox reaction) between CCP and hydroperoxide has already taken place upon the formation of Compound ES.

Reduction of Compound ES with 2 moles of ferrocytochrome c generates the original enzyme rapidly. It has not been possible to detect the formation of the one-equivalent, ferrocytochrome c-reduced intermediate of Compound ES and to determine the rate constants of reactions of Compound ES with first and second moles of ferrocytochrome c individually. However, using ferrocyanide as a reductant, it has been possible to examine the mechanism of reduction of Compound ES to the original enzyme in detail[13]. Comparison of optical and EPR titrations shows that the reaction of Compound ES with ferrocyanide in a range from pH 5 to 8 is biphasic and strongly supports a mechanism in which two one-equivalent intermediates are at rapid equilibrium:

$$Fe(IV) \rightleftharpoons Fe(III)-R^*$$

where Fe(IV) and Fe(III) are ferryl and ferric heme irons and R^* is a protein free radical. Optical and EPR parameters thus far available are not sufficient to identify the chemical nature of the protein group responsible for R^*. However, the spontaneous decay of Compound ES results in destruction of several amino acid residues[14]. At pH 4 and 8, tryosine and tryptophan are the residues principally affected. Thus, it is possible that one of these residues may be responsible for the formation of the free radical, R^*, in Compound ES (Fe(IV)-R*) and its one-equivalent reduced form (Fe(III)-R*). Low-temperature magnetic susceptibility[15] and Mossbauer spectroscopic[16] data are consistent with assumption that Compound ES contains Fe(IV).

On reaction with a stoichiometric amount of hydroperoxide, catalase and horseradish peroxidase are converted to a green colored intermediate, Compound I[17]. The
chemical nature of Compound I has been extensively debated since its discovery by
Theorell[18]. Recently, Dolphin et. al.[19] have demonstrated that upon one-equivalent
oxidation several metalloporphyrins are converted to stable porphyrin πcation
radicals, the absorption spectra of which possess the spectral characteristics
of Compound I, namely, a decreased Soret π-π^* transition and an appearance of
the 620–670-nm absorption bands. Since Moss et. al.[20] proposed the presence of
Fe(IV) in Compound I of horseradish peroxidase from Mossbauer spectroscopic measurements, it is attractive to describe Compound I as Fe(IV)-P^*, where P^* is a porphyrin
π-cation radical.Then, Compound I and Compound ES become isoelectronic. Both
contain Fe(IV) and a radical: the former as a porphyrin radical (P^*) and the latter
as a protein radical (R^*). Then the reaction cycles of horseradish and cytochrome
c peroxidases may be compared as shown in Fig. 2.

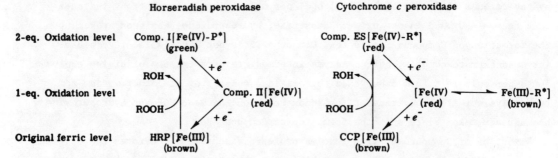

Fig. 2. Reaction cycles of horseradish and cytochrome c peroxidases.

   Mode of Interaction with Cytochrome c. The formation of a reversible Michaelis-
Menten-type complex of the enzyme and ferrocytochrome c can be postulated from
initial steady-state kinetics of the cytochrome c peroxidase reaction[21]. Since
cytochrome c peroxidase and cytochrome c are acidic and basic proteins, respectively, their interaction may be governed principally by electrostatic attraction.
This assumption is further supported by the fact that several polycations which
reversibly and irreversibly bind cytochrome c peroxidase inhibit its enzymic
activity in competition with ferrocytochrome c[21,22].

   The 220-MHz proton-NMR spectrum of horse heart ferricytochrome c is well characterized by the presence of two low-field methyl resonances at 35 and 32 ppm and
two other relatively sharp high-field methyl resonances at +2.4 and +2.7 ppm relative to the standard 2,2-dimethyl-2-silopentane-5-sulfonate reference at 25° [23].
The two low-field resonances at 35 and 32 ppm have been assigned to two methyl
side chains of the heme group at positions 8 and 3, respectively[24]. In the presence of equimolar amounts of cytochrome c peroxidase, the linewidths of these
two low-field resonances broaden from 20 to ~100 Hz. This is accompanied by a
decrease in the separation between these two resonances from 3.1 to 2.25 ppm,
as the consequence of mutual shifts of the 35 and 32 ppm resonance to up- and

down-field directions, respectively. NMR titrations of ferricytochrome c with cytochrome c peroxidase as a function of either the resonance linewidths or $\Delta\nu$ give a
stoichiometry of 1:1 to confirm the formation of a stoichiometric complex between
these two macromolecules[25]. The observed approximately fourfold broadening of the
resonance linewidths is consistent with the expected change in the tumbling correlation time of the whole molecule upon a 1:1 complexation. When cytochrome c is
present in an excess over cytochrome c peroxidase, a time-averaged NMR spectrum of
free and complexed ferricytochrome c rather than a simple superposition of two distinct spectra is observed. This indicates that the association and dissociation
rates for the peroxidase-cytochrome c complex must be much greater than the reverse
of the frequency separation between complexed and free states. It is possible to
set a lower limit of 200 sec^{-1} on the dissociation rate. It is further observed that
the linewidths of the resonances in the cytochrome c peroxidase-cytochrome c complex
are not sensitive to changes in the electronic structure (high-spin or low-spin) and
consequently also the electronic relaxation time of the heme iron of cytochrome c
peroxidase upon complexing with fluoride and cyanide. These observations indicate
that the heme group of cytochrome c is a considerable distance from the heme iron
of the enzyme in the complex (>25 Å, assuming an electronic relaxation time of the
heme iron of cytochrome c peroxidase of approximately 10^{-10} sec).

The fluorescence studies of the interaction of cytochrome c with the anilino-
napthalene sulfonate-apoenzyme and protoporphyrin-apoenzyme complexes provide another
line of evidence[3] in support of the above-mentioned conclusion. Both fluorescence
steady-state and lifetime titrations of these fluorescence-labeled apoenzymes with
ferro- and ferri-cytochrome c indicates the formation of a 1:1 complex, the
affinity for ferricytochrome c being less than that for ferrocytochrome c. From the
phosphorescence and fluorescence quenching, the distance between the emitter (a
fluorescence label) and the quencher (the heme of cytochrome c) can be calculated by
assuming that no direct electronic interaction exists between them, and that the
quenching is derived from the Forster-type energy transfer to the heme. The distance
from the apoenzyme-bound emitter to the heme group of cytochrome c is estimated to
be 19 Å and 14 Å for the anilinonaphthalene sulfonate and protoporphyrin emitter,
respectively. The latter should correspond to the heme-heme distance in the complex
of the holoenzyme and cytochrome c. These NMR and fluorescence data preclude the
electron transfer mechanism through a direct contact of two prosthetic groups. The
possibility of electron transfer via the polypeptide chains or quantum mechanical
tunneling must be seriously considered.

The elucidation of the mode of interaction between cytochrome c peroxidase and
cytochrome c is not only essential in our understanding of the reaction mechanism
of this enzyme but also provides important clues for formulating a general mechanism
of electron transfer in biological systems such as mitochondrial and microsomal electron transfer systems. The question of whether or not the direct contact between two

prosthetic groups is a prerequisite for the electron tranfer processes in biological reactions is long-standing and yet to be answered. This problem becomes experimentally approachable by the use of the cytochrome c peroxidase-cytochrome c couple.

ACKNOWLEDGEMENTS

This work has been supported in part by research grants, PCM 77-00811 from the National Science Foundation and HL 14508 from the National heart, Lung, and Blood Institute.

REFERENCES

1. Yonetani, T. (1967) in Methods in Enzymology, Estabrook, R.E. and Pullman, M.E., eds., Vol. X, Academic Press, New York, page 336.

2. Yonetani, T. (1967) J. Biol. Chem. 242, 5008.

3. Leonard, J. and Yonetani, T. (1974) Biochemistry 13, 1465.

4. Yonetani, T. (1976) in The Enzymes, Boyer, P.D., ed., Academic Press, New York, p. 346.

5. Yonetani, T., Yamamoto, H., Erman, J.E., Leigh, J.S., Jr., and Reed, G. (1972) J. Biol. Chem. 247, 2447.

6. Larsson, L.O., Skoglund, U., Kierkegaard, P., and Yonetani, T. (1977), to be published.

7. Yonetani, T., Wilson, D.F., and Seamonds, B. (1966) J. Biol. Chem. 241, 5347.

8. Morishima, I., Inubushi, T., Iizuka, T., and Yonetani, T. (1977), to be published.

9. Chance, B. (1949) Arch. Biochem. Biophys. 24, 389.

10. Yonetani, T. (1965) J. Biol. Chem. 240, 4509.

11. Yonetani, T., Schleyer, H., and Ehrenberg, A. (1966) J. Biol. Chem. 241, 3240.

12. Yonetani, T. (1966) J. Biol. Chem. 241, 2562.

13. Coulson, A.F.W., Erman, J.E., and Yonetani, T. (1971) J. Biol. Chem. 246, 917.

14. Coulson, A.F.W. and Yonetani, T. (1972) Biochem. Biophys. Res. Commun. 48, 391.

15. Iizuka, T., Kotani, M., and Yonetani, T. (1968) Biochim. Biophys. Acta 167, 257.

16. Lang, G., Spartalian, K., and Yonetani, T. (1976) Biochim. Biophys. Acta 451, 250.

17. Chance, B. (1951) Adv. Enzymology 12, 153.

18. Theorell, H. (1941) Enzymologia 10, 250.

19. Dolphin, D., Felton, R.H., Borg, D.C., and Fajer (1970) J. Amer. Chem. Soc. 92, 743.

20. Moss, T.H., Ehrenberg, A., and Bearden, A.J. (1969) Biochemistry 8, 4159.

21. Yonetani, T. and Ray, G.S. (1966) J. Biol. Chem. 241, 700.

22. Mochan, E. and Nicholls, P. (1970) Biochim. Biophys. Acta 216, 80.

23. Wuthrich, K. (1961) Proc. Nat. Acad. Sci. U.S. 63, 1071.

24. Redfield, A.G. and Gupta, R.K. (1971) Cold Spring Harbor Symposium Quantum Biology 36, 405.

25. Gupta, R.K. and Yonetani, T. (1973) Biochim. Biophys. Acta 292, 502.

ELECTRONIC STATES OF HEME IN CYTOCHROME OXIDASE

Graham Palmer, Toni Antalis, Gerald T. Babcock*, Lucia Garcia-Iniguez, Michael Tweedle*, Lon J. Wilson and Larry E. Vickery*. Departments of Biochemistry and Chemistry, Rice University, Houston, Texas and the Laboratory of Biodynamics, University of California, Berkeley, California.

ABSTRACT

Magnetic Circular Dichroism (MCD), Electron Paramagnetic Resonance (EPR) and optical spectroscopy, together with variable temperature magnetic susceptibility measurements have been used to characterize the electronic states of cytochromes a and a_3. The spectroscopic data supports a model which has cytochrome a low-spin in the resting and fully reduced species, while cytochrome a_3 is high-spin in these two redox states. The absence of a g=6 EPR resonance in the resting enzyme is explained as a consequence of an antiferromagnetic exchange interaction between cytochrome a_3^{3+} (high-spin, S=5/2) and a copper species (Cu_u^{2+}, S=1/2). These assignments are fully supported by magnetic susceptibility data over the temperature range 7-200°K and estimates of the exchange integral and the axial and rhombic zero-field splitting parameters of the coupled spin-system have been obtained. Susceptibility measurements on the cyanide-treated enzyme reinforce our confidence in the accuracy of the assignment described above: the magnitude of the exchange integrals are consistent with the presence of an imidazole bridge between cytochrome a_3 and Cu_u.

An analysis of optical, EPR and MCD spectra taken throughout reductive titrations suggests a linkage in the potentials of the two cytochromes such that a change in oxidation state of one of the hemes lowers the potential of the other by ca 135 mV. In these titrations high-spin species are detected which account for 0.25 spins/oxidase. Evidence is presented to indicate that at least some of these signals can be attributed to cytochrome a^{3+} which has undergone a temporary low-spin to high-spin transition in the course of the titration.

INTRODUCTION

Despite the incisive and imaginative experiments which first defined the nature of cytochrome oxidase almost forty years ago, the mechanism whereby this enzyme effects the oxidation of 4 moles of ferricytochrome c with the concommitant reduction of oxygen to water is still one of the vigorously investigated areas of contemporary Biochemistry.

Keilin and Hartree[1] clearly defined the presence of two chromophores in the enzyme. One of these they named cytochrome a and stipulated that it was unreactive with common heme ligands: the second they called cytochrome a_3 and deduced that it was the site of reaction of compounds such as azide, carbon monoxide, cyanide and, by inference, oxygen.

The introduction of the names cytochromes $\underline{a}$ and $\underline{a}_3$ engendered a long and unproductive debate[2] over whether or not there were in fact two discrete hemoproteins present in oxidase, for only one species of heme, namely heme $\underline{a}$, could be isolated from purified preparations of the enzyme.

Today it is agreed that cytochrome oxidase is an oligomeric protein which contains heme $\underline{a}$ in two sites, and it is the differences endowed on heme $\underline{a}$ by the differences in each site that give rise to the entities which are named cytochrome $\underline{a}$ and $\underline{a}_3$ (it is important to appreciate that the nomenclature is a little atypical and it should not be inferred that cytochrome $\underline{a}$ and $\underline{a}_3$ are intrinsically separable though with an improved ability to manipulate hydrophobic proteins even this might prove possible).

We now know that iron is not the only transition metal present in the enzyme, but that the full complement comprises two copper ions in addition to the two iron ions. Thus, in the simplest terms the reaction might be represented:

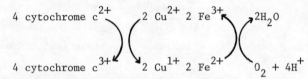

$$4 \text{ cytochrome c}^{2+} \quad 2 \text{ Cu}^{2+} \; 2 \text{ Fe}^{3+} \quad 2H_2O$$

$$4 \text{ cytochrome c}^{3+} \quad 2 \text{ Cu}^{1+} \; 2 \text{ Fe}^{2+} \quad O_2 + 4H^+$$

In pursuing the understanding of the mechanism whereby cytochrome oxidase functions, it is essential to be able to describe the chemical states of these redox centers as completely as possible, for a mechanistic approach in the absence of discrete structural information is unrewarding. Historically the principal tool employed in characterizing the protein has been optical spectroscopy (eg., [3]), but this technique is highly limited because of the absence of any useful analysis of the optical spectra of heme $\underline{a}$ derivatives. The introduction[4] of electron paramagnetic resonance (EPR) spectroscopy was a major development which led first of all to the acceptance of Cu^{2+} as a bona fide component of the enzyme[4], then to the discovery that some of the heme[5] (now known to be 50%[6,7]) is in the low-spin form, and finally that high-spin heme signals were present as transient species during oxidation-reduction reactions though these signals were not to be found in either fully oxidized or fully reduced enzyme[5].

The characterization of the electronic state of the hemes is an important objective for two reasons. Firstly, the spin-state is a consequence of the coordination of the central metal ion by the pyrolle nitrogens of the porphyrin, the amino acid residues of the protein and binding of exogenous ligands by the metal ion. Thus, both iron-protein and iron-substrate and/or inhibitor interactions can be investigated. Secondly, the chemical reactivity of the heme is related to its spin-state, low-spin heme proteins having a primarily electron transfer function, while high-spin heme protein participate in reactions which involve ligand binding by heme.

EPR is unfortunately a rather limited technique in this regard, for its use is restricted to the ferric valence states (although high-spin ferrous ion is paramagnetic (S=2) its EPR is rarely observed in any system, and we are not aware of any report of EPR signals from this state in a biochemical system). Ideally, we would like a spectroscopic probe which is responsive to all four of the common spin-valence states of iron (high and low-spin Fe^{3+}, high and low-spin Fe^{2+}). A method that we have found to have considerable value in this regard is Magnetic Circular Dichroism, the differential absorption of right and left circularly polarized light (i.e. circular dichroism) which is induced in a material when it is immersed in a magnetic field (A good introductory review of this technique is provided by Schatz and McCaffery[8]). In particular, as I hope to show the combined application of EPR and MCD, provides a powerful synergistic tool in the characterization of this enzyme[9,10].

RESULTS AND DISCUSSION

MCD of Cytochrome Oxidase and its derivatives

The MCD spectrum of isolated cytochrome oxidase and that of an equal concentration of bis-imidazole heme a, is shown in Fig. 1[9]. There are four notable features

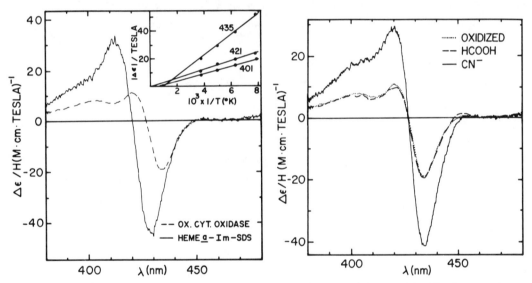

Fig. 1. MCD spectra of isolated cytochrome oxidase, and bis-imidazole heme a in sodium dodecyl sulfate. The concentrations of both compounds are identical, on a heme basis. From Ref. 9, with permission.

Fig. 2. MCD spectra of isolated cytochrome oxidase together with the derivatives formed on incubation with excess sodium formate or potassium cyanide. From Ref. 9, with permission.

about these spectra. 1) They resemble markedly the MCD spectra of conventional low-spin heme proteins such as cytochrome c and b_5[11]. 2) The amplitude of the spectrum varies with temperature, being linear with T^{-1}. This is as expected if

223

the MCD originates from a paramagnetic state such as a low-spin heme[8]. 3) The amplitude of the oxidase spectrum is only one-half of that of a comparable concentration of heme $\underline{a}$, implying that only one-half of the heme in oxidase is low-spin, and 4) The zero-crossing of the MCD spectrum is at 427 nm, thus identifying an electronic band at 427 nm in the optical spectrum as the origin of the MCD. As the absorbance maximum of the enzyme is close to 420 nm one thus concludes that the observed spectrum in the Soret region has two contributions, one centered at 427 nm and the second at higher energies $\underline{eg}$. 414 nm, with the observed Soret band the composite of these two individual components. In a rather elegant analysis of the optical spectra Vanneste[3] has resolved these two contributions to the Soret region and shown that the heme $\underline{a}$ which does not bind ligands (=cytochrome $\underline{a}$ by definition) has its Soret maximum at 427 nm. We can thus conclude that the species that exhibits the MCD in the resting enzyme is cytochrome $\underline{a}$, that it is a low-spin ferriheme complex and that it comprises 50% of the total heme present in the enzyme.

The most obvious question that now arises is the chemical nature of the 50% of the heme that is underdetectable by MCD. From the relationship between MCD intensity and heme spin-states first demonstrated by Vickery $\underline{et}$ $\underline{al}$[11], the most obvious possibility is that the missing heme is in the high-spin state, for high-spin hemes have very low MCD intensity. This conclusion is reinforced by the results shown in Fig. 2. Addition of the weak-field ligands formate or fluoride to the enzyme does not change the shape or intensity of the MCD[9], though the optical spectrum shows a clear shift to the blue. On the other hand the strong-field ligand, cyanide, produces a two-fold increase in the intensity of the MCD with no change in lineshape and the maximum in the optical spectrum shifts to 427 nm[9]. This behavior is consistent with the interpretation that cytochrome $\underline{a}_3$ and the cytochrome $\underline{a}_3 \cdot$formate complex is high-spin while the cytochrome $\underline{a}_3 \cdot$CN complex is low-spin, and reinforces the assignments that cytochromes $\underline{a}$ and $\underline{a}_3$ are low- and high-spin respectively.

When oxidase is reduced the MCD spectrum changes dramatically with a large increase in intensity and an overall change of sign of the various features (Fig. 3[9]): the amplitude of this new spectrum is also temperature dependent demonstrating that, at least in part, the spectrum originates from a paramagnetic center. In fact, the observed spectrum is a composite of two contributions. This can be demonstrated in a number of ways. The simplest is to record the spectrum of the fully reduced enzyme$\cdot$carbon monoxide complex ($\underline{a}^{2+}\underline{a}_3^{+2} \cdot$CO) (Fig. 4). The validity of this demonstration rests on the assumption that the MCD of $\underline{a}_3^{2+} \cdot$CO has only small intensity. That this is indeed the case, is documented by a comparison of the MCD of the resting enzyme ($\underline{a}^{+3}\underline{a}_3^{+3}$) and that of the mixed valence$\cdot$CO compound ($\underline{a}^{+3}\underline{a}_3^{+2} \cdot$CO): these are extremely similar (Fig. 5) which requires that $\underline{a}_3^{+2} \cdot$CO make only a small

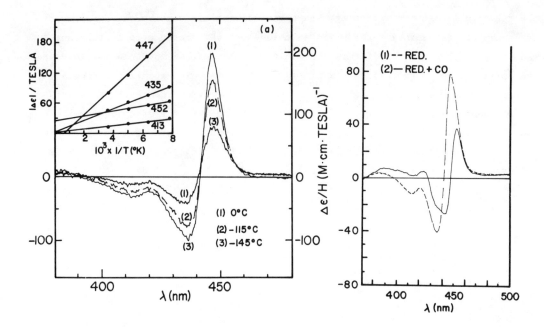

Fig. 3.　MCD spectrum of reduced cytochrome oxidase.　From Ref. 9, with permission.

Fig. 4.　MCD spectrum of the reduced oxidase·CO complex.　From Ref. 9, with permission.

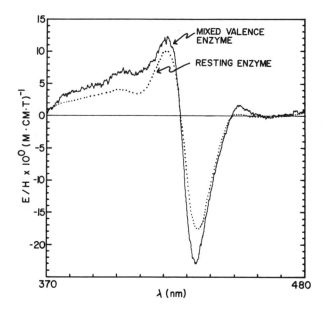

Fig. 5.　Comparison of the MCD spectrum of the resting (oxidized) cytochrome oxidase with that of the complex of the mixed-valence enzyme and carbon monoxide.

contribution to the MCD intensity. An alternative is to observe the spectrum of the enzyme in the aerobic steady state achieved in the presence of a reductant (ascorbate + TMPD) and the terminal inhibitor, formate (Fig. 6). In this system the formal state of the oxidase is $\underline{a}^{2+} \cdot \underline{a}_3^{+3} \cdot HCOOH$. As high-spin cytochrome $\underline{a}_3^{3+}$ has small MCD, the observed spectrum is dominated by cytochrome $\underline{a}^{+2}$.

We thus conclude that Figs. 4 and 6 represent the spectrum of cytochrome $\underline{a}^{+2}$, and that the difference in MCD between eg. fully reduced enzyme and the fully reduced enzyme·CO complex should yield the contribution of cytochrome $\underline{a}_3^{2+}$. This difference spectrum (Fig. 7) has an intensity appropriate for 50% of the heme; it is very similar to that observed with deoxyhemoglobin[11,12] implying that cytochrome $\underline{a}_3^{+2}$, like deoxyhemoglobin, is a high-spin ferrous heme compound.

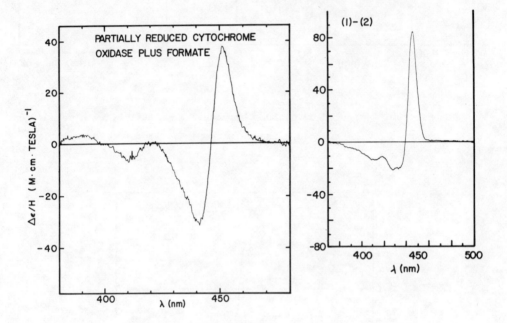

Fig. 6. MCD spectrum of cytochrome oxidase taken during the aerobic steady-state in the presence of excess formate, From Ref. 9, with permission.

Fig. 7. Computed MCD spectrum of reduced cytochrome $\underline{a}_3$ (Reduced enzyme minus reduced enzyme·CO compound), From Ref. 9, with permission.

Our results show that cytochrome $\underline{a}$ is low-spin in both fully oxidized and fully reduced enzyme whereas cytochrome $\underline{a}_3$ is high-spin in these two states. The various spectra of the several components of cytochrome oxidase are summarized in Fig. 8; of particular note is the fact that the zero-crossing of cytochrome $\underline{a}^{+2}$ occurs at the wavelength of maximum amplitude of cytochrome $\underline{a}_3^{+2}$ (447 nm) a fact which is of value in the analysis of the titration behavior reviewed later.

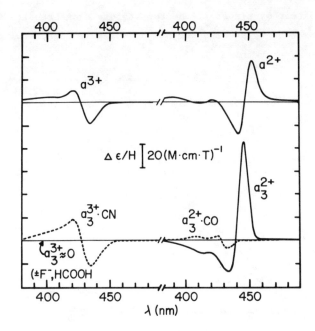

Fig. 8. Summary of the MCD spectra of the individual cytochromes and their derivatives. From Ref. 10, with permission

Variable Temperature Magnetic Susceptibility Studies

These assignments seem to have a substantial problem for one would have anticipated that the high-spin ferricytochrome a_3 would have been readily detected in
the EPR spectrum of the oxidized enzyme, whereas in fact, this species exhibits
only a very small EPR signal in the g=6 region. To rationalize this inconsistency
we have proposed[13] that cytochrome a_3 is antiferromagnetically coupled to one of
the two copper ions (Cu_u) present in the protein: this coupled system has a
ground state of S=2 and an excited state with S=3 to higher energy by an amount
3J, where J is the exchange coupling constant. The existence of a magnetic coupling
between cytochrome a_3 and Cu had been made earlier by Van Gelder and Beinert[5] to
explain the presence of a high-spin signal at intermediate levels of reduction.

The most penetrating test of the assignments just described in a measurement
of the magnetic susceptibility of the enzyme and its derivatives. Several such
measurements have been reported in the literature[14,15,16,17] and we have previously
described[13] how these data are as consistent with the above scheme as with any
obvious competing alternative. For reasons which are discussed in full elsewhere[18]
we have repeated and extended these magnetic measurements over a wide temperature
range, 7-250°K[18].

The experimental quantity of interest is $\mu^2\text{eff}$, the effective magnetic moment
squared. This can be defined as $1/3(g_x^2 + g_y^2 + g_z^2) \cdot S(S+1)$, in units of β^2. For
cytochrome a^{+3} and Cu_d^{+2}, S=1/2 and with the g-values reported by Aasa et al[6] the
values of μ^2_{eff} are 4.0 and 3.2 respectively. The magnetic susceptibility of the

resting enzyme is linear over most of the temperature range studied (Fig. 9) with a slope which corresponds to a value for μ^2_{eff} of 31.5 ± 1. Subtracting the

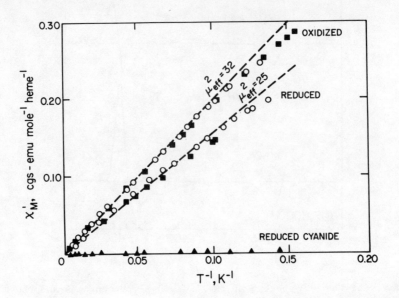

Fig. 9. The temperature variation of the magnetic susceptibility of oxidized, reduced and reduced cyanide derivatives of cytochrome oxidase. The lines shown along side the data have no theoretical significance and are for reference purposes only. From Ref. 18, with permission.

contributions of cytochrome $\underline{a}$ and Cu_d we find the contribution of cytochrome $\underline{a}_3$ plus Cu_u to be 24.3. Consideration of all possible alternatives (Table 1, Ref. 18) immediately establishes that only two electronic configurations fit the experimental result: 1) a strongly-coupled antiferromagnetic pair comprising high-spin cytochrome $\underline{a}_3^{3+}$ and Cu_u^{2+}, or 2) a strongly coupled ferromagnetic pair of Cu_u^{2+} and intermediate-spin (S=3/2) cytochrome $\underline{a}_3^{3+}$. The latter has been discounted as being improbable for several persuasive reasons[18]. Thus the paramagnetic susceptibility has contributions from three terms:

$$\chi_{total} = \chi_a + \chi_d + \chi_{pair} \tag{1}$$

due to cytochrome $\underline{a}$, Cu_d and the cytochrome $\underline{a}_3$-Cu_u pair. The first two terms are due to S=1/2 species and give simple linear contributions to a Curie plot (Fig. 9). The χ_{pair} term can introduce a variety of complications but the linearity of the data over most of the temperature range studied establishes $-J>200$ cm^{-1}; consequently, the complications that can arise from an exchange interaction can be ignored and attention focused on the possible non-linearities which seem to be present at the lowest temperatures studied.

The S=2 spin state of the antiferromagnetic pair will split in zero-field under the influence of axial and rhombic components of the ligand field. The magnitude of the splitting (D) due to the axial components is expected to be 5-10 cm^{-1},[19]

228

while the effect (E) of the rhombic contribution is expected to be small, <u>ca</u>. .2 cm^{-1},[18]. Thus in computing equation 1, we have used formulas for χ_{pair} which accommodate these small splittings in the S=2 set of states. The results of such a calculation are shown in Fig. 10. As expected, the data is quite linear over

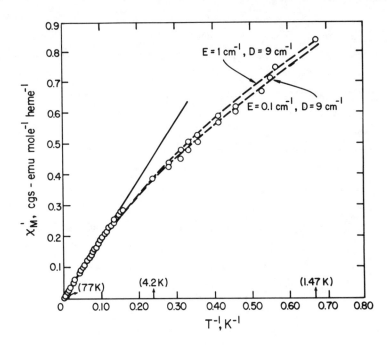

Fig. 10. A theoretical fit to the susceptibility of the oxidized enzyme using equation 1 and the values of the zerofield splitting shown. The data taken below 4.2° was kindly provided by Moss <u>et al</u>[20]. From Ref. 18, with permission.

most of the temperature range with a slope which is independent of the magnitude of the zero-field splitting parameters (for plausible values of D and E). Below 7°K however, the slope of the calculated line decreases as kT approaches D and E. For the values of D and E shown the slope of the data below 4.2°K is about one-half of its value at high temperatures. Furthermore, the presence of the contributions from cytochrome $\underline{a}^{+3}$ plus Cu_d^{+2} together with the admixture of $M_s=2$ states into the lowest lying $M_s=0$ component minimize the usual curvature observed when kT∿D. Fig. 10 also show a set of susceptibility data obtained independently by Moss <u>et. al</u>.[20] which falls closely on the computed lines. The ability of equation 1 to satisfactorily account for two apparently quite different results obtained in two different temperature ranges by two independent laboratories must be taken as rather persuasive evidence in favor of the model we have adopted.

In the absence of added ligands reduced cytochrome oxidase is highly paramagnetic (Fig. 9). The data is linear over most of the temperature range with a slope which yields a value of 24 for μ_{eff}^2. This value is readily accounted for by recognizing that Cu_d, Cu_u and cytochrome $\underline{a}$ are all diamagnetic in their reduced form while cytochrome $\underline{a}_3$ is proposed to be high-spin ferrous (S=2) for which $\mu_{eff}^2 = 24-29$, depending on the magnitude of the orbital contribution. These data are consistent with the MCD result (Fig. 7) and there seems to be no reason to question the interpretation that cytochrome $\underline{a}_3$ is the only paramagnetic species in the reduced enzyme.

Incubation of the reduced enzyme with cyanide should convert cytochrome $\underline{a}_3$ to the diamagnetic state (low-spin ferrous) and in agreement with this the susceptibility of the reduced enzyme is found to be very small ($\mu_{eff}^2<1$) (Fig. 9).

An additional penetrating test of the model is provided by the compound obtained by incubating the oxidized enzyme with cyanide. From the MCD data cytochrome $\underline{a}_3$ should now be present as a low-spin (S=1/2) species. We would, therefore, expect that the cytochrome $\underline{a}_3 \sim Cu_u$ binuclear center will have modified properties since the antiferromagnetic interaction is now between two S=1/2 centers with a total spin of zero in the ground state and an excited state with S=1. Thus, χ_a and χ_d will be unchanged and χ_{pair} should be replaced with the classic expression for two interacting S=1/2 centers. The observed susceptibility of the oxidase-cyanide complex is essentially linear below 50°K but above this temperature the susceptibility decreases rapidly (Fig. 11). The limiting slopes at the high and low temperature

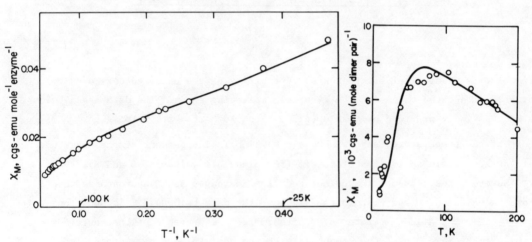

Fig. 11. Magnetic Susceptibility of the oxidized oxidase-cyanide complex. The solid line was calculated as described in the text. From Ref. 18, with permission.

Fig. 12. The difference susceptibility. Oxidized enzyme-cyanide complex minus (cytochrome $\underline{a}^{+3}$ plus Cu_d^{+2}). The solid line is the computed contribution of the exchange coupled pair $\underline{a}^{+3} \cdot CN \sim Cu_u^{+2}$ with $-J \sim 40cm^{-1}$. From Ref. 18, with permission.

extremes are $\mu_{eff}^2 = 15.1$ and 7.8 respectively which agree closely with the prediction for $J\hbar kT^{18}$. The best fit to the data (Fig. 11) is obtained with $-J=38.5$ cm^{-1}. The agreement is emphasized in Fig. 12 which shows the susceptibility obtained after subtracting the contributions of cytochrome $\underline{a}$ and Cu_u from the total susceptibility values together with the computed susceptibility due to the binuclear center[18].

The susceptibility results we have obtained are compelling evidence for the correctness of the model under investigation and thus raises the question of the pathway for the antiferromagnetic interaction between cytochrome $\underline{a}_3$ and Cu_u. We are attracted to the possibility that the imidazole anion fulfills this role by functioning as a bridging ligand between the two metal ions. The evidence to

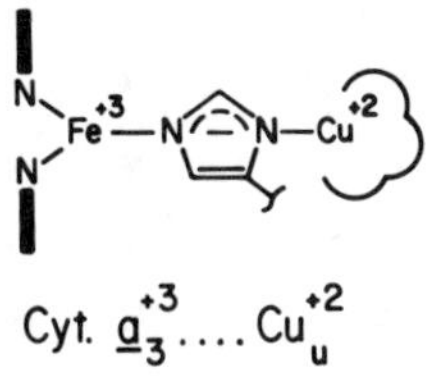

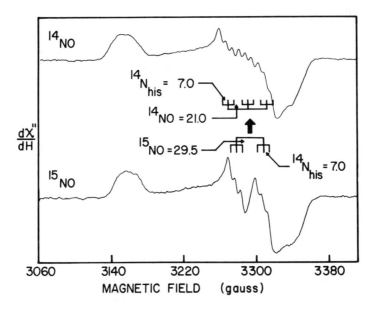

Cyt. $\underline{a}_3^{+3} \ldots Cu_u^{+2}$

Fig. 13. Assumed structure for the cytochrome $\underline{a}_3$-Cu_u binuclear center. From Ref. 18, with permission.

support this proposal is admittedly circumstantial but nonetheless valid. First of all, use of the heme probe NO clearly establishes the presence of a N atom in the ligating position trans to the NO i.e. in the proximal heme site. The EPR spectra of the reduced cytochrome oxidase·NO complex (Fig. 14) clearly demonstrate hyperfine

Fig. 14. EPR spectra of the reduced oxidase · ^{14}NO and ^{15}NO compounds. The stick spectra represent the analysis for the [15]NO derivative and the subsequent prediction for the [14]NO derivative. T. Antalis and G. Palmer, (unpublished).

coupling to two N atoms. When $^{14}NO(I_N=1)$ is employed a nine-line hyperfine signature is observed at g_y: this has been interpreted as a major hyperfine coupling with the [14]NO (a_N = 21 gauss) and a minor hyperfine coupling (a_N = 7.0 gauss) to the ^{14}N imidazole[21]. Proof of this interpretation is obtained by the use of [15]NO: the EPR spectrum of the oxidase [15]NO compound exhibits six hyperfine lines at g_y, arranged as a doublet of triplets. The major splitting is now 29 gauss, increased by precisely the ratio of the g-factors of ^{15}N to [14]N : the minor splitting is unchanged as required. These data establish the proximal ligand to be Nitrogen, though of course, does not prove that histidine is the donor amino acid: nevertheless, it should be remembered that histidine is by far and away the most common amino acid at this axial coordination site in hemeproteins. The data does however establish that structures invoking oxygen atoms[22] in the bridging positions are incorrect.

The second piece of data favoring histidine is the magnitude of the exchange interaction: the extremely strong antiferromagnetic coupling that is observed points to a facile path for the transfer of electron spin between the two metal ions: the aromatic imidazolate anion is well suited for this role, and the magnitude of the exchange interaction in the oxidase·CN complex is quite comparable to that measured in imidazole bridged copper-copper pairs[23]. Finally, we can point to the fact that histidine has already been established as a ligand bridging the copper and zinc components of superoxide dismutase[24] and that copper and cobalt analogs of this protein exhibit strong exchange interactions[25].

Behavior of Cytochrome oxidase and its derivatives during Reductive Titrations

Of course cytochrome oxidase is not confined to the limiting states having just been described. On the contrary, one must anticipate 1e, 2e, 3e and 4e reduced states of the protein and the characterization of these states by both stoichiometric and potentiometic oxido-reduction techniques has been an area of substantial activity and diametrically conflicting interpretations[26,27,28].

Thus we have seen the low-spin species responsible for the g=3 resonance attributed to both cytochrome $\underline{a}^{+3}$[7] and cytochrome $\underline{a}_3^{+3}$,[29] while the interpretation of the potentiometic data has depended on allosteric linkage of the redox potentials of cytochromes $\underline{a}$ and $\underline{a}_3$[28] or alternatively on the allosteric linkage of the optical properties of these two redox centers with little modification of heme redox potentials[26].

Our contributions to this area have relied upon the comparison of the optical, EPR and MCD spectra of cytochrome oxidase as the enzyme is progressively reduced by the addition of small increments of a chemical reductant[10].

In the prototype experiment (Fig. 15), no inhibitors are present. Among the experimental observables are the optical transitions in the Soret (443 nm) the

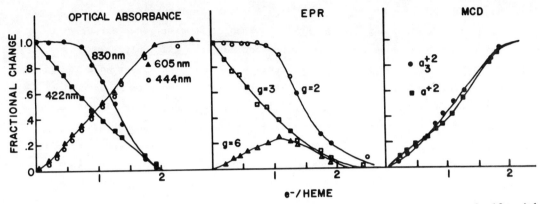

Fig. 15. Reductive titrations of cytochrome oxidase under Argon. From Ref. 10, with permission.

visible (605 nm) and the near IR (830 nm) regions, the EPR resonances at g=3 (cyto-chrome $\underline{a}^{+3}$), g=2 (Cu_d^{+2}) and g=6 and the contributions to the MCD of cytochromes $\underline{a}^{+2}$ and $\underline{a}_3^{+2}$ obtained from deconvolution of the MCD spectra in the 440-460 nm region. There are a number of notable features to the data. Firstly, the absor-bance at 443 nm and 605 nm, the EPR (g=3) and the MCD contributions from cyto-chromes $\underline{a}^{+2}$ and $\underline{a}_3^{+2}$ change continuously throughout the entire titration. These data cannot be explained by the assumption that cytochromes $\underline{a}$ and $\underline{a}_3$ have signifi-cantly different potentials for this would require that the spectral changes due to one of the hemes be complete before those due to the other. Rather, the data clearly indicates that the potentials of $\underline{a}$ and $\underline{a}_3$ are similar. Secondly, one sees that the reduction of Cu_d^{+2} (g=2, 830 nm) is characterized by a lag lasting for the first half of the titration, during which cytochromes $\underline{a}$ and $\underline{a}_3$ become approximately 50% reduced and then proceeds rapidly during the latter half of the titration to-gether with the remainder of the cytochromes. These data require that the two hemecenters have potentials about 100mV more positive than Cu_d in the early part of the titration while towards the end of the titration cytochromes $\underline{a}$ and $\underline{a}_3$ have potentials about 35 mV more negative than Cu_d. The simplest explanation for this observation invokes a linkage between the two cytochromes as illustrated in the following scheme[28,30], where E_H and E_L are the midpoints potentials for the

$$\text{SCHEME} \quad 1 \quad a^{+3} a_3^{+3} \underset{E_H}{\overset{E_H}{\rightleftharpoons}} \begin{matrix} a^{+3} a_3^{+2} & E_L \\[1ex] a^{+2} a_3^{+3} \end{matrix} \overset{E_L}{\rightleftharpoons} a^{+2} a_3^{+2}$$

high-potential and low-potential forms of the two cytochromes. Taking the midpoint potential of Cu_d as 250 mV[31], E_H and E_L are 350 and 215 mV respectively. The

233

difference in potential $E_H - E_L$ is then the linkage in potentials and is proportional to the mutual free energy of coupling[32]. Thus, in the oxidized enzyme the
reduction of cytochrome $\underline{a}$ and $\underline{a}_3$ are equally probable; however, once either cytochrome is reduced, the reduction of the other becomes more difficult. In this
representation the contributions of cytochromes $\underline{a}$ and $\underline{a}_3$ to the absorbance changes
are actually irrelevant and the sigmoidal Minnaert plots[33] observed in potentio-
metric experiments[26] are a consequence of the equivalence and linkage in the potentials and not of widely different potentials for $\underline{a}$ and $\underline{a}_3$ which would lead to
stoichiometric behavior totally different to that which is observed experimentally[10].

With the basic assumption that the fate of cytochrome $\underline{a}^{+3}$ is simply the capture
of an electron with reduction to cytochrome $\underline{a}_3^{+2}$ then the data should satisfy the
relationship:

$$\text{cytochrome } \underline{a}^{+3} \ (g=3) + \text{cytochrome } \underline{a}_3^{+2} \ (\text{MCD}) = 1$$

that is, the quantity of cytochrome $\underline{a}^{+3}$ measured by the g=3 EPR signal should be
the complement of the quantity of cytochrome $\underline{a}^{+2}$ measured by MCD. This is, in fact,
not the case, the MCD data consistently predicting more cytochrome $\underline{a}^{+3}$ than is
measured by the EPR method. The magnitude of the discrepancy between the two
sets of measurements is maximum at the midpoint of the titration (Fig. 16 left):

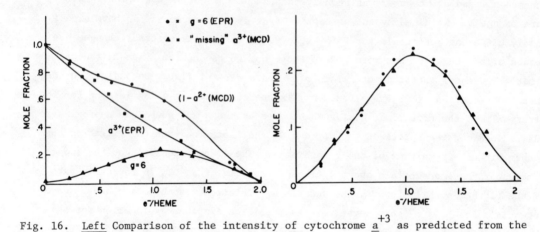

Fig. 16. <u>Left</u> Comparison of the intensity of cytochrome $\underline{a}^{+3}$ as predicted from the
MCD results $(1-\underline{a}^{+2}(\text{MCD}))$ with that actually observed by EPR. <u>Right</u> Comparison of
the quantity of "missing $\underline{a}^{+3}$" with the intensity of the g=6 high-spin signal.

this difference we call the "missing $\underline{a}^{+3}$". There is a striking correspondence
between the appearance and disappearance of the g=6 high-spin ferric EPR signal
and the missing $\underline{a}^{+3}$ (Fig. 16 right), which leads us to believe that during the
redox event part of the cytochrom $\underline{a}^{+3}$ undergoes a low-spin to high-spin transition,
triggered perhaps by a redox event at cytochrome $\underline{a}_3$. It should be emphasized that
the EPR at g=6 is heterogeneous with contributions from at least two species, one
of them rhombic and the other axial: in view of the data which Dr. Beinert will
review in the next presentation, it seems plausible that the axial species

originates from cytochrome $\underline{a}$.

In titrations performed in the presence of CO the observed behavior is quite different[10]. In particular, we observe that the EPR amplitude at g=3 and the intensity of the MCD trough at 432 nm (Fig. 17) does not change until the enzyme is more than 50% reduced. Indeed the MCD spectrum of half-reduced enzyme (and that of the mixed-valence enzyme (Fig. 5)) is not markedly different from that of the fully oxidized protein. CO, of course, has a very high affinity for high-spin ferrous hemes: it binds avidly to cytochrome $\underline{a}_3^{+2}$ and the binding energy specifically raises the potential of $\underline{a}_3$ so that the sequence for electron addition is now:

$$\underline{a}^{+3}\underline{a}_3^{+3} \longrightarrow \underline{a}^{+3}\underline{a}_3^{+2}\cdot CO \longrightarrow \underline{a}^{+2}\underline{a}_3^{+2}\cdot CO$$

i.e. the stabilization of cytochrome $\underline{a}_3^{+2}$ by carbon monoxide confines the redox sequence to the upper branch of Scheme 1.

Alternatively, titrations performed in the presence of sodium azide might be expected to proceed by the lower path of Scheme 1 for azide binds avidly to

$$\underline{a}^{+3}\underline{a}_3^{+3}\cdot N_3 \longrightarrow \underline{a}^{+2}\underline{a}_3^{+3}N_3 \longrightarrow \underline{a}^{+2}\underline{a}_3^{+2}$$

cytochrome $\underline{a}_3^{+3}$ and stabilization of the ferric form of the hemoprotein will drive its potential more negative than cytochrome $\underline{a}$. Our data (Fig. 18) is consistent with this expectation. The EPR at g3 disappears rapidly during such a titration and the MCD spectrum of half-reduced enzyme shows that cytochrome $\underline{a}^{+2}$ is substantially reduced. The EPR still contains a low-spin resonance, with $g_z = 2.9$; this is most satisfactorily assigned to the cytochrome $\underline{a}_3^{+3}\cdot N_3$ species for the MCD results show that cytochrome $\underline{a}_3$ is still largely oxidized past the midpoint of

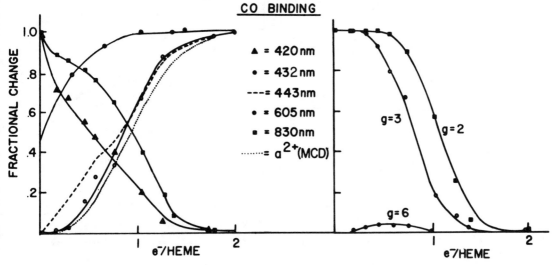

Fig. 17. Reductive titration of cytochrome oxidase under CO. From Ref. 10, with permission.

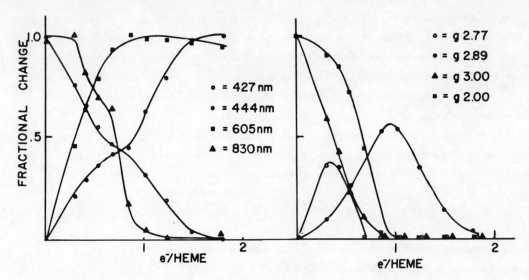

Fig. 18. Reductive titration of cytochrome oxidase under Argon and in the presence of excess sodium azide. (G. Palmer, L. Garcia-Iniquez, G.T. Babcock and L.E. Vickery; (unpublished results)).

the titration. Rather similar results are obtained in the presence of formate which also stablizes a_3^{+3} (Fig. 19). Again the EPR of cytochrome a^{+3} is eliminated and the typical MCD of cytochrome a^{+2} is fully developed by the midpoint of the titration

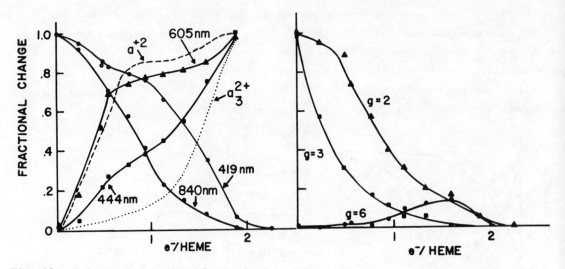

Fig. 19. Reductive titration of cytochrome oxidase under Argon and in the presence of excess sodium formate. (G. Palmer, L. Garcia-Iniquez, G.T. Babcock and L.E. Vickery, (unpublished results)).

and yet, for instance, the optical changes in the Soret region are only half-complete at this point indicating only 50% reduction of heme; a small g=6 signal appears during the latter half of this experiment which we interprete as the high-spin cytochrome a_3·HCOOH complex.

The optical changes observed in these experiments are consistent with the assignment that cytochrome $\underline{a}$ contributes approximately 80% to the changes in the alpha-band at 605 nm and about 50% to the changes in the reduced Soret at 443 nm while cytochrome $\underline{a}_3$ contributes the remainder and that the extinction changes due to cytochrome $\underline{a}$ are insensitive to the state of ligation of $\underline{a}_3$. With these extinction coefficients together with Scheme 1 one can adequately account for much of the potentiometric data obtained with this enzyme[29].

It should be appreciated that Scheme 1 is in fact only the simplest approximation to the real situation. The antiferromagnetic system, $\underline{a}_3^{+3} \cdot Cu_u^{+2}$, appears to function as a n=2 redox system for, at most, only small amounts of possible one-electron reduced (e.g. $\underline{a}_3^{+2}Cu_u^{+2}$, $\underline{a}_3^{+3}Cu_u^{+1}$) states are observed. Thus the free energy of coupling which is responsible for the linkage is between cytochrome $\underline{a}$ on the one hand, and the cytochrome $\underline{a}_3 \cdot Cu_u$ pair on the other. As a consequence, whereas the potential of cytochrome $\underline{a}$ will undergo a decrease of 135 mvolts (n=1) when $\underline{a}_3 \cdot Cu_u$ is reduced, the potential of $\underline{a}_3 \cdot Cu_u$ will only be decreased by 68 mvolts (n=2) when cytochrome $\underline{a}$ is reduced, assuming that the coupling energy is partitioned equally between $\underline{a}_3$ and Cu_u. In the extreme that the coupling energy is restricted to only one of the two metal ions present in the binuclear center then the n=2 redox pair may well be converted to a n=1, two-fold, system. The significance of this potential asymmetry in coupling and its implication for both static and kinetic experiments are currently being evaluated.

ACKNOWLEDGEMENTS

This research was supported in part by NIH Grant GM-21337, NIH Postdoctoral fellowship HL-02052 (to GTB), grants from the Welch Foundation, C-636 (GP) and C-627 (LW) and by the U. S. Energy Research and Development Association. Liquid Helium was obtained from the Helium Liquidification facility operated under Navy Contract at Rice University.

*GTB is currently at Michigan State University, East Lansing, Michigan, MT at the Department of Chemistry, Stanford University, Stanford, California and LEV at the Department of Physiology, University of California, Irvine, California.

REFERENCES

1. Keilin, D. and Hartree, E.F., (1939) Proc. R. Soc. Lond. B. Biol. Sci. 127, 167-191.

2. Oxidases and Related Redox Systems (King, T.E., Mason, H.S. and Morrison, M.), Wiley Intescience, 1964, Vol. 2.

3. Vanneste, W. (1966) Biochemistry 5, 838-848.

4. Beinert, H., Griffiths, D.F., Wharton, D.C., and Sands, R.H. (1962) J. Biol. Chem. 237, 2337-2346.

5. Van Gelder, B.F. and Beinert, H. (1969) Biochim. Biophys. Acta, 189, 1-24.

6. Aasa, R., Albracht, S.P.J., Falk, K.-E., Lanne, B., and Vanngard (1976) Biochim. Biophys. Acta, 422, 260-272.

7. Hartzell, C.R., and Beinert, H. (1976) Biochim. Biophys. Acta, 423, 323-338.

8. Schatz, P.N., and McCaffery, A.J. (1969) Q. Res. Chem. Soc. 23, 552-

9.   Babcock, G.T., Vickery, L.E. and Palmer, G. (1976) J. Biol. Chem. 251, 7901-7919.

10. Babcock, G.T., Vickery, L.E. and Palmer, G. (1978) J. Biol. Chem. 253, in press.

11. Vickery, L.E., Nozawa, T. and Sauer, K. (1976) J. Am. Chem. Soc. 98, 343-350.

12. Treu, J.J. and Hopfield, J.J. (1975) J. Chem. Phys. 63, 613-623.

13. Palmer, G., Babcock, G.T. and Vickery, L.D. (1976) Proc. Nat. Acad. Sci. U.S.A. 73, 2206-2210.

14. Ehrenberg, A. and Yonetani, T. (1961) Acta, Chem. Scand. 15, 1071-1080.

15. Ehrenberg, A. and Vanneste, W. (1968) 19th Colloqium der Gesellschaft fuer Biologische Chemie Mosbach, (Springer, New York), pp.121-124.

16. Tsudzuki, T. and Okunuki, K. (1971) J. Biochem. 69, 909-922.

17. Falk, K.E., Vanngard, T. and Angstrom, J. (1977) FEBS Letters 75, 23-27.

18. Tweedle, M., Wilson, L., Babcock, G.T., Garcia-Iniguez, L., and Palmer, G. (1978) submitted.

19. Brackett, G., Richards, P. and Caughey, W. (1971) J. Chem. Phys. 56, 4383-4401.

20. Moss, T.H., Beinert, H. and King, T.E. (1977) personal communication.

21. Blokzihl-Homan, M.F. and Van Gelder, B.F. (1971) Biochim. Biophys. Acta, 234, 493-498.

22. Wilson, D.F., Erecinska, M., Lindsay, J.G., Leigh, J.S., Jr., and Owen, C.S. (1975) Proc. 10th FEBS Meeting Elsevier, Amsterdam 1976, p. 195-210.

23. Kolks, G., Frihart, C.R., Rabinowitz, H.N. and Lippard, S.J. (1977) J. Am. Chem. Soc. 99, 5804-5806.

24. Moss, T.H. and Fee, J.A. (1975) Biophys. Biochem. Res. Comm. 66, 799-808.

25. Richardson, J.S., Thomas, K.A., Rubin, B.H. and Richardson, D.C. (1975) Proc. Nat'l. Acad. Sci. U.S.A. 72, 1349-1353.

26. Wilson, D.F., Lindsay, J.G. and Brocklehurst, E.S. (1972) Arch. Biochem. Biophys. 256, 277-286.

27. Leigh, J.S., Jr., Wilson, D.F., Owen, C.S. and King, T.E., (1974) Arch. Biochem. Biophys. 160, 476-486.

28. Mackey, L.N., Kuwana, T. and Hartzell, C.R. (1973) FEBS Letters 36, 326-329.

29. Wikstrom, M.K.F., Harmon, H.J., Ingledew, W.J. and Chance, B. (1976) FEBS Letters 65, 259-276.

30. Wilson, D.F. and Leigh, J.S. (1972) Arch. Biochem. Biophys. 150, 154-163.

31. Nicholls, P. and Chance, B. (1974) in Molecular Mechanisms of Oxygen Activation (Hayaishi, O. ed) pp. 479-534, Academic Press, New York.

32. Wilson, D.J., Erecinska, M., and Owen, C.S. (1976) Arch. Biochem. Biophys. 175, 160-1728.

33. Weber, G. (1972) Biochemistry 11, 864-878.

34. Minneart, K., (1965) Biochim. Biophys. Acta, 110, 47-56.

CHARACTERISTICS OF HIGH SPIN COMPONENTS
AND ALTERNATIVE OXIDIZED STATES IN CYTOCHROME c OXIDASE

HELMUT BEINERT, ROBERT W. SHAW AND RAYMOND E. HANSEN

Institute for Enzyme Research, University of Wisconsin, Madison, Wisc. 53706 (U.S.A.)

ABSTRACT

Cytochrome c oxidase was reduced by an equivalent amount of ferrocytochrome c in the presence of a small excess of ascorbate and rapidly reoxidized by ferricyanide or porphyrexide. According to Beinert and Shaw [Biochim. Biophys. Acta, 462 (1977) 121], this procedure makes cytochrome a_3^{3+} detectable by EPR in the form of a rhombic high spin signal (g $\sim$ 6;2) and thus makes it possible to study reactions of cytochrome a_3^{3+} directly. By this approach, the responses of a_3^{3+} to O_2 and ferrocytochrome c and to ligands such as CO, CN^-, S^- and N_3^- were studied with the use of multiple rapid mixing techniques.

INTRODUCTION
Components of cytochrome c oxidase

It is now quite generally accepted that in cytochrome c oxidase, four different metal components can be distinguished, that these components are involved in the four-electron reduction of dioxygen and that the minimal functional unit contains at least one set of these four metal components. Two of these are designated as cytochromes, viz., a and a_3, while two are copper ions. Cytochrome a_3 has been defined as the component that, in its reduced state, binds CO, a reaction reversed by light, and by analogy O_2. This leaves cytochrome a as the component that is relatively inert towards most ligands. In the reduced form, no more than 50% of the total heme binds CO.

Search for distinguishing features of components

Earlier work. Attempts to distinguish the two cytochromes by spectrophotometry were at best partly successful when ligands were present, but remained largely inconclusive in unliganded forms because of the overlap of spectral features and the lack of knowledge of molar absorptivities of the pure components. Great expectation was therefore placed in EPR spectroscopic investigations, since, in principle, this technique has the potential of detecting all four metal components in the oxidized state and of furnishing information on quantities involved, oxidation states and ligand effects. As is well known today, this did not materialize.

Abbreviations: l.s., low spin; h.s., high spin; AR, species of enzyme obtained on anaerobic reoxidation of reduced enzyme; RO, resting, oxidized enzyme; Cu_d (Cu_u) EPR-detectable (undetectable) copper species; a, a_3, c, cytochromes a, a_3 and c.

The only straight-forward interpretation of the available EPR spectra is that not more than one heme (in the l.s. form) and one of the copper components are detectable in the oxidized resting enzyme by conventional EPR spectroscopy. As has been discussed in more detail in the preceding presentation of this symposium[1] this very EPR "silence" of one heme and one copper ion contains valuable information concerning the mutual disposition of the components of the enzyme.

During the early EPR studies on the enzyme it was observed that, as the signal of the l.s. heme of the enzyme disappeared during reduction, a ferric h.s. heme became detectable at intermediate oxidation states[2]. This raised the question as to whether the ferric h.s. heme which, paradoxically, was emerging during reduction, was due to that heme component which is EPR silent in the resting enzyme or to the same component that is seen as l.s. heme in the resting form and disappears on reduction.

Conflicting interpretations were presented concerning the identity of these h.s. signals[2-7]. A major complicating factor was and still is that not one but a number of such signals appear depending on a variety of conditions[6], and it is even likely that a second type of l.s. signal, which appears particularly at elevated pH, originates from the same components(s) that produce(s) the h.s. signal[7].

Recent work. Since the earlier EPR work, cytochrome oxidase has been studied in a number of states of oxidation and of liganding and situations have been found in which the EPR silent cytochrome becomes EPR detectable. The most unambiguous cases involve liganded states, so that the information derived from these studies is not directly applicable to the ligand-free enzyme. Cytochrome a_3 can be detected by EPR as a_3^{2+}-NO [8], a_3^{3+}-CN$^-$ [9] and a_3^{3+}-S^{2-} [10], in the last two cases in the ferric l.s. state. Cytochrome a_3 can, however, also be detected in the absence of ligands, when the enzyme is reduced and rapidly reoxidized anaerobically by ferricyanide or other chemical oxidizing agents, but not O_2[11]. In this case largely rhombic ferric h.s. heme signals are seen, which, together with the l.s. heme present, can account for up to 80% of the total heme present in the enzyme. Since the l.s. heme shows kinetic and ligand binding properties generally attributed to cytochrome a, the bulk of the h.s. signals must originate from a_3 under these conditions.

A second approach which in all probability (see below) produces high spin signals from a_3 is that of dissociating the a_3^{2+}-CO complex by light. If this is done in the presence of ferricyanide, rhombic h.s. signals are seen in the presence of fully developed l.s. signals[12,13]. The quantities of heme represented in these h.s. signals were, however, not reported to be of a magnitude to make this case entirely convincing[11,12]. While in the mentioned situations there is either little or no doubt that both cytochromes can be detected in the enzyme, the search for the silent copper component has been generally unsuccessful.

If, as has been discussed in more detail by Dr. Palmer[1], a_3 and Cu_u interact antiferromagnetically so as to make them undetectable by EPR, one might expect that the copper component should become detectable when the interaction is broken up so that the silent heme component is converted to a detectable form. This has not been observed. There is one report in which this is claimed for the sulfide liganded form, but the data presented thus far are not convincing[14]. It has also been proposed that one of the copper components is in the cuprous state[15] and the observed supposed copper signal arises from sulfur ligands of the metal[16].

It has long been known that cytochrome oxidase has a broad absorption band centered at 655 nm[17]. This band is very conspicuous in reflectance spectra, inasmuch as they emphasize long wavelength absorptions. Attention has been drawn to the pattern of appearance and disappearance of this band and its correlation with the behavior of other features in optical and EPR spectra[4]. From its position in the spectrum, it could be due to copper or h.s. heme. However, since formate enhances this absorption band[11] and formate is known as one of the strongest ligands of the h.s. heme, it is most likely that the 655 nm absorption band is related to heme. The 655 nm band is absent in the reduced form and returns, on reoxidation with O_2, at the earliest reproducible times ($\sim$ 5 msec). It is absent whenever strong rhombic ferric h.s. heme signals are present. If what we think we have demonstrated in the experiments on reoxidation of reduced oxidase with ferricyanide[11] is accepted, namely, that $\underline{a}_3$ is represented in these rhombic h.s. signals, then ferric $\underline{a}_3$ per se does not exhibit the 655 nm absorption. Since Cu_d and the l.s. heme (cytochrome $\underline{a}$) are reoxidized under these conditions, neither one of these components can give rise to this absorption band. Thus the 655 nm band must be due to a form of the EPR silent heme (cytochrome $\underline{a}_3$) different from that detectable by its rhombic ferric h.s. signal, namely a form present in the resting oxidized enzyme, and this form may be the spin coupled h.s. heme—copper species which was just discussed by Dr. Palmer[1].

Alternative oxidized forms of the enzyme

There have been indications of different kinds that cytochrome $\underline{c}$ oxidase can exist in a number of alternative oxidized forms[18-21]. This has mainly been concluded from shifts of light absorption maxima in the Soret or α-region[18-20] and from differences in oxidation-reduction kinetics[21] and ligand binding[22] with preparations of different history. While in the past considerable attention had been given to the so-called "oxygenated" form[18-20,23-25] (which does not appear to be what the name implies), more recently interest has concentrated on activated species such as the "pulsed" enzyme of Antonini et al.[21]. Our observations on a form or forms of the enzyme generated by anaerobic reoxidation of the previously reduced enzyme are germane to this topic[11]. We have observed two states of the enzyme, in which at least the detectable metal components, namely both hemes and one of the copper ions, are in the oxidized state and which can be distinguished from the resting oxidized form by differences in the presence of a rhombic ferric h.s. heme EPR signal and the 655 nm absorption. In the work to be described below, we have investigated the reactivity of one of these forms with oxygen and ferrocytochrome c and with ligands and have studied their responses toward pH changes. The approaches, materials and techniques used were those described in previous publications from this laboratory[4-7,11].

RESULTS
Species observed

As described above, the principal species (AR) with which we will be concerned is that produced by anaerobic reoxidation of the previously reduced enzyme by agents such as ferricyanide and porphyrexide. This species has a lifetime of a few seconds (maximal development at $\sim$ 1 sec), and is characterized by the absence of absorption at 655 nm and an intense rhombic ferric h.s. heme signal (extrema at g=6.34 and 5.44), which is maximal in the presence of an amount of cytochrome c approximately equivalent to that of the total heme in the oxidase. Observations made during the reaction of species AR with O_2 (Fig. 1) show that AR is not directly converted to the resting oxidized enzyme (species RO), namely, the rhombic h.s. signal disappears within 10 msec on mixing with O_2, whereas the 655 nm absorption is restored to its full intensity only within seconds to minutes. Thus, there must be an intermediate form that lacks both the rhombic high spin signal of species AR and the 655 nm band of species RO[11].

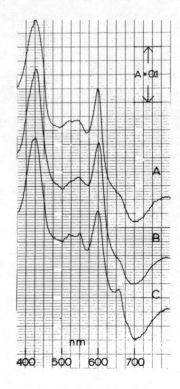

Fig. 1. Optical reflectance spectra at -175°C of cytochrome c oxidase, previously reduced with NADH, in various states of reoxidation. A, reoxidized anaerobically for 1.2 sec by 4 mM ferricyanide; B, as in A and then exposed to O_2 for 2 sec; C; reoxidized for 6 msec immediately with O_2. Cytochrome c oxidase[6], 1 mM, in 10 mM cacodylate of pH 7.2 was reduced for 20 hrs at 2° with a 4 fold molar excess of NADH in the presence of 10 μM cytochrome c. The enzyme was then rapidly mixed with equal volumes of reactants as specified above, and rapidly frozen[5]. All samples were subjected to two mixing steps so that the final concentrations of enzyme were equal.

242

As is evident from the foregoing, the possibilities of studying the properties of these transient species are limited. We have not yet studied properties of the second transient species which arises on exposure of species AR to O_2.

While it is possible by rapid reaction techniques using multiple mixing arrangements to study effects of ligands, oxidants and reductants, it is very difficult to ascertain the oxidation state of the enzyme in these transient species. Thus, we are unable to say at this time whether species AR contains zero or one reducing equivalent per minimal active unit (i.e. per 2 hemes and 2 copper ions). It is obvious from EPR and optical spectroscopy that both hemes and Cu_d are $\geq$ 80% oxidized but the state of Cu_u is uncertain. Since we obtain species AR with ferricyanide (E_o' = 420 mV) as well as with porphyrexide (E_o' = 720 mV)[26] we are inclined to assume that this species contains zero reducing equivalents and can, on this basis, be considered as an alternative oxidized form.

Reaction with O_2 and ferrocytochrome c

As mentioned in the preceding section, species AR reacts very rapidly ($t_{1/2}$ < 10 msec) with O_2. It was also pointed out, that we are not certain whether any oxidation is involved in this reaction. In addition to the argument brought forth above, namely that we would expect a high potential oxidant such as porphyrexide to have withdrawn all electrons from the metal components of the oxidase, it is hard to rationalize that O_2 should accept electrons from a species (AR) in which $\underline{a}_3$ is in the oxidized state. Thus, it would seem that O_2 exerts an allosteric effect on species AR, transforming it into the oxidized resting form RO via the intermediate form mentioned above. In this intermediate form, $\underline{a}_3^{3+}$ and Cu_u must already be anti-ferromagnetically coupled, since the ferric h.s. heme EPR signal has disappeared and the alternative, viz. disappearance of the signal through reduction of $\underline{a}_3$, is most unlikely under the prevailing conditions. Some rearrangement in the enzyme, which leads to the resting form RO with pronounced absorption at 655 nm, has, however, not yet taken place. Thus far, it has remained an enigma, why in species AR Cu_u does not become detectable. As shown in previous work[5,11] however, it has been found that the signal, generally attributed to Cu_d, reaches, in species AR, the highest intensities ever observed. This signal usually accounts for 35-40% of the copper determined by chemical analysis, but can, in the state of species AR, account for ~50% of the copper. If Cu_u had an EPR signal very similar to that of Cu_d, we may not be able to distinguish contributions from Cu_u in the signal we observe. On the other hand, the very EPR silence of Cu_u under the conditions mentioned above, when the silent heme becomes detectable, seems to give support to interpretations which postulate states other than that of Cu^{2+} for Cu_u[15,16,27].

Species AR also reacts rapidly with ferrocytochrome c. While significant rhombic ferric h.s. heme signals appear only within seconds during reduction of species RO by ferrocytochrome $\underline{c}$[4,5], the rhombic signals of species AR disappear within 10 msec on addition of ferrocytochrome $\underline{c}$. This was shown in an experiment involving 4 syringes and 3 separate mixing steps. Enzyme, prereduced with NADH, was reoxidized with excess porphyrexide, the excess of porphyrexide was removed by NADH, a very slow reductant of the oxidase, and then ferrocytochrome $\underline{c}$ was added. Thus, form AR has properties akin to the "activated" form of the enzyme described by Rosén et al.[28], except that in our experiments the response of the rhombic species with ferrocytochrome $\underline{c}$ was more rapid and more extensive, presumably because under the conditions of Rosén et al. less of the active species (presumably AR or a related form) was present.

Reaction with ligands

If, as we have concluded, $\underline{a}_3$ is predominantly present in an EPR detectable h.s. form in species AR, the rhombic ferric h.s. heme species may be expected to undergo the reactions with ligands typical of a h.s. heme species and previously reported for $\underline{a}_3$ for a number of conditions[29-31].

Carbon monoxide. Although one might not expect much information from exposing $\underline{a}_3^{3+}$ to CO, our experiments on the interaction of species AR with CO have given an important clue. When species AR was generated by porphyrexide and CO was then mixed in in a second step, the rhombic ferric h.s. heme signal did not immediately decrease in intensity but the rhombic splitting increased (cf. Fig. 2A,B) close to that previously observed[6] with relatively weak signals seen in the presence of CO (extrema at g=6.54 and 5.25). The strongest signals of the rhombic species observed in the presence of CO had been seen on dissociation of the $\underline{a}_3^{2+}$-CO complex by light in the presence of ferricyanide[11,12], and it was suggested that these signals were due to $\underline{a}_3^{3+}$[11,12]. Our observations on exposure of species AR to CO strongly support this interpretation. Thus, it seems that CO, like O_2, can exert an allosteric effect on the enzyme in the state of species AR which leads to an increase in the rhombicity of $\underline{a}_3^{3+}$. We are now tempted to ask: would one expect either heme in its h.s. form to show this increase in rhombicity or is it not likely to be more specific, that is, specific for $\underline{a}_3$, with which it has, in fact, been observed? If we consider this increase in rhombicity as a specific marker of $\underline{a}_3$, we may now examine other rhombic ferric h.s. heme signals observed with cytochrome $\underline{c}$ oxidase. The most important of these, unquestionably, is that observed on reduction with ferrocytochrome $\underline{c}$ (extrema at g=6.41 and 5.37)[4-7]. As reported previously, this rhombic signal responds within a few msec to CO (cf. Fig. 2C, D.), assuming increased rhombicity typical for the presence of CO[6,12]. Taken together, these observations obviously raise the question whether the rhombic h.s. species observed on reduction with cytochrome $\underline{c}$ is not also due to $\underline{a}_3^{3+}$. This h.s. species, and for that matter all h.s. heme species for which measurements are available, have been found to have relatively low oxidation-reduction midpoint potentials[3,13]. These include species generated in the presence of mediators by reduction with dithionite or by reoxidation with ferricyanide. Considerations of midpoint potentials have played a con-

Fig. 2. EPR spectra of cytochrome c oxidase samples exposed to the following conditions: A, 0.8 mM enzyme reduced anaerobically with a slight excess of ascorbate in the presence of an equimolar quantity of cytochrome c, reoxidized anaerobically with 6 mM porphyrexide for 1.2 sec and then exposed anaerobically to a saturated solution of CO in 10 mM cacodylate, pH 7.2, for 100 msec. B, (control) as A except that CO was replaced by N_2. C, cytochrome c oxidase partly reduced anaerobically with cytochrome c to maximal development of h.s. signals at g=6 (cf. 6,7). D, as C, but then mixed with a saturated solution of CO for 6 msec. Equal

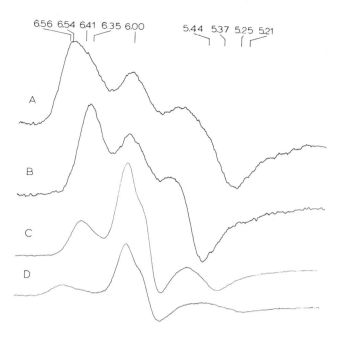

volumes were mixed and rapidly frozen (cf. 5). The concentrations given refer to the initial concentrations in the syringes in terms of total heme a. The conditions of EPR spectroscopy were: power, 2.7 mwatt; modulation frequency and amplitude, 100 kHz and 8 G, resp; scanning rate, 200 G per min; time constant 0.5 sec; and temperature 13K.

siderable role in attempts to assign the observed EPR signals to cytochromes a and a_3, respectively[3,4]. In view of the observations just reported we may recall the arguments of Malmström[32], who pointed out that there was no compelling reason why a_3^{3+} had to be a high potential heme species.

Cyanide. It is known that the resting enzyme in form RO reacts but slowly with cyanide[9,33-35]. The reaction with the reduced enzyme, the "oxygenated" form[23] and enzyme during turnover is more rapid[9]. Under such conditions the a_3 heme can be detected in the form of a_3^{3+}-CN^- as a l.s. ferric heme signal at g=3.58. This signal is not observed on mere addition of cyanide to oxidase[9,10].

When we produced species AR by reoxidation of reduced oxidase with ferricyanide or porphyrexide, 17 mM cyanide (final concentration) eliminated the rhombic EPR signal of a_3 by over 95% in 6 msec. About 10% of the axial signal persisted at this time. A signal of l.s. a_3^{3+}-CN^- appeared. The quantitative relationships between these signals are yet to be determined. There was no significant change in the l.s. ferric heme signal representing cytochrome a, except for a small shift of g-values as previously reported[6]. The results of these experiments are in line with our interpretation given above and in ref. (11), namely that the rhombic h.s. ferric heme species represents $\underline{a}_3^{3+}$. As present in form AR of the oxidase, $\underline{a}_3^{3+}$ readily reacts with cyanide in contrast to $\underline{a}_3^{3+}$ in form RO.

While we have concentrated thus far on the rhombic h.s. species, experiments of the kind reported here may also be able to give information on the identity of the axial h.s. species which, to some extent, is always present together with the rhombic one. The present experiments, as well as those using sulfide (see below) in which the axial species also disappeared rapidly by ∿90% on reaction with cyanide, a ligand typical for $\underline{a}_3$, make it plausible that the axial species may also be due to $\underline{a}_3^{3+}$, but in an environment different from that characteristic for the appearance of the rhombic species.

Sulfide. Sulfide is a strong ligand of $\underline{a}_3^{3+}$, but again, partial reduction has to precede formation of the EPR detectable $\underline{a}_3^{3+}$-S^{2-} (g=2.54; 2.23;1.87)[10,14]. Exposure times of the oxidase in form RO of ∿1 min were found optimal for formation of the $\underline{a}_3^{3+}$-S^{2-} signal. At this time there is considerable reduction of all components of the enzyme including $\underline{a}$, so that the assignment of the signals to heme species may become ambiguous. We have exposed form AR of the oxidase, generated by reoxidation with ferricyanide or phorphyrexide, to 7 and 13 mM Na_2S, respectively, and observed complete disappearance of the rhombic h.s. signals and >90% of the axial signals within 8 msec. Strong signals of $\underline{a}_3^{3+}$-S^{2-} appeared immediately. Since, as the EPR spectra showed, there was still unreacted ferricyanide or porphyrexide present at this stage, $\underline{a}$ underwent no significant changes in these experiments. The conclusions from these experiments are analogous to those drawn above from experiments involving cyanide and thus support what was said there.

Azide. As shown previously[10] azide is a weaker ligand of $\underline{a}_3^{3+}$ than cyanide and sulfide. Experiments in which oxidase in the form AR was exposed to 30 mM azide for up to 100 msec showed at best 40% disappearance of the h.s. species. Resonances typical for the azide species (cf. 2,6) were observed. Again, since there was excess oxidant present, no major change in the signal representing $\underline{a}$ was observed, except for a small shift in g-values similar to that seen in addition of cyanide or even chloride[6]. These experiments thus confirm observations reported previously[10], namely that azide is the weakest ligand in the series, cyanide, sulfide and azide. The experiment is also of interest with a view to the assignment of the previously reported[2,6] azide induced signals at g=2.9, 2.2 and 1.67. It has been suggested that azide binds to cytochrome $\underline{a}$ [13,36]. Our experiments, however, suggest strongly that it is the $\underline{a}_3^{3+}$-N_3^- complex which is represented in the observed l.s. EPR signals. Another observation made in the experiments with azide deserves mentioning. Frequently a weak l.s. signal at g=2.6; 2.2; 1.86[6] is seen at states when h.s. signals at g=6 appear, particularly at elevated pH. We have, therefore, suggested that this signal represents a l.s. form of the same heme that is manifested in the h.s. signal at g=6;2 [6]. This minor l.s. signal was also present before addition of azide but was immediately deleted on addition of azide, indicating that azide reacts faster with this l.s. form of $\underline{a}_3^{3+}$ than with the h.s. form represented at g=6;2.

Response to pH changes

As shown previously[5-7], the intensity and shape of h.s. ferric heme signals of cytochrome $\underline{c}$ oxidase observed at states of partial reduction with cytochrome $\underline{c}$ depend on pH. At pH ∿9 practically no h.s. signals are seen, at pH ∿ 7.5, a mixture of axial and rhombic species is observed as shown above, while at pH ∿ 6 relatively intense axial signals appear, which differ in shape from those seen at neutral pH[6,7]. It was, therefore, of interest to examine the response of the h.s. species in form AR to changes in pH. This form was generated as usual by anaerobic reoxidation of reduced oxidase in the presence of cytochrome $\underline{c}$ and then subjected to a pH jump by mixing in anaerobically from a third syringe an appropriate buffer (glycine, pH 10.5 or acetate, pH 4.0), so that the average oxidation state of the components of the system should not change. pH jump experiments from neutral to low pH had previously been done on enzyme that was partly reduced by cytochrome $\underline{c}$[7] and it was then observed that there was a very rapid (< 6 msec) equilibration between axial and rhombic components, with axial increasing and rhombic decreasing. The same observation was made with the h.s. signals of species AR, i.e., exposure to low pH elicits the same response from the largely rhombic h.s. signal of species AR as it does from the h.s. signal of the partly reduced enzyme, namely, a shift of signal intensity from the rhombic to the axial species with decreasing pH. When a jump to pH 9.2 was carried out, all h.s. ferric heme signals were abolished within 25 msec, the earliest observation time chosen. It was of particular interest in these experiments whether, as the h.s. signals disappeared, signals of the minor l.s. species (g=2.6; 2.2; 1.8) would appear. This species generally has been observed at states of partial reduction and at elevated pH[6,7] and a relationship to the h.s. signal at g=6;2 had been suggested. However, 25 msec after the pH jump from 7.2 to 9.2, these l.s. signals were minimal and no larger than in the control sample at pH 7.2. After 3 minutes exposure to pH 9.2, however, substantial signals of the minor l.s. species had emerged. Experiments in which the pH jump preceded the reoxidation gave similar results.

Although the strict quantitative relationships among species in these experiments have yet to be worked out, three observations seem to us particularly significant. 1.) The axial and rhombic h.s. species are in rapid equilibrium with respect to pH changes. As pointed out previously[4,5], they are not in rapid equilibrium with respect to oxidation-reduction. This can be understood, if it is considered that any functional unit has a h.s. heme in either the rhombic or the axial form and that oxido-reduction between units is not rapid, whereas a pH change does not involve electron exchange between units. 2.) The response of the h.s. signals observed in the partly reduced enzyme to low pH is the same as that in the anaerobically reoxidized enzyme (form AR), namely a shift of intensity from rhombic to axial. This

247

recalls the observation communicated above, that the response of both species to CO is the same, namely an increase in rhombicity of the rhombic components. Both, observations 1.) and 2.) add weight to the suggestion, that the h.s. signals in the enzyme partly reduced by cytochrome c and in form AR originate from the same component of the enzyme, viz., a_3^{3+}. 3.) The jump to a alkaline pH showed that the h.s. component whose EPR signal is abolished on alkalinization, does not immediately appear in the form of the minor l.s component at g=2.6; 2.2; 1.8. These experiments therefore, do not furnish direct evidence for the relationship of the h.s. signals at g=6;2 and the minor l.s. signals, although these l.s. signals do appear at later times. Apparently, intermediate EPR silent states are involved.

CONCLUSIONS

We may first state the assumptions which we make in interpreting our experimental results and arriving at the conclusions below:

1.) The major l.s. ferric heme signal at g=3; 2.2;1.5 of the resting oxidized enzyme represents cytochrome a under all conditions.

2.) The double integration procedure of the h.s. ferric heme signal at g=6;2 according to Aåsa et al.[37] and Hartzell and Beinert[6], using ferrimyoglobin as standard is correct and does not have a systematic error exceeding +20%.

We consider both of these assumptions justified.

On this basis, the first conclusion following from our work is that we have a means of studying by EPR the properties of a_3^{3+} in the "free" state, so to speak, and that this cytochrome is a h.s. heme compound which undergoes the ligand reactions that one would expect of a h.s. heme compound and from what has been known about cytochrome a_3 from earlier work.

However, perhaps the most important consequence of the work reported here and in one of our previous publications[11] is in our opinion the necessity for critical reappraisal of previous assignments of the h.s. ferric heme signals around g=6 and g=2 observed in cytochrome c oxidase. Whereas on the basis of early work in our laboratory[2], it was proposed that these signals represent a_3^{3+} in more recent work[4-7], we adduced evidence and pointed to the possibility that a may be represented in both the l.s. and h.s. signals and a_3^{3+} may be EPR silent under most conditions. Evidence, however, contradicting this was already reported in ref. (5). The quantitative accounting of heme species communicated in ref. (11) and the experiments reported here have now convinced us that the original proposal of Van Gelder and Beinert[2] was correct, at least for some of the rhombic h.s. ferric heme species that can be observed, if not for all of them, viz., that they represent a_3^{3+}. Evidence was brought forth above that this may also be true for some of the axial h.s. species observed. Wilson and Leigh[3] have assigned the h.s. signals to a, largely on the basis of considerations of the midpoint oxidation-reduction potentials which they measured. Cytochrome a_3^{3+} was postulated to be the high potential heme component of

248

cytochrome $\underline{c}$ oxidase while the h.s. signals observed during oxidoreductive titrations with cytochrome $\underline{c}$ or with dithionite and ferricyanide in the presence of potential mediators clearly belong to the low potential heme[3,7]. If the criterion introduced above, namely the increase in rhombicity of h.s. ferric heme signals in the presence of CO should indeed be a valid one for the recognition of $\underline{a}_3^{3+}$, then it would be permissible to generalize and conclude that $\underline{a}_3^{3+}$, in states in which it is detectable by the h.s. ferric heme signals at g ∿ 6 and g ∿ 2, is indeed a component of the enzyme of low midpoint potential. This possibility has been pointed out by Malmstrbm[32]. The recognition of $\underline{a}_3^{3+}$ as a low potential component in the oxidase at states where $\underline{a}_3^{3+}$ is detectable by its h.s. signal has no implications for the mid-point potential of $\underline{a}_3^{3+}$ in the EPR silent state of the resting oxidized enzyme or other forms where $\underline{a}_3^{3+}$ is undetectable.

Comparing the alternative oxidized forms of the oxidase, viz., AR, to those des-cribed previously[18-25], we come to the conclusion that AR cannot be identical to the "oxygenated" species, since, to our knowledge, the "oxygenated" form shows the 655 nm absorption and no significant h.s. signal at g ∿ 6;2. However, it seems likely that form AR is related to the species described by Rosén et al.[28], although under their conditions, the timing for formation of AR was probably not optimal. Form AR is presumably not identical with the "pulsed" enzyme of Antonini et al.[21], since these authors reoxidized the reduced enzyme with an O_2 pulse, which more likely led to the "oxygenated" species. The precise relationship between these alternative forms of the oxidized enzyme (if indeed it is completely oxidized in all these forms) and their possible significance in the biological function of the enzyme are a subject for future research.

ACKNOWLEDGEMENTS

This work was supported by a research grant (GM-12394) and a Research Career Award (5-K06-GM-18492) to H. B. from the Institute of General Medical Sciences, National Institutes of Health, U.S.P.H.S.

REFERENCES

1. Palmer, G., this volume.

2. Van Gelder, B. F. and Beinert, H. (1969) Biochim. Biophys. Acta, 189, 1-24.

3. Wilson, D. F. and Leigh, J. S., Jr. (1972) Arch. Biochem. Biophys., 150, 154-163.

4. Hartzell, C. R., Hansen, R. E. and Beinert, H. (1973) Proc. Nat. Acad. Sci. (USA), 70, 2477-2481.

5. Beinert, H., Hansen, R. E. and Hartzell, C. R. (1976) Biochim. Biophys Acta, 423, 339-355.

6. Hartzell, C. R. and Beinert, H. (1974) Biochim. Biophys. Acta, 368, 318-338.

7. Hartzell, C. R. and Beinert, H. (1976) Biochim. Biophys. Acta, 423, 323-338.

8. Blokzijl-Homan, M. F. J. and Van Gelder, B. F. (1971) Biochim. Biophys. Acta, 234, 493-498.

9. DerVartanian, D. V., Lee, I. Y., Slater, E. C. and Van Gelder, B. F. (1974) Biochim. Biophys. Acta, 347, 321-327.

10. Wever, R., Van Gelder, B. F. and DerVartanian, D. V. (1975) Biochim. Biophys. Acta, 387, 189-193.

11. Beinert, H. and Shaw, Robert W. (1977) Biochim. Biophys. Acta, 462, 121-130.

12. Wever, R., Van Drooge, J. H., Van Ark, G. and Van Gelder, B. F. (1974) Biochim. Biophys. Acta, 347, 215-223.

13. Leigh, J. S., Jr., Wilson, D. F. and Owen, C. S. and King, T. E. (1974) Arch. Biochem. Biophys., 160, 476-486.

14. Seiter, C. H. A., Angelos, S. G. and Perreault, R. A. (1977) Biochem. Biophys. Res. Commun., 78, 761-765.

15. Hu, V. W., Chan, S. J. and Brown, G. S. (1977) Proc. Nat. Acad. Sci. (USA), 74, 3821-3825.

16. Peisach, J. and Blumberg, W. E. (1974) Arch. Biochem. Biophys., 165, 691-708.

17. Yakushiji,   E. and Okunuki, K. (1941) Proc. Imp. Acad., 17, 38-40.

18. Tiesjema, R. H., Muijsers, A. O. and Van Gelder, B. F. (1972) Biochim. Biophys. Acta, 256, 32-42.

19. Muijsers, A. O., Tiesjema, R. H. and Van Gelder, B. F. (1971) Biochim. Biophys. Acta, 234, 481-492.

20. Orii, Y. and King, T. E. (1976) J. Biol. Chem., 251, 7487-7493.

21. Antonini, E., Brunori, M., Colosimo, A., Greenwood, C. and Wilson, M. T. (1977) Proc. Nat. Acad. Sci. (USA), 74, 3128-3132.

22. Brittain, T. and Greenwood, C. (1976) Biochem. J., 155, 453-455.

23. Sekuzu, I., Takemori, S., Yonetani, T. and Okunuki, K. (1959) J. Biochem., 46, 43-49.

24. Wharton, D. C. and Gibson, Q. H. (1968) J. Biol. Chem. 243, 702-706.

25. Lemberg, R. and Mansley, G. E. (1966) Biochim. Biophys. Acta, 118, 19-35.

26. Kuhn, R. and Franke, W. (1935) Ber.Dtsch. Chem. Ges., 68, 1528-1536.

27. Beinert, H. (1966) in The Biochemistry of Copper, Peisach, J., Aisen, P. and Blumberg, W. E. eds., Academic Press, New York, pp. 213-234.

28. Rosén, S. Brändén, R., Vänngård, T. and Malmström, B. G. (1977) FEBS Letters, 74, 25-30.

29. Nicholls, P. and Chance, B. (1974) in Molecular Mechanisms of Oxygen Activation, Hayaishi, O. ed., Academic Press, New York and London, pp. 479-534.

30. Caughey, W. S., Wallace, W. J., Volpe, J. A. and Yoshikawa, S. (1976) in The Enzymes 3rd Edition part C, Boyer, P.D., ed., pp. 299-344.

31. Lemberg, M. R. (1969) Physiol. Revs., 45, 48-121.

32. Malmström, B. G. (1973) Q. Rev. Biophys., 6, 389-431.

33. Van Buuren, K. J. H., Zuurendonk, P. F., Van Gelder, B. F. and Muijsers, A. O. (1972) Biochim. Biophys. Acta, 256, 243-257.

34. Van Buuren, K. J. H., Nicholls, P., and Van Gelder, B. F. (1972) Biochim. Biophys. Acta, 256, 258-276.

35. Nicholls, P., Van Buuren, K. J. H., and Van Gelder, B. F. (1972) Biochim. Biophys. Acta, 275, 279-287.

36. Wilson, D. F., Lindsay, J. G., Brocklehurst, E. S. (1972) Biochim. Biophys. Acta, 256, 277-286.

37. Aasa, R., Albracht, S. P. J., Falk, K-E., Lanne, B. and Vänngård, T. (1976) Biochim. Biophys. Acta, 422, 260-272.

THE EFFECTS OF CROSS-LINKING ON CYTOCHROME c AND CYTOCHROME OXIDASE

MAURICE SWANSON, LARS CHR. PETERSEN AND LESTER PACKER
Membrane Bioenergetics Group, Lawrence Berkeley Laboratory and the Department of
Physiology-Anatomy, The University of California, Berkeley, California 94720 U.S.A.

ABSTRACT

The terminal portion of the mitochondrial electron transport chain may be arranged such that cytochrome c_1 is able to reduce cytochrome c bound to its high affinity site on Complex IV. This model suggests that c might possess some type of rotational mobility while situated on cytochrome aa_3, and that cytochrome oxidase is arranged as a stable array of inter-associated polypeptides. In the present study these questions have been approached by cross-linking purified cytochrome c and cytochrome oxidase with bifunctional alkylating agents.

Amidination of cytochrome c with dimethylsuberimidate, but not with a mono-functional control, methylacetimidate, renders cytochrome c a less efficient electron donor for oxidase. This has been found by gel electrophoresis studies to be due to formation of oligomers. Since monofunctional amidination of lysyl residues does not interfere with the cytochrome c reaction with oxidase, this implies that surface charge interactions are still strong enough after introduction of bulky groups (up to 6 Å chain length) to permit electronic communication between cytochrome c and oxidase.

Cytochrome oxidase is a complex of polypeptides and lipids. Treatment of the complex with monofunctional reagents slightly inhibits, whereas cross-linking markedly restricts, activity indicating that certain conformational variations are essential for maximum activity. Indeed 5-10% more lysyl amino groups are exposed for reaction with monofunctional alkylating reagents in reduced as compared to oxidized oxidase. Gel electrophoresis studies indicate that polypeptide I is the most inaccessible and presumably hydrophobic species as it is not cross-linked to other polypeptides, whereas polypeptides II through VII readily become cross-linked together to form an 87,000 molecular weight aggregate. These results indicate that the cytochrome oxidase complex functions without the necessity for long-range motion and that it requires only a small degree of polypeptide mobility.

INTRODUCTION

Cytochrome c oxidase (ferrocytochrome c: oxygen oxidoreductase, EC 1.9.3.1) is a transmembrane complex of seven polypeptides (Table I)[1-17] which apparently displays very limited rotational diffusion within the mitochondrial inner membrane[18]. Labeling studies of the inner membrane of beef heart mitochondria using impermeable surface probes such as p-diazonium benzene [^{35}S] sulfonate[3] have established the relative dis-position of these polypeptides in the plane of the membrane with subunit II exposed on the cytoplasmic side as the probable binding site for cytochrome c[12]. In order to resolve the functional requirements of cytochrome oxidase at the molecular level more precise structural information on the role of each polypeptide in the complex is required, and as indicated in Table I, this information is still scarce.

Spin-labeling studies of both cytochrome $\underline{c}$[19], and the polypeptides of cytochrome oxidase[9] have indicated that certain types of conformational changes occur within these molecules depending on their local environment and redox state. However, the importance of the mobility of the individual polypeptides within complex IV in terms of function and the specific types of interactions cytochrome $\underline{c}$ and oxidase must undergo in order for effective electron transfer to occur has not been elucidated. Margoliash and coworkers[15,20] have established the electrostatic nature of the binding of cytochrome $\underline{c}$ to cytochrome $\underline{aa}_3$ and kinetic studies[20] suggest that two binding sites of high ($K_D=0.028\mu M$) and low ($K_D=1.2\mu M$) affinity exist on each cytochrome oxidase molecule. These workers have also calculated that the reaction of cytochrome $\underline{c}$ and $\underline{aa}_3$ must have an "off" constant of $1.5\text{-}30s^{-1}$ depending on the ionic strength of the medium and the resultant rate constants. This indicates that cytochrome $\underline{c}$ bound to the oxidase is capable of being reduced by cytochrome $\underline{c}_1$[20]. This study involves the effects of mono- and bifunctional alkylating reagents on various kinetic parameters of the cytochrome $\underline{c}$-cytochrome $\underline{aa}_3$ system. Since these reagents may be used to induce particular types of molecular perturbations both within the enzyme and cytochrome $\underline{c}$ we have attempted to correlate these structural alterations to changes in enzymic activity. These studies provide information on the relative importance of electrostatic and stearic factors on cytochrome $\underline{c}$ oxidase activity, and indicate that a specific structural arrangement of closely associated polypeptides is required within the enzyme for maximal electron transfer.

METHODS

The abbreviations used in this text are: MA, methylacetimidate; DMA, dimethyl-adipimidate dihydrochloride; DMS, dimethylsuberimidate dihydrochloride; TMPD, N,N,N',N'-tetramethylphenylenediamine dihydrochloride.

Preparations. Fresh and frozen beef heart mitochondria, kindly provided by Dr. T. P. Singer and H. Tisdale, were prepared by the method of Blair[21]. Cytochrome oxidase was isolated from these mitochondria basically as described by Capaldi and Hayashi[22], however modifications were required in order to obtain reproducible results[23]. The cholic acid (Sigma Chemicals Co.) was recrystallized from 95% ethanol, and deoxycholic acid (Gold Label) was obtained from the Aldrich Chemical Company. Cytochrome oxidase, purified from Keilin-Hartree particles by the method of Van Buren[24] and enzymatically characterized by Nicholls et al[25], was used for comparative studies. Results with Keilin-Hartree cytochrome oxidase preparations yielded identical results to those observed with the oxidase purified by the other procedure. Identical results were also obtained if the high molecular weight contamination of the cytochrome oxidase preparations was eliminated by the gel filtration method of Briggs et al[7]. Sepharose 4B was purchased from Pharmacia Chemicals.

TABLE 1

BEEF HEART CYTOCHROME c OXIDASE

Subunit	Molecular Weight[1,2]	Membrane Disposition[3,4]	Polar Amino[5,7] Acids (%)	Nearest Neighbors[8]	Characteristics
native enzyme I - VII	122,500 (~101,000 - 145,000)	transmembrane complex	—	—	functional unit: dimer (4 Cu^{++}, 4 Fe^{++})[17]; conformations: reduced: open[9-11], oxygenated: intermediate, oxidized: compact
I	35,500 (33,000 - 43,000)	intramembrane [40,000 (I)][a]	35.5	not cross-linked	hydrophobic
II	22,000 (19,000 - 27,000)	cytoplasmic [22,500 (II)]	44.7	V	hydrophobic-hydrophilic; cytochrome c binding[12]
III	20,500 (19,000 - 25,000)	II & III unresolved	39.9	V	hydrophobic; N-ethyl maleimide reactivity (pH 7.0)[2]
IV	16,500 (13,800 - 17,200)	matrix [15,000 (III)]	48.6	VI	hydrophilic
V	12,400 (6,000 - 13,000)	intramembrane [11,200 (IV)]	47.8	II, III, VII	hydrophilic; possible heme a binding subunit (11,600)[13]
VI	8,800 (6,700 - 11,200)	cytoplasmic [9,800 (V)]	49.7	IV	hydrophilic; N-ethyl maleimide reactivity (pH 7.0)[2]
VII	6,800 (3,400 - 9,000)	cytoplasmic [7,300 (VI)]	53.7	V	hydrophilic
cytochrome c	12,400	—			binding to aa$_3$: primarily electrostatic[14,15]; binds to III in yeast[16]; binds to II in beef heart[12]

For the cytochrome c row, Nearest Neighbors: II[12].

a. Original nomenclature of Eytan, Carroll, Schatz and Racker (1975).

Imidoester Treatment. Stock solutions of imidoesters were prepared as previously described by Tinberg et al[26] in 0.133 M triethanolamine-chloride at pH 9.0. The bifunctional reagents, DMA or DMS, were used as 5-40mM incubation solutions and the
monofunctional reagents, MA or MBI as 10-80 mM solutions. Reduced oxidized cytochrome
oxidase (1.23 - 2.0 mg/ml) or cytochrome c (6.5 mg/ml) preparations were incubated at
the above imidate concentrations as described[23]. The degree of amidination resulting
from imidoester treatment was determined by following the loss of free amino groups by
a fluorometric assay using fluorescamine by a modification of the Bohlen et al[27]
procedure by placing samples in 1% sodium dodecyl sulphate[28]. Fluorescense was
measured in a Perkin-Elmer MPF-44A apparatus at 390 nm excitation/475 nm emission.
Alkylating reagents and fluorescamine were obtained from Pierce Chemical Co.

After bifunctional amidination, separation of monomeric and oligomeric cytochrome
c was accomplished by gel filtration[23]. Cytochrome c concentration was calculated
using a $\Delta\varepsilon_{540-550}$ = 20mM^{-1}cm^{-1} [20].

Gel Electrophoresis. Sodium dodecyl sulphate polyacrylamide gel electrophoresis was
performed in a BIO-RAD Model 220 dual vertical slab cell as previously described[23].

Protein determinations were made as follows: mitochondrial and crude oxidase
protein was determined by the rapid biuret procedure of Van Buren[24]; the protein of
purified cytochrome oxidase was determined by the Lowry et al method[29] using bovine
serum albumin as a standard. During isolation of the oxidase in fractions where
ammonium sulphate was present, protein was determined by the Lowry method after acid
precipitation to avoid interference with the assay.

Assays. Two assays of cytochrome c oxidase activity were employed. A spectrophotometric procedure was utilized to enzymatically characterize the preparations
while polarographic determinations were used for routine analysis. The spectrophotometric assays were performed in 0.05M potassium phosphate at pH 7.0 in 0.5%
Tween-80 at 25°C by following the oxidation of ferro-cytochrome c[6]. Initial molecular
activities ($MA_{o_{max}}$), determined according to Vanneste et al[30], were in the range of
250-300, and thus the activity of cytochrome oxidase employed in these studies are
in the range of the highest reported[7]. Heme a concentrations, calculated using a
$\Delta\varepsilon_{605-630}$nm = 13.5mM^{-1}cm^{-1} [24], were in the range of 8.0-11.3nmoles heme a/mg.
protein. Heme a_3 concentration was determined from difference spectra of reduced
minus reduced + CO using $\Delta\varepsilon_{428.5-445}$ = 148mM^{-1}cm^{-1} [30]. Polarographic assays of cytochrome oxidase activity were carried out in a medium containing 12.5mM potassium
ascorbate, 0.5mM TMPD, 12.5μm cytochrome c, 67mM potassium phosphate, pH 7.4, and
0.5% Tween-80. Polarographic and spectrophotometric assays give equivalent results.

RESULTS
CYTOCHROME c - Cross-linking of cytochrome c with DMS results in the loss of monomers
and the formation of oligomers, and when the cross-linked material is subjected to
chromatography purified monomer and oligomer fractions can be recovered. These
fractions were then tested for activity with cytochrome oxidase. Native cytochrome c

254

monomers or the monomers recovered from the cross-linked sample, as well as cytochrome $\underline{c}$ amidinated with the monofunctional reagent MA, exhibit substantially the same kinetics. However, cross-linked mixtures of monomers and oligomers of cytochrome $\underline{c}$ or oligomer-enriched fractions show markedly inhibited cytochrome oxidase activities, the decrease in V_{max} (from 1.0 to 0.3, arbitrary units) is accompanied by a substantial decrease in K_m (from 10.0 to 2.4μM cytochrome $\underline{c}$). In other experiments, Scatchard plots revealed the K_D for high-affinity binding of $\underline{c}$ to oxidase using c-depleted rat liver mitochondria[31] is substantially the same for native and oligomerized cytochrome $\underline{c}$ being 0.030 μM and 0.035 μM, respectively. Amidination or cross-linking of cytochrome $\underline{c}$ does not alter its spectral characteristics and anaerobic potentiometric[31] titrations of native, MA-treated, and DMS-treated cytochrome $\underline{c}$ all yield $E_{M_{7.0}}$ values of 285 mV.

Several explanations may account for a change in reactivity of oligomeric cytochrome $\underline{c}$: decreased binding, a decrease in effective substrate concentration, or the inability of the larger aggregates to interact effectively at the binding site on the enzyme. The first possibility seems ruled out by the binding experiments. Double reciprocal plots using various fixed concentrations of oligomeric $\underline{c}$ demonstrates that the cross-linked molecule is a partial competitive inhibitor of native cytochrome $\underline{c}$ with a Ki, calculated from slope replots, of 3.5μM. This suggests that reduction in effective substrate concentration is not the main inhibitory factor. It may also be mentioned that the introduction of bulky groups onto cytochrome $\underline{c}$ by the use of the monofunctional reagents MA and MBI does not alter cytochrome $\underline{c}$'s ability to reduce the oxidase when up to 90% of the exposed lysine residues of cytochrome $\underline{c}$ have been modified. Indeed, some slight enhancement of activity has been observed. These results confirm the importance of electrostatic factors in the interaction of cyto-chrome $\underline{c}$ with oxidase which demonstrates this contribution to binding dominates over molecular size; in other words a 10-15% increase in the molecular weight of cytochrome $\underline{c}$ does not affect its ability to transfer electrons to cytochrome aa_3.

Since oligomeric cytochrome $\underline{c}$ can interact with, but cannot effectively reduce, the enzyme, this suggests that binding site stearic constraints alter activity.
CYTOCHROME OXIDASE: The preparations of cytochrome oxidase utilized in this investi-gation exhibit one of the most clear SDS gel electrophoretic patterns thus far reported for this enzyme (Fig. 1). Following cross-linking, enzyme oligomers are more readily formed with the longer chain-length DMS (11.5Å) as compared to DMA (8.5Å). No perturbation in the gel electrophoresis pattern is observed with samples highly amidi-nated (65%) with MA.

Some information can be gained on nearest neighbors of polypeptides from such studies: most of the polypeptides are cross-linked while polypeptide I is completely unaffected, suggesting its lysine residues, which are able to react with DMS, are not within 11.5Å of the lysine residues of adjacent polypeptides of the oxidase. The major predominating band that appears in highly cross-linked samples has an aggregate molecular weight of approximately 87,000. This is precisely the molecular size that

CYTOCHROME OXIDASE

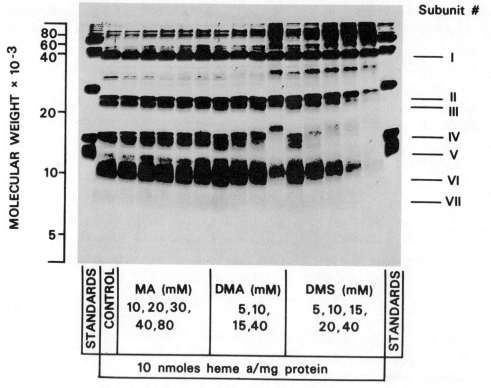

Fig. 1   SDS gel electrophoresis pattern of purified beef heart mitochondrial
cytochrome oxidase. Subunit molecular weights are: (1) 35,500;
(II) 22,000; (III) 20,500; (IV) 16,500; (V) 12,400; (VI) 8,800;
(VII) 6,800. Each channel contains 30 µg of protein.

would be expected if subunits II through VII were cross-linked to one another.

Treatment of cytochrome oxidase with MA or DMS,results in very extensive reaction
with free amino groups (Table II). However, the bifunctional reagents inhibit enzyme
activity much more extensively. This is particularly evident when low concentrations
of reagents are used. Monofunctional amidination of cytochrome oxidase, which results
in the introduction of bulky groups onto exposed lysine residues of the molecule, also
reduces molecular activity. Since at equivalent extents of amidination the cross-
linking reagents are more inhibitory, a second conclusion derived from these studies is
that the introduction of molecular mobility restraints on subunit interactions reveals
a necessity for polypeptide movement for electron transport in Complex IV. The percentage inhibition of activity induced by DMS was identical throughout a wide incubation concentration range of oxidase (0.01 → 2.5 mg/ml) and thus intermolecular cross-
linking of cytochrome $\underline{aa}_3$ is not responsible for the observed decline in activity.

It has been reported that cytochrome oxidase exhibits different conformations in
its oxidized and reduced state[9]. We have found that the reduced form of the enzyme is
more extensively amidinated and inhibited than the oxidized form, an effect which is
only seen with the monofunctional reagent. The small molecular size of MA must

256

therefore allow it to be permeable and capable of reacting with amino groups that are exposed in the reduced but not oxidized form of the enzyme.

Other studies have been made with 60% delipidated cytochrome oxidase preparations in which 75% of the phosphatidylethanolamine has been extracted[8], and substantially the same pattern of inactivation is observed following amidination and cross-linking. Hence it does not appear that lipid-protein coupling is involved in the inhibition of the enzyme.

When cytochrome c is present with oxidase during the modification procedures, there is a 5% greater degree of inhibition with the bifunctional reagents. This could be due to fixing cytochrome c tightly at its binding site restricting the necessary inter-action that native cytochrome c must undergo with the electron donor, Complex III in the respiratory chain or in this case ascorbate plus TMPD.

The data summarized in Table II show that DMS (11.5Å), at the highest concentrations used, is about 5 times more inhibitory than MA (2.5Å) and about 4 times greater than MBI (4.5Å). This greater degree of inhibition correlates with its cross-linking and not with its amidination per se.

TABLE II CHEMICAL MODIFICATION OF OXIDIZED CYTOCHROMES c AND OXIDASE

[Imidoester]	(mM)	Free amino Groups (% decrease)	Gel Protein (% excluded)	Enzyme Activity (% remaining)
MA	10	12	< 1	99
	80	66	< 1	55
MBI	10	30	< 1	100
	80	54	< 1	47
DMS	5	20	5	62
	40	64	23	11

DISCUSSION

By using purified cytochrome oxidase and cytochrome c for chemical modification studies, it has been possible to construct a more detailed model for the inter-actions of cytochrome c oxidase polypeptides with each other and with cytochrome c.

The main findings upon which this model is based are enumerated in Figure 2. These are: (1) oligomerization of cytochrome c inhibits its ability to reduce cyto-chrome oxidase while monofunctional amidination or intramolecular coupling does not; (2) cross-linkage of cytochrome c to oxidase results in a slight increase in inhibi-tion over that observed in the absence of cytochrome c; (3) polypeptides II through VII are readily cross-linked together to form an aggregate of molecular weight ~ 87,000; (4) polypeptide I is resistant to cross-linking to other oxidase poly-peptides; (5) intramolecular cross-linking of cytochrome oxidase even at low levels severely inhibits its enzymatic activity while amidination _per se_ does not.

Charge and conformational factors are both important for cytochrome _c_ oxidase activity. Margoliash and coworkers[14,15,32,33] have made some specific modifications of cytochrome _c_. The 4-nitrobenzo-2-oxa-1,3-diazole derivative of lysine -13 of cytochrome _c_ and the bis-phenylglyoxal derivative of arginine -13 from Candida krusei are reduced by succinate-cytochrome _c_ reductase, but the apparent K_m of the cytochrome _c_ oxidase reaction is increased 5-10 fold. These workers have also shown, using selective trifluoracetylation of certain lysines of cytochrome _c_, that the elimination of positive charges and not side group stearic interference is responsible for inhibiting cytochrome oxidase when lysines -13,-25 and possibly -27,-72,-79 are derivatized. In the present study it is quite clear that the reactivity of cytochrome _c_ with oxidase is unaffected by amidination with the bulky groups of MA or MBI. Even when 17 of the 19 lysines, or 90%, of the exposed groups are reacted, activity is unchanged. This treatment increases the molecular size of cytochrome _c_ about 15% while charge is conserved. If cytochrome _c_ is modified by inter-molecular cross-linking to form oligomeric cytochrome _c_, only then is activity inhibited.

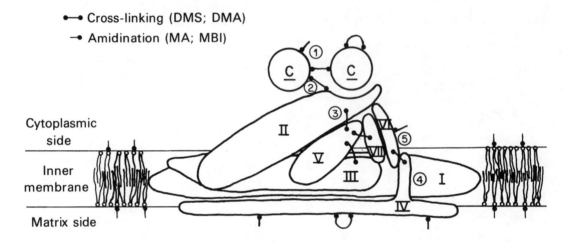

Fig. 2   Scheme showing the arrangement of cytochrome _c_ oxidase in the inner membrane of mitochondria. The numbers given in the figure correspond to points enumerated in text.

The function of cytochrome _c_ in electron transport is to transfer electrons from complex III to complex IV and thus it must interact efficiently with both of these complexes. How many molecules of cytochrome _c_ are involved in this interaction and whether complex III, cytochrome _c_, and complex IV are organized as a closely-knit functional unit is unknown. Presumably the membrane environment and the specific structure of the proteins determine their disposition in the membrane plane. If a closely-knit functional unit exists, must cytochrome _c_ alternatively bind and associate to these complexes in a reversible manner? Kinetic data suggests that for cytochrome oxidase located in the inner membrane of mitochondria there are a small number of tight and a larger number of loose binding sites for cytochrome c[31]; the former may be the common site of interaction with complex III and IV. The inhibition of cytochrome oxidase activity when oligomeric cytochrome c is the substrate suggests that stearic factors are exceedingly important in the interaction. Alternatively, inhibition might be caused by slowing diffusion of cytochrome c onto and off of oxidase if dissociation is required for reduction of the oxidized c. It is important to note the oligomerized cytochrome c is a partial competitive inhibitor of native cytochrome c, and thus oligomerized cytochrome c inhibits binding of the native cytochrome probably to the same site on the enzyme. Although cytochrome oxidase is severely inhibited by cross-linking, if the enzyme is cross-linked in the presence of cytochrome c, inhibition of oxidase activity is only 5% greater. Thus covalent attachment of c to oxidase does not appreciably affect the inhibitory pattern suggesting that the cytochrome c is bound in the vicinity of its high affinity binding site and that cytochrome c dissociation from that site is not required.

Cross-linking studies reveal that the introduction of structural mobility restraints into individual polypeptides of the cytochrome oxidase complex markedly inhibit the ability of this molecule to transfer electrons to reduce oxygen. The cross-linking studies also reveal nearest neighbor relationships between the individual subunits (Figure 2). Subunit I is presumed to be the most hydrophobic of the polypeptides, based on compositional studies[5,7], labeling studies[3], and the inability to cross-link this polypeptide to any of the other polypeptides of the complex. Alternatively, the cross-linking reagents may not be permeable to the interior of the complex to react with this polypeptide or the lysines[6] of polypeptide I may not be within 11.5Å of a reactive free amine on another polypeptide of the complex. The other polypeptides, II through VII, can all be cross-linked together. Since the total molecular weight of the individual polypeptides II through VII is 87,000, this further indicates that only one copy of each of the higher molecular weight polypeptides is present in cytochrome oxidase. Since the chain length of the effective cross-linking reagent is 11.5Å or less, these polypeptides in the complex are closer to one another than this distance. The importance of conformation of the individual polypeptides towards activity is also revealed by the 5% greater ability of monofunctional reagents to react with exposed amino groups when the enzyme is in the reduced state.

ACKNOWLEDGMENT

This research was supported by the Department of Energy.

REFERENCES

1. Capaldi, R.A., Bell, R.L. and Branchek, T. (1977) Biochem. Biophys. Res. Commun., 74, 425-433.
2. McGeer, A., Lavers, B. and Williams, G.R. (1977) Can. J. Biochem., 55, 988-994.
3. Eytan, G.D., Carroll, R.C., Schatz, G. and Racker, E. (1975) J. Biol. Chem., 250, 8598-8603.

4. Ruben, G.C., Telford, J.N. and Carroll, R.C. (1976) J. Cell Biol., 68, 724-739.

5. Capaldi, R.A. and Vanderkooi, G.(1972) Proc. Natl. Acad. Sci. U.S.A., 69, 930-932.

6. Downer, N.W., Robinson, N.C. and Capaldi, R.A. (1976) Biochemistry, 15, 2930-2936.

7. Briggs, M., Kamp, P.-F., Robinson, N.C. and Capaldi, R.A. (1975) Biochemistry, 14, 5123-5128.

8. Briggs, M. and Capaldi, R.A. (1977) Biochemistry, 16, 73-77.

9. Dasgupta, U. and Wharton, D. (1977) Arch. Biochem. Biophys., 183, 260-272.

10. Kornblatt, J.A., Kells, D.I.C. and Williams, G.R. (1975) Can. J. Biochem., 53, 461-466.

11. Kornblatt, J.A. (1976) Fed. Proc., 35, 1598.

12. Bisson, R., Azzi, A., Gutweniger, H., Montecucco, C., Colonna, R. and Zanotti, A. (1977) Abs. 7, Internat. Sym. Membrane Bioenerget., Spetsai, Greece.

13. Yu, C.A. Yu, L. and King. T.E. (1977) Biochem. Biophys. Res. Commun. 74, 670-676.

14. Staudenmayer, N., Ng, S., Smith, M.B. and Millett, F. (1977) Biochemistry 16, 600-604.

15. Margoliash, E., Ferguson-Miller, S., Tulloss, J., Kang, C.H.,Feinberg, B.A., Brautigan, D.L. and Morrison, M. (1973) Proc. Natl. Acad. Sci. U.S.A., 70, 3245-3249.

16. Birchmeier, W., Kohler, C.E. and Schatz, G. (1976) Proc. Natl. Acad Sci. U.S.A., 12, 4334-4338.

17. Robinson, N.C. and Capaldi, R.A. (1977) Biochemistry, 16, 375-380.

18. Kunze, U. and Junge, W. (1977) FEBS Letters, 80, 429-434.

19. Azzi, A., Tamburro, A.M., Farnia, G. and Gobbi, E. (1972) Biochem. Biophys Acta, 256, 619-624.

20. Ferguson-Miller, S., Brautigan, D.L. and Margoliash, E. (1976) J. Biol. Chem., 251, 1104-1115.

21. Blair, P. (1965) Methods Enzymol., 10, 78-81.

22. Capaldi, R.A and Hayashi, H. (1972) FEBS Letters, 26, 261-263.

23. Swanson, M., Petersen, L.C. and Packer, L. (1977) Arch. Biochem. Biophys., submitted

24. Van Buren, K.J.H. (1972) Ph.D. Thesis, University of Amsterdam, Gerja Waarland.

25. Nicholls, P., Petersen, L.C., Miller, M. and Hansen, F.B. (1976) Biochim. Biophys. Acta, 449, 188-196.

26. Tinberg, H.M., Nayudu, P.R.V. and Packer, L. (1976) Arch. Biochem. Biophys., 172, 734-740.

27. Bohlen, P., Stein, S., Dairman, W. and Udenfriend, S. (1973) Arch. Biochem. Biophys., 155, 213-220.

28. Tinberg, H.M., Lee, C. and Packer, L. (1975) J. Supramol. Struct., 3, 275-283.

29. Lowry, O.H., Rosebrough, N.J., Farr, A.L. and Randall, R.J. (1951) J. Biol. Chem., 193, 265-275.

30. Vanneste, W.H., Ysebaert-Vanneste, M. and Mason, H.S. (1974) J. Biol. Chem., 249, 7390-7401.

31. Erecinska, M. (1975) Arch. Biochem. Biophys., 169, 199-208.

32. Smith, L., Davies, H.C., Reichlin, M. and Margoliash, E. (1972) J. Biol. Chem., 248, 237-243.

33. Brautigan, D.L. and Ferguson-Miller, S. (1976) Fed. Proc., 35, 1598.

COPPER ENZYMES

ELECTRON DISTRIBUTION AMONG THE REDOX SITES OF *RHUS* LACCASE AND ITS REACTION WITH O_2 AND H_2O_2

I. PECHT, M. GOLDBERG, S. WHERLAND, AND O. FARVER

Department of Chemical Immunology, The Weizmann Institute of Science, Rehovot, Israel

ABSTRACT

The possible formation of a peroxy intermediate in the catalytic reduction process of dioxygen by *Rhus* laccase has been examined. In the reaction of the partially reduced enzyme with O_2, an intermediate was identified which spectroscopically and chemically is similar to the peroxy laccase species produced by the reaction of oxidized laccase with H_2O_2. The same type of spectral changes were obtained in an O_2 containing solution of laccase, by adding limiting amounts of reductant or under steady state conditions. Since the electron distribution pattern among the three redox sites of laccase determines the pathway of reaction with O_2, this distribution pattern was studied for several reductants and found to vary depending on the redox potential of the donor. A model, involving the coupling-uncoupling transition of the type 3 copper-pair site, is proposed to explain the observed behavior.

INTRODUCTION

The mechanism of dioxygen reduction to water by oxidases has attracted significant interest because it is both a crucial step in the biological energy conversion process and chemically a very interesting, multiple electron transfer reaction[1,2,3]. Much effort was invested in the search for intermediates of the dioxygen reduction by the "blue" copper oxidases[3]. We have previously shown that a stable peroxy derivative of *Rhus vernicifera* laccase can be prepared specifically by the reaction of the native, oxidized enzyme with one molecule of hydrogen peroxide[4]. To explore the significance of this peroxy species to the O_2 reduction catalyzed by laccase, we also studied its possible formation in the reaction of partially reduced oxidase with O_2 or by reducing the enzyme in the presence of oxygen (including steady state conditions). Indeed, the results support the notion that the peroxy-laccase is an intermediate in the reaction of O_2 with the enzyme during the catalytic cycle.

The formation of this intermediate by the reaction of the partially reduced laccase with dioxygen was found to depend on the electron distribution pattern in laccase molecules. It could be correlated with the concentration of fully reduced type 3 site. Thus the mode of electron distribution among the laccase molecules and within its different redox sites is decisive for the reaction with dioxygen. It is noteworthy that in reductive titrations of laccase using different electron donors, we found the electron distribution pattern among the enzyme's sites to depend on the reductant employed. The redox potential of the reductant was found to be the parameter determining the electron distribution. This observation implies a quasi-equilibrium situation which is interpreted in terms of the coupling-uncoupling of the type 3 copper site.

MATERIALS AND METHODS

Laccase was prepared from acetone powder of the lacquer of *Rhus vernicifera* supplied by Saito Company, Tokyo, Japan. A modified version of Reinhammar's[5] procedure was employed. Instead of applying the filtrate from step 1 of the latter procedure directly to a CM-Sephadex C-50 column, it was first dialyzed extensively against 0.01 M potassium phosphate buffer at pH 6.0, thus removing a large amount of low-molecular weight yellow dyestuff, and then passed through a 7 x 12 cm DEAE-Sephadex A-50 column equilibrated with the same buffer. This procedure eliminated most of the yellow pigment. The spectroscopic, EPR and catalytic properties of the enzyme were in agreement with those reported earlier[4,5]. The A_{280}/A_{615} ratio was 14.6. The concentration of the protein was determined from its absorbance at 615nm, using $\varepsilon = 5700$ $M^{-1}cm^{-1}$.

1,4-benzohydrozuinone (benzohydroquinone) was purified by sublimation. 2,5-dimethyl-1,4-benzohydroquinone (xylohydroquinone) and 2,3,5,6-tetramethyl-1,4-benzo-hydroquinone (durohydroquinone) were prepared by reduction of the corresponding quinones in ethanol by sodium borohydride. All these compounds were purified by sublimation immediately before use. Due to the low water solubility of the methyl substituted hydroquinones, their stock solutions were made up in 50% ethanol/water. However, the final ethanol concentration in the laccase solution never exceeded 3%. The hexamineruthenium(II) chloride was prepared by reducing a weighed amount of the corresponding Ru(III) complex with amalgamated Zn in a phosphate buffer, pH 7.0 and I = 0.22 M, containing 5×10^{-3} M EDTA while passing argon through the solution. After approximately 30 minutes, the reduction was complete, and this stock solution was used directly for the reductive titrations. The concentration of hexammineru-thenium(II) ion ($Ru[NH_3]_6{}^{2+}$), usually 3×10^{-3} M, was determined by reduction of Cu(II)-bathocuproinesulphonate. All experiments with $Ru(NH_3)_6{}^{2+}$ were carried out in an all glass system.

All other chemicals were of analytical grade and used without further purification. Laccase solutions were made up in 0.1 M potassium phosphate buffer (pH 7.0, I = 0.22 M, doubly distilled water). The range of protein concentrations in the reductive titrations extended from 3×10^{-5} to 1.2×10^{-4} M. EDTA concentrations varied from 0 to 4×10^{-4} M without having any effects on the results.

Absorption and difference absorption spectra were recorded on a Cary model 118 spectrophotometer at 20°C and EPR spectra on a Varian spectrometer model E-3 at liquid nitrogen temperature. The anaerobic titration procedures and the method of data analysis have been described earlier[4,6]. It was found that optimal eliminat-tion of O_2 was achieved by bubbling purified Ar or N_2 through the buffer for 15 minutes prior to adding the concentrated enzyme. The fraction of oxidized type 2 sites was calculated from the overall stoichiometry and the measured amounts of reduced type 1 and type 3 copper ions. All experimental errors therefore accumulate in these computed values for the type 2 site, leading to a considerably enhanced data scatter.

The simulations of the titrations were generated by calculating the concentration of each of the 16 possible species (0 to 4 electrons in 4 sites) linked by 15 equilibrium constants. These were defined via reduction potentials, assuming the formalism of the Nernst equation. First the concentrations of the species were obtained by assuming a certain electroehcmical solution potential, and then the reduction equivalents taken up were calculated from the concentrations weighted by the number of electrons in each species. Care was taken to maintain path independence of the free energy change for total reduction. All calculations and plots were produced by a BASIC program running on an HP 2100 computer.

RESULTS AND DISCUSSION

Introduction of dioxygen into a continuously stirred solution of laccase previously partially reduced by two reduction equivalents of ascorbate, causes a fast change in the absorption spectrum of the enzyme (Figure 1). This transient absorption decays within a few minutes to form a new spectrum which is stable for many hours (up to 12 hours) and has striking similarity to that of the peroxy-laccase in terms of its shape and extinction at about 330nm (Figures 2 and 3). It differs from it in showing an additional broad band of low intensity between 400 and 540nm. This broad band does show up in the peroxy-laccase formed by reacting the reduced enzyme with H_2O_2 but not when formed from the oxidized laccase and H_2O_2[4] (Figure 3). This stable intermediate can be titrated reductively (with ascorbate) under anaerobic conditions and the course of this titration is similar to that of the reduction of the peroxy-laccase: first the extra absorption at 337nm is reduced while no change takes place in the visible range of the spectrum. As further reductant is added the absorption at 330 continues to decrease and that of the type 1 site (at 614nm) start only gradually to decrease. The broad band at 400-540nm decays only in the last phase of the litration. The stoichiometry of the titration is consistent with a situation where all the laccase molecules which initially contained two reduction equivalents, form a peroxy intermediate upon reaction with the dioxygen. The formation of the above stable intermediate was also observed when other reductants were employed (e.g. $Ru(NH_3)_6^{2+}$ or benzohydroquinone) yet some correction had to be made for the contributions made by their respective oxidation products (cf. Figure 2).

It is significant that upon mixing an aerobic solution of laccase with a solution containing 1 to 2 reducing equivalents of ascorbate, a sequence of spectral events is observed (Figures 1 and 2), which is very similar to that described above for reacting partially reduced enzyme with O_2; again a transient absorption is

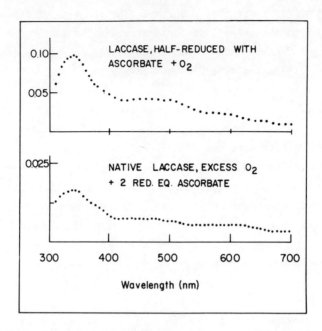

Figure 1. Transient spectra obtained immediately upon mixing half reduced laccase (fifty μ M) with O_2 or O_2 saturated oxidized laccase with limiting reduction equivalents. The data are the difference spectrum between the transient and the stable intermediates.

formed decaying to a stable species which displays characteristic differences with respect to the oxidized native enzyme. However, the intensity of the stable intermediate (minus oxidized) difference spectrum obtained by the above procedure, even under optimal conditions (found by varying amounts of reductant added), was lower than that obtained by reacting half reduced laccase with excess O_2. Similar stable difference spectra were also produced by reacting with O_2 an anaerobic solution of laccase treated previously with excess reductant (5-6 equivalents).

The site of interaction between laccase and peroxide has been proposed to be the type 3 copper pair[4]. This was deduced mainly by excluding the possible binding to type 2 copper as no effect of F- ions on the reaction with H_2O_2 was found (F⁻ is known to specifically bind to the latter site[3]). Furthermore, the fluoride binding to laccase type 2 site and the EPR spectrum of this copper (II) ion were unaffected by equimolar concentrations of H_2O_2. The type 1 Cu site, which is known to be inaccessible to the solvent[2] is also found by us to be unaffected by H_2O_2 and is therefore excluded from being the site of interaction. Thus it is the type 3 site which remains the most probably locus of this interaction. This in line with the idea that the type 3 copper ions pair is the reduction site of dioxygen[3,6]. The above experiments, where the spectral features of the peroxy-laccase species could also be generated by reacting partially reduced laccase with O_2 , further corroborate the latter concept. It is also in line with the analogy between the behavior of the type 3 like copper sites in hemocyanine and in laccase[7].

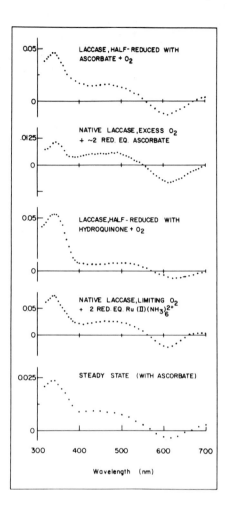

Figure 2. Spectra of the stable intermediate obtained by the different approaches
indicated in the figure. The data are the
difference in absorbance between the intermediate and the fully oxidized enzyme.
Laccase concentrations were in the range of
50-70 μM except for the steady state experiments where it was 14 μM and [ascorbate]
$< [O_2]_0 = 2.5 \cdot 10^{-4}$ M.

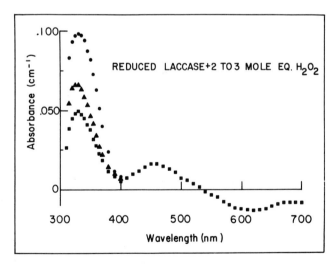

Figure 3. Difference spectra between the product of the reaction of fully (ascorbate) reduced laccase with H_2O_2 and fully oxidized laccase ■ 2.35, ▲ 2.7, and
● 3.0 molar equivalents of H_2O_2.

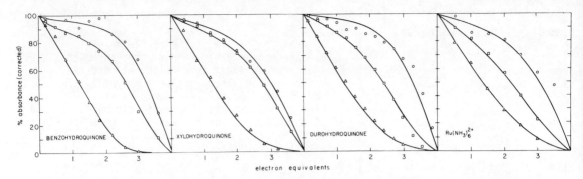

Figure 4. Anaerobic spectrophotometric titration of (about 1 x 10⁻⁴ M) laccase by different reductants: corrected fractional absorbance (A-Ared)/(Aox-Ared) at 330nm Δ; at 614nm ▢ and calculated fraction of type 2 Cu oxidized o. The fully extended line represents the simulated titrations based on the model described in the text.

The reaction of dioxygen with laccase was shown to depend critically on the distribution of reduction equivalents among and within the protein molecule's sites. Whereas the fully reduced species reoxidize fast and without resolvable intermediates at room temperature[2,3], it is the partially reduced species which react to form the intermediates described above. Therefore we examined systematically the electron distribution pattern in the enzyme solution and revealed, to our surprise, that major differences may be found in it, depending on the reductant used. In Figure 4, the anaerobic reductive titrations of laccase by four different electron donors are plotted showing the degree of reduction of the three sites of the enzyme throughout the course of titrant addition. From an examination of these data it becomes evident that major differences are found in the electron distribution patterns. These differences are even better resolved in a double logarithmic plot made following the Nernst formalism (Figure 5). Thus the slope of the line obtained varies between 1 and 2 according to the reductant employed. The only correlation that can

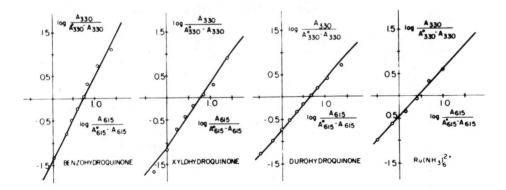

Figure 5. A double logarithmic plot of the results of the titrations shown in Figure 4. A_{330} and A_{614} are (A-Ared) for 330 and 614 nm respectively and A_{330}° and A_{614}° are (Aox-Ared) for the respective wavelength. The lines are a result of the simulation according to the model proposed in the text.

be found is with the first oxidation potential of the reagents used. The stronger reductants (like durohydroquinone and $Ru(NH_3)_6^{2+}$, $E^{0'}$ = 355 and 51 mv, respectively) lead to a slope of 1 whereas the milder produce a value of 2 (benzohydroquinone or ferrocyanide ions[5]; E^{0i} = 480 and 420 mv, respectively). Interesting are the results obtained with reductants of intermediate driving force which also lead to an intermediary value of the slope (e.g. xylohydroquinone).

The dramatic dependence of the electron distribution pattern on the employed reductant is further illustrated in a titration where two different reductants are used sequentially. Thus when benzohydroquinone is used to titrate laccase to approximately half reduction and then $Ru(NH_3)_6^{2+}$ is used to continue it to completion, the reduction pattern in each half is the same as that observed for independent titration with each of the reagent separately. This is clearly illustrated in the double logarithmic plot where a sharp break in the slope is observed (Figure 6). Also the presence of benzoquinone as a mediator (approximately 40 µM) in laccase solutions titrated with $Ru(NH_3)_6^{2+}$ had no effect on the electron distribution pattern. Very important is the observation that several redox cycles of the enzyme made prior to the reductive titration led to results identical to those obtained with the native enzyme.

From these data it emerges that the variation in electron distribution patterns cannot stem from certain chemical or structural features of the substrates reductants, nor in characteristics of their electron transfer mechanism. Reductants as different in their properties as benzohydroquinone and $Fe(CN)_6^{4-}$ ions or durohydroquinone and $Ru(NH_3)_6^{2+}$ ions lead respectively to rather similar titrations.

Before any attempt is made to formulate a model explaining the above results it is important to recognize that these patterns of reduction must involve a quasi-equilibrium state. If a true thermodynamic equilibrium was to be attained in the systems described, a single, reductant independent electron distribution pattern should always be observed. That this is not the case is amply illustrated above. An allosteric control of the enzyme by direct binding of one of the electron donors-substrates finds no support in our results. Allosteric control could therefore be expected from the degree of reduction of the different redox sites of the enzyme.

We have tried to formulate a possible mechanism that underlies this behavior of laccase. One model which is based on thermodynamic arguments, assumes that it is the difference in free energy between the donor couple and the redox sites of the enzyme which determines the eventual electron distribution pattern. An alternative model assumes a kinetic control based eventually on a linear free energy correlation between the rate of electron transfer and the driving force of the reaction.

In the first model the type 3 site is considered as a two electron acceptor with formally a varying degree of positive cooperativity. This degree is determined by the relative stability of the half reduced state (single electron reduced) of the type 3 site and can be expressed by the "interaction potential," defined as half the difference between the potentials of the second and the first reduction steps of this site. This formalism was used to simulate the electron distribution patterns in a quantitative manner as illustrated in Figures 4 and 5. In order to carry out these simulations, the relation between the extinction coefficient of the type 3 site and its oxidation and coupling had to be expressed. The best fit was obtained when an equal contribution to the extinction by each of the two Cu(II) was assumed. The overall agreement between the experimental data and the simulation using the above assumptions is good. Somewhat larger discrepancies are found only for the data of the type 2 Cu site. These stem from the fact that these points are obtained only indirectly from balancing the stoichiometry. This formal analysis of the data can be rationalized in terms of the following molecular mechanism: 1) The capacity to produce the half reduced state of the initially fully coupled type 3 site depends on the redox potential of the reductant. Only the relatively strong reductants are capable of producing the half reduced state in appreciable concentrations. 2) The process of generating the half reduced state of the type 3 site leads to its uncoupling and thereby conversion into a pair of single electron acceptors. 3) Inherent in this model is the postulate that the uncoupled state of site 3 is a metastable one which does not relax within the timescale of the titrations. A more detailed analysis of these results and alternative models that were considered are reported elsewhere[8].

The relationship between the different states (of reduction and coupling) of the type 3 site and its proposed role as reaction site with dioxygen are evidently complicated by the quasi-equilibrium situation. A systematic study of the electron uptake from laccase, i.e. its oxidation pathways, is now carried out in a similar way to that described above for the reduction. Results obtained using O_2 and H_2O_2 have

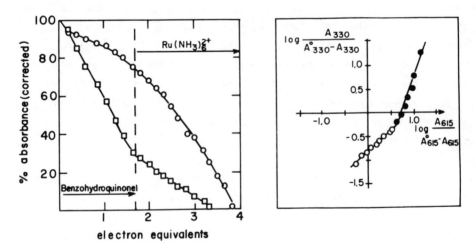

Figure 6. Anaerobic sequential titration of laccase by benzohydroquinone followed by $Ru(NH_3)_6^{2+}$ ions. Left: the switch from benzohydroquinone to $Ru(NH_3)_6^{2+}$ is indicated by the broken line at ~ 1.7 electron equivalents. Corrected fractional absorbance at 330nm □ at 614nm ○. Right: double logarithmic plot of the above data ● by benzohydroquinone, ○ by $Ru(NH_3)_6^{2+}$.

also shown differences in the electron distribution pattern throughout the titrations[4].

REFERENCES

1. Caughey, W. S., Wallace, W. J., Volpe, J. A. and Yoshikawa, S. (1976) The Enzymes, Boyer, P. D. ed., Academic Press Inc., New York, 13, pp. 302.

2. Holwerda, R. A., Wherland, S. and Gray, H. B. (1976) Ann. Rev. Biophys. Bioengin., 5, 363.

3. Malmström, B. G., Andreasson, L.-E. and Reinhammar, B. (1975) The Enzymes, Boyer, P. D. ed., Academic Press Inc., New York, 12, pp. 507.

4. Farver, O., Goldberg, M., Lancet, D. and Pecht, I. (1976) Biochem. Biophys. Res. Comm., 73, 494.

5. Reinhammar, B. (1970) Biochem. Biophys. Acta, 275, 245.

6. Pecht, I., Farver, O. and Goldberg, M. (1977) Bioinorganic Chemistry II: Adv. Chem. Ser., Am. Chem. Soc. Washington D. C., 162, pp. 179.

7. Thamann, T. J., Loehr, J. S. and Loehr, T. M. (1977) J. Am. Chem. Soc., 99, 4187.

8. Farver, O., Goldberg, M., Wherland, S. and Pecht, I., Submitted for Publication.

MECHANISMS OF SUPEROXIDE DISMUTASES*

J. A. FEE and G. J. McCLUNE

Biophysics Research Division and Department of Biological Chemistry, The University
of Michigan, Ann Arbor, Michigan 48109 U.S.A.

Superoxide dismutases are metal containing proteins which catalyze the reaction

$$2 \ O_2^- + 2 \ H^+ \rightarrow H_2O_2 + O_2 \qquad\qquad [1]$$

The hypothesis that superoxide is an extremely dangerous cytotoxin which is quickly
eliminated by the action of these proteins has been vigorously popularized by Fridovich and his co-workers[1,2]. Unfortunately, since the unequivocal demonstration that
superoxide can accumulate to some degree during the action of xanthine oxidase by
Bray's group[3] and the discovery of biological catalysis of Reaction 1 almost a decade
ago[4], not a single reaction of superoxide in aqueous solution which can be seriously
considered as posing a threat to the living cell has been unequivocally documented!
Thus, while the importance of superoxide in biological systems remains conjectural,
it is a fact that superoxide is formed in many *in vitro* and in some *in vivo* processes[5] and the elucidation of its chemistry is necessary to understand its potential
biological roles. Within this context we are primarily interested in the reactivity
of superoxide toward complexes of Cu, Fe and Mn being motivated by the desire to
understand the chemical mechanism whereby these ions catalyze Reaction 1.

The purposes of this article are to present a simple, general reaction scheme for
metal ion catalyzed dismutation in general, and describe some of our recent results
on the iron containing protein from *E. coli* B.

The Spontaneous Dismutation Reaction

Investigations into the catalysis of superoxide dismutation should be predicated
on an understanding of the self-dismutation reaction, i.e., the uncatalyzed process.

Many workers have studied this reaction, and the process has been broken down into
partial reactions occurring in different pH ranges:

$$HO_2 + HO_2 \xrightarrow{k_1} H_2O_2 + O_2 \qquad\qquad [2]$$

$$HO_2 + O_2^- \xrightarrow{k_2} H_2O_2 + O_2 + OH^- \qquad\qquad [3]$$

$$O_2^- + O_2^- \xrightarrow{k_3} H_2O_2 + O_2 + 2OH^- \qquad\qquad [4]$$

*
Supported by the U. S. Public Health Service Grant, GM-21519.

$$HO_2 \underset{}{\overset{K_a}{\rightleftharpoons}} O_2^- + H^+ \qquad [5]$$

The general expression for k_{obsd} is

$$k_{obsd} = \frac{(k_1 + k_2 K_a/[H^+])}{(1 + K_a/[H^+])^2} \qquad [6]$$

where $k_1 = 7.61 \pm 0.55 \times 10^5 \ M^{-1}sec^{-1}$, $k_2 = 8.86 \pm 0.43 \times 10^7 \ M^{-1}sec^{-1}$ and $k_3 <$ $0.3 \ M^{-1}sec^{-1}$; $pK_a = 4.75 \pm 0.08$ (Ref. 6). Qualitatively, these data show that k_{obsd} decreases by a factor of 10 for each unit increase in pH above the pK_a, and they strongly suggest that O_2^- never reacts with itself to dismute but must react with its conjugate acid, HO_2. A more general statement is that the reaction is specific acid actalyzed and the transition state contains two molecules of superoxide and one proton. Since general acid catalysis by water does not occur even in strongly basic solution[6] the proton must be closely associated with one of the superoxide ions[7], and we presume this to be the one which becomes peroxide. In other words, super-oxide dismutation is specific acid catalyzed, and one can conclude that an important feature of the transition state is that charge neutralization on the putative per-oxide anion must be nearly complete. We believe that a lack of charge neutraliza-tion constitutes the kinetic barrier which prevents superoxide ion from being the powerful oxidant toward organic substances portended by its thermodynamic potential[8]. Superoxide is, however, an effective oxidant of metal ions, thus it would seem in these reactions neutralization of charge occurs in a metal-peroxide complex.

Catalyzed Dismutation

There are three distinct types of superoxide dismutases all of which were dis-covered in Professor Fridovich's laboratory. For our purposes, they are character-ized by the metal ions required for catalysis; Cu, Fe and Mn. The properties of these proteins have been reviewed elsewhere[9,10].

It is important to realize that the protein is not necessary for catalysis. Thus, several small complexes of Cu, Fe and Mn have been shown to have dismutase activity ranging from very weak to more effective than that of the protein complexes[8]; this implies that the inherent catalytic activity must rest with the metal ion itself.[a] The simplest question is then: What are the structural requirements necessary in any complex ion of Cu, Fe or Mn which will confer superoxide dismutase activity?

Foregoing complete documentation of the development of these ideas (Cf. Ref. 10 for a preliminary disçussion), we conjecture that metal ion catalyzed superoxide

---

[a]. Because of this fact, the possibility cannot be excluded that the superoxide dis-mutase activity displayed by these proteins is a trivial one unrelated to other yet undiscovered physiological functions.

dismutation occurs according to the basic scheme given below:

$$M^{n+}\text{-}OH_2 + O_2^- \rightleftharpoons [M^{n+}\text{-}OH_2 \cdots O_2^-]_{os} \tag{7}$$

$$[M^{n+}\text{-}OH_2 \cdots O_2^-]_{os} \rightleftharpoons [M^{n+}\text{-}O_2^-\text{---}(H_2O)]_{is} \tag{8}$$

$$[M^{n+}\text{-}O_2^-\text{---}(H_2O)]_{is} \rightleftharpoons M^{(n-1)+}\text{-}O_2 \cdots H_2O \tag{9}$$

$$M^{(n-1)+}\text{-}O_2 \cdots H_2O \rightleftharpoons M^{(n-1)+}\text{-}OH_2 + O_2 \tag{10}$$

$$M^{(n-1)+}\text{-}OH_2 + O_2^- \rightleftharpoons [M^{(n-1)+}\text{-}OH_2 \cdots O_2^-]_{os} \tag{11}$$

$$[M^{(n-1)+}\text{-}OH_2 \cdots O_2^-]_{os} \rightleftharpoons [M^{(n-1)+}\text{-}O\text{-}O^- \cdots H\text{-}O_H]_{is} \tag{12}$$

$$[M^{(n-1)+}\text{-}O\text{-}O^- \cdots H\text{-}O_H]_{is} \overset{H_2O}{\rightleftharpoons} [M^{n+}\text{-}O\text{-}O\text{-}H \text{---} (H_2O)] \tag{13}$$

$$[M^{n+}\text{-}O\text{-}O\text{-}H \text{---} (H_2O)] \overset{H_2O}{\rightleftharpoons} M^{n+}\text{-}OH_2 + H_2O_2 + OH^- \tag{14}$$

The essential features of this scheme are:

1. The metal ion is able to exist in at least two valence states.

2. Both M^{n+} and $M^{(n-1)+}$ have open coordination position(s) where O_2^- enters into direct contact with the metal as a typical liganding species.

3. All electron transfer processes take place within the $M\text{-}O_2^-$ complexes. The scheme suggests at least three points where rate limitation may occur:

a. Diffusion or encounter control. In this situation, reactions [7] and [11] are rate limiting. This appears to apply to the Zn/Cu protein (Table I), and to aquo Cu^{2+} (Cf. Ref. 8).

b. Ligand exchange. Since water off-rates are very high for Cu^+, Cu^{2+}, Fe^{2+}, Mn^{2+}, Mn^{3+}, and Co^{2+} it is unlikely that exchange limitations will occur for these ions. By contrast the water off-rate for Fe^{3+} is moderately slow and for Co^{3+} it is extremely slow (Table I). Thus, for Fe-EDTA catalysis ligand exchange is rate limiting both at the level of $H_2O\text{-}Fe^{3+}\text{-}EDTA$ and for the $[EDTA\ Fe\text{-}O_2]$ peroxo complex[11]. The kinetic inertness of Co^{3+} complexes almost certainly precludes a Co catalyst of superoxide dismutation, and none has been reported.

In the context of the scheme it would appear that monovalent anionic ligands which act as inhibitors of superoxide dismutases do so by forming complexes which have relatively slow dissociation rates.

c. Electron transfer limited. As described below, iron superoxide dismutase appears to form an O_2^- complex which does not undergo rapid electron transfer.

TABLE I

SOME REPRESENTATIVE EXCHANGE RATES FOR METAL BOUND WATER

Metal Ion	Complex	Rate[a]	Reference
Cu^+		Rapid	
Cu^{2+}	Aquo	2×10^8	18
Fe^{2+}	Aquo	3×10^6	18
	EDTA	$\sim 10^{7[b,c]}$	27
Fe^{3+}	Aquo	0.6	18
	-OH	4×10^2	18
	Met hemoglobin	$< 10^3$	20
	EDTA	$\sim 10^3 - 10^{4[b,c]}$	11
Mn^{2+}	Aquo	3×10^7	18
	EDTA	4.4×10^8	21
	NTA	1.5×10^9	21
Mn^{3+}	Aquo	$10^4 - 10^{5[b]}$	22,23
	-OH	$10^8 - 10^{9[b,c]}$	23
Co^{2+}	Aquo	4×10^5	19
	EDTA	$2 \times 10^{6[b]}$	24
	$(NH_3)_2 (H_2O)_4$	6.5×10^7	25
	tmc[d]	4.2×10^4	26
Co^{3+}	Aquo	10^{-5}	18

(a) Sec^{-1}

(b) $M^{-1}Sec^{-1}$

(c) Deduced from substitution rates

(d) 1,4,8,11-tetramethyl-1,4,8,11-tetraazacyclotetradecane

In order to test the hypothesis put forward in reactions [7-14], we must obtain the following:

a. The overall rate of the catalytic process.

b. The rates of the reactions.

$$i \quad O_2^- + M^{n+} \rightarrow M^{(n-1)+} + O_2 \qquad [15]$$

$$ii \quad O_2^- + M^{(n-1)+} \overset{H^+}{\rightarrow} M^{n+} + H_2O_2 \qquad\qquad [16]$$

c. Unequivocal evidence for open coordination positions on both M^{n+} and $M^{(n-1)+}$ forms.

d. The H_2O off-rates for both valence states of the metal.

e. Kinetic evidence for the complexes $M - O_2^-$ and $M - OOH$ and their respective rates of decomposition.

f. Finally, it would be necessary to trap the intermediate complexes and examine their physical properties.

A summary of available results relevant to these objectives is presented in Table II for the various superoxide dismutases and for Fe-EDTA. While the details of the individual systems must be worked out in greater detail, on the whole, the extant results are consistent with the proposed reaction scheme.

Catalysis of O_2^- Dismutation by iron superoxide dismutase (FeSD) from *E. Coli* B.

The remainder of this chapter will be devoted to our recent work on the mechanism of FeSD as deduced from the decay profiles of O_2^- in the presence of FeSD.

In a recent paper[12] concerned with the anion binding properties of $Fe^{3+}SD$ it was proposed that the active site of this protein consisted of an iron atom having two open coordination positions only one of which was necessary for superoxide dismutase activity. It was shown that anions such as N_3^- and F^- could bind to one of these without loss of activity; that F^- could bind to the second site with loss of activity, but N_3^- could not. Indeed it was concluded from both steady state procedures for measuring activity and by our stopped-flow procedure[13] that F^- inhibited the protein but only at relatively high concentrations ($\sim$30 mM for half inhibition) and that N_3^- did not inhibit even at quite high concentrations in the stopped-flow experiments. In our hands, the traditional steady-state assay systems were poorly behaved in the presence of azide and gave no clear cut evidence for N_3^- inhibition. Therefore, in spite of the reports of Asada, et. al.[14] and Yamakura[15] who utilized steady state methods and concluded that N_3^- was an inhibitor of FeSD from *Plectonema boryanum* and *Ps. ovalis* we maintained that N_3^- was not an inhibitor.

We have now examined the FeSD catalyzed dismutation reaction in greater detail and wish to present evidence to support the following new conclusions:

(1) O_2^- forms a complex with FeSD having a K_d of $\sim$65 μM, and this complex is devoid of dismutase activity.

(2) F^- inhibition of dismutase activity becomes apparent only at low O_2^- concentrations, $\overset{\sim}{<}$ 50 μM with a K_I of $\sim$50 mM.

(3) Similarly N_3^- inhibition of FeSD activity becomes apparent only at low O_2^- concentrations, $\overset{\sim}{<}$ 50 μM, with a K_I of $\sim$10 mM.

TABLE II

SOME PROPERTIES OF CATALYSTS OF SUPEROXIDE DISMUTATION

Property	Catalyst[a]			Fe Protein		Fe EDTA	
	Cu^{2+}	Cu/Zn Protein	Cu^+	Fe^{3+}	Fe^{2+}	Fe^{3+}	Fe^{2+}
Overall catalytic rate							
k_{cat} $(M^{-1}sec^{-1})$		$\sim 2 \times 10^9 (28)^b$			$\sim 4 \times 10^8 (16)^c$		$\sim 10^6 (11)^c$
k_1 $(M^{-1}sec^{-1})$	$\sim 2 \times 10^9$						
k_2 $(M^{-1}sec^{-1})$			$\sim 2 \times 10^9$				
Number of open coordination positions	1 (29,30)		1 (31)	2 (12)		1 (32)	1 (27)
H_2O off-rate (sec^{-1})	$< 10^6$	(29)				$< 10^3$	10^5
Evidence for							
M – Superoxide	– (28)				+(This work)		
M – Peroxide			– (28)			+ (11)	

(a) The mangano protein is not included here as this is the least studied protein and the one pulsed radiolysis study indicated an unusual complexity of the O_2^- decay curve (33). The overall catalytic rate is $10^8 - 10^9$ $M^{-1}sec^{-1}$.

(b) References are given in parentheses.

(c) pH dependent.

278

RESULTS AND DISCUSSION

The evidence that O_2^- at concentrations above $\sim$100 μM has a pronounced inhibitory effect on the rate of superoxide dismutation is shown in Fig. 1. At O_2 concentrations $\sim < 50$ μM the log A_{275} vs. t plots are nominally linear. However, as the initial O_2^- concentration is raised above 50 μM the plots show two distinct portions. In Fig. 1 the dashed lines were computed to best fit the experimental data using the scheme

$$O_2^- + FeSD \underset{}{\overset{K_{O_2^-}}{\rightleftharpoons}} FeSD\text{-}O_2^- \qquad [17]$$

$$O_2^- + Fe^{3+}SD \overset{k_1'}{\to} Fe^{2+}SD + O_2 \qquad [18]$$

$$O_2^- + Fe^{2+}SD \underset{2H^+}{\overset{k_2'}{\to}} Fe^{3+}SD + H_2O_2 \qquad [19]$$

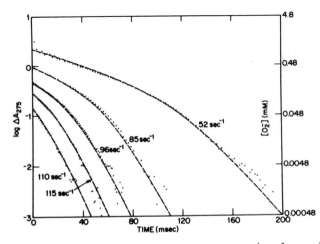

Fig. 1. Stopped-flow decay traces of O_2^- in the presence of 0.225 μM [Fe] iron superoxide dismutase at pH 8.5, 0.25 mM pyrophosphate buffer. The initial concentration at O_2^- was varied from $\sim$ 1 mM (upper) to 0.06 mM (lower). The experimental procedure[12] and instrumentation [13] were as described previously. The dashed lines were best fit using a Runge-Kutta[17] procedure for the solution of the differential equation given in the text. The second-order rate constant for spontaneous dismutation under these conditions is 1.7×10^4 M^{-1} sec^{-1}. The value of K_{O_2} - was fixed and k_{cat} was changed to give the best fit. The values in the figure correspond to the "best fit" k_{cat} times the concentration of iron associated with the protein.

Reactions [18] and [19] are based on the work of Lavelle et al[16],

$$-\frac{d[O_2]}{dt} = k_s \, [O_2^-]^2 + \frac{k_{cat}[E]_{tot}[O_2^-]}{1 + [O_2^-]/K_{O_2^-}} \qquad [20]$$

FeSD $-O_2^-$ has no activity, and $k_{cat} = 2k_1' x \, k_2'/(k_1' + k_2')$. While the fit to the experimental data is quite good using the individual rate constants given in the legend we do not understand why the rate at low O_2^- concentrations depends on the initial concentration of the O_2^-. These subtleties are still under investigation. What is clear from these data however is that O_2^- can act as an inhibitor of the FeSD by forming a reversible complex with the catalyst characterized by $K_{O_2^-} \sim 65$ μM. The nature of this complex is uncertain.

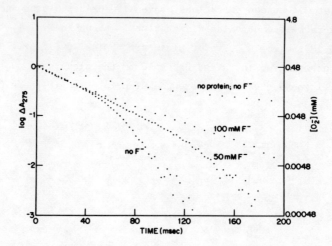

Fig. 2. Effects of fluoride on stopped-flow decay traces of O_2^- in the presence of iron superoxide dismutase. Conditions were as described in the legend to Fig. 1.

The effects of F^- on the rate of O_2^- dismutation in the presence of FeSD are shown in Fig. 2. It can be seen that F^- has no effect on the initial rate of the reaction which accounts for about 70% of the initial O_2^- concentration. In a somewhat arbitrary fashion, two apparent first order constants are taken from these plots. The initial portion is termed k_1 and the most linear part of the faster phase is termed k_2.

Fig. 3 shows how $k_2/[Fe]$ varies with F^- concentration; k_1 is unchanged even at 100 mM F^-. The data show a scatter of approximately 20% but are quite reproducible

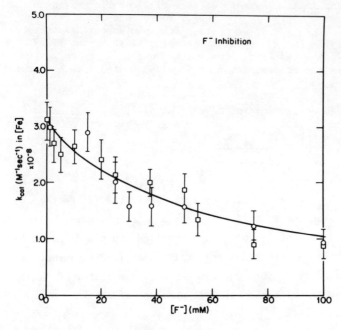

Fig. 3. Variation of k_{cat} (=$k_2/[Fe]$) with the concentration of fluoride. See text for details. The solid line is calculated using an inhibition constant K_F^-, of 50 mM.

within this limit. The decrease in $k_2/[Fe]$ is consistent with an inhibition constant K_F- of 50 mM which is similar to the value of 33 mM reported from an earlier steady state measurement and to the binding of the second F^- to the Fe^{3+} form of the protein, 42 mM[12]. The binding of the first F^- ($K_F- \sim 2$ mM) is apparently not involved (see below).

Azide showed a behavior similar to F^- (Fig. 4) except at concentrations above approximately 30 mM where the apparent value of k_1 begins to decrease. Fig. 5 shows the variation of $k_1/[Fe]$ and $k_2/[Fe]$ with azide concentration. It can be seen that $k_2/[Fe]$ decreases with an apparent K_{N_3-} of 10 mM while $k_1/[Fe]$ is constant out to 20 mM after which it decreases. At high N_3^- concentrations ~ 100 mM the decay curve approaches the spontaneous dismutation curve suggesting a complete loss of activity. K_{N_3-} bears no apparent relation to the binding constant of N_3- to the oxidized protein (see below).

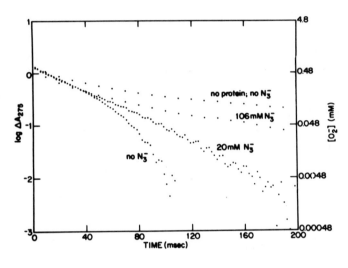

Fig. 4. Effects of azide on stopped-flow decay traces of O_2^- in the presence of iron superoxide dismutase. Conditions were as described in the legend to Fig. 1.

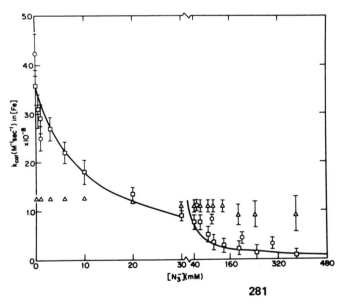

Fig. 5. Effect of azide on k_{cat} ($K_1/[Fe]$), $\triangle$; and k_{cat} ($k_2/[Fe]$) for two different preparations at iron dismutase, O and $\square$. The initial concentration of O_2^- in all experiments was 0.45 mM. The solid line was calculated using an inhibition constant, $K_{N_3}-$, of 10 mM.

These observations account for the disparate conclusions of our group and Asada et. al.[14] and Yamakura[15] concerning azide inhibition. Since, in our earlier studies[12], we only considered the first portion of the decay curve and were not convinced that corrections for N_3^- interference in the steady state assays could be properly made, our conclusion that N_3^- is not an inhibitor of FeSD was incorrect. Azide inhibits FeSD with an apparent K_I of 10 mM at concentrations of $O_2^- < K_{O_2^-}$.

The inhibition patterns have been tentatively modeled with the kinetic scheme

$$FeSD + I \rightleftharpoons I\text{-}FeSD \quad ; \quad K_I \tag{21}$$

$$FeSD + O_2 \rightleftharpoons O_2^-\text{-}FeSD \quad ; \quad K_{O_2}$$

$$Fe^{3+}SD + O_2^- \xrightarrow{k_1'} Fe^{2+}SD + O_2$$

$$Fe^{2+}SD + O_2^- \xrightarrow[2H^+]{k_2'} Fe^{3+}SD + H_2O_2$$

Assuming I-FeSD and O_2^--FeSD are mutually exclusive and have no catalytic activity

$$-\frac{d[O_2]}{dt} = k_s[O_2^-]^2 + \frac{k_{cat}[FeSD]_{tot}[O_2^-]}{1 + \dfrac{[O_2^-]}{K_{O_2^-}} + \dfrac{[I]}{K_1}} \tag{22}$$

When $[I]/K_I \ll [O_2^-]/K_{O_2}$- inhibition is due solely to O_2^- accounting for the invariability of the initial portion of the decay curves under these conditions. Fits to the experimental data (comparable to Fig. 1) are obtained using this scheme.

TABLE III

SUMMARY OF CONSTANTS

Dissociation or Inhibition Constant	F^-	N_3^-	O_2^-
Site 1	1.9×10^{-3}	1.3×10^{-3} [a]	-
Site 2	4.2×10^{-2}	-	-
Inhibition	3.3×10^{-2}, 5.0×10^{-2}	1.0×10^{-2}	6.5×10^{-5}

[a] Independent of pH (7.5-8.5) and DMSO (0-5%).

It is interesting to compare the inhibition patterns with the binding of the inhibitors to the ferric form of the protein in terms of the two site model proposed earlier[12]. The relevant constants are summarized in Table III. While O_2^- is the most effective inhibitor it does not seem possible to measure its binding to $Fe^{3+}SD$, and it is not known whether inhibition occurs at the oxidized or reduced level. Azide inhibition shows no correlation with binding to $Fe^{3+}SD$. Thus, at the concentrations of N_3^- where half-inhibition is evident all the $Fe^{3+}SD$ should exist as the mono-azido form – at least at low O_2^- concentration. It is possible that N_3^- is inhibiting by binding to the reduced form of the iron, but we have no independent evidence for this. Based on the correlation of the binding and inhibition constants for F^- it is possible that this anion inhibits by binding to the second site on the $Fe^{3+}SD$. Thus, K_I corresponds to the process $F^- -Fe^{3+}SD \underset{}{\overset{+F^-}{\rightleftharpoons}} 2F^- -Fe^{3+}SD$.

Many questions remain about the mechanism of FeSD catalysis; in particular, we still do not know the rate limiting step.

REFERENCES

1. McCord, J.M., Keele, Jr., B.B. and Fridovich, I. (1971) Proc. Nat'l. Acad. Sci. 68, 1024-1027.

2. Fridovich, I. (1975) American Scientist 63, 54-59.

3. Bray, R.C. and Knowles, P.F. (1968) Proc. Roy. Soc. A 302, 351-353.

4. McCord, J.M. and Fridovich, I. (1969) J. Biol. Chem. 244, 6049-6055.

5. Michelson, A.M., McCord, J.M. and Fridovich, I. (eds.) (1978) Superoxide and Superoxide Dismutases, Academic Press, In press.

6. Bielski, B.H.J. and Allen, A.O. (1977) J. Phys. Chem. 81, 1048-1050.

7. Bender, M.L. (1971) Mechanisms of Homogeneous Catalysis From Protons to Proteins, Wiley Interscience, New York, pp. 19-144.

8. Fee, J.A. and Valentine, J.S. (1978) in Superoxide and Superoxide Dismutases, Michelson, A.M., McCord, J.M. and Fridovich, I. (eds.), Academic Press, In press.

9. Fridovich, I. (1974) Adv. Enzymol. 41, 35-97.

10. Fee, J.A. (1978) in Superoxide and Superoxide Dismutases, Michelson, A.M., McCord, J.M. and Fridovich, I. (eds.), Academic Press, In press.

11. McClune, G.J., Fee, J.A., McClusky, G.A. and Groves, J.T. (1977) J. Am. Chem. Soc. 99, 5220-5222.

12. Slykhouse, T.O. and Fee, J.A. (1976) J. Biol. Chem. 251, 5472-5477.

13. McClune, G.J. and Fee, J.A. (1976) FEBS Letters 67, 294-298.

14. Asada, K., Yoshidawa, K., Takahashi, M-A., Maeda, Y. and Enmanji, K. (1975) J. Biol. Chem. 250, 2801-2807.

15. Yamakura, F. (1976) Biochim. Biophys. Acta 422, 280-294.

16. Lavelle, F., McAdam, M.E., Fielden, E.M. and Roberts, P.B. (1977) Biochem. J. 161, 3-11.

17. Boyce, W.E. and DiPrima, R.C. (1969) Elementary Differential Equations and Boundary Value Problems, John Wiley and Sons, New York, pp. 353-357.

18. Eigen, M. and Wilkins, R.G. (1965) Adv. Chem. Ser. 49, 55-67.

19. Eigen, M. and DeMaeyer, L. (1963) in Technique of Organic Chemistry, Weissberger, A. (ed.) Interscience Publishers, New York, p. 1042.

20. Pifat, G., Maricit, S. and Grandju, S. (1973) Biopolymers 12, 905-920.

21. Zetter, M.S., Grant, M.W., Wood, E.J., Dodgen, H.W., and Hunt, J.P. (1972) Inorg. Chem. 11, 2701-2706

22. Davies, G., Kirschenbaum, L.J. and Kustin, K. (1968) Inorg. Chem. 7, 146-154.

23. Dibler, J., (1969) Z. Phys. Chem. 68, 64-78.

24. Jones, J.P. and Margerum, D.W. (1969) Inorg. Chem. 8, 1486-1490.

25. Chmelnick, A.M. and Fiat, D. (1967) J. Chem. Phys. 47, 3986-3990.

26. Meier, P., Merbach, A., Burki, S. and Kaden, T.A. (1977) Chem. Commun. 36-37.

27. Woodruff, W.H. and Margerum, D.M. (1974) Inorg. Chem. 13, 2578-2585.

28. Fielden, E.M., Roberts, P.B., Bray, R.C., Lowe, D.J., Mautner, G.N., Rotilio, G., and Calabrese, L. (1974) Biochem. J. 49-60.

29. Gaber, B.P., Brown, R.D., Koenig, S.H. and Fee, J.A. (1972) Biochim. Biophys. Acta 271, 1-5.

30. Richardson, J.S., Thomas, K.A., Rubin, B.H. and Richardson, D.C. (1975) Proc. Nat'l. Acad. Sci. 72, 1349-1353.

31. Fee, J.A. and Ward, R.L. (1976) Biochem. Biophys. Res. Commun 71, 427-437.

32. Lind, M.D., Hamor, M.J., Hamor, T.A. and Hoard, J.L. (1964) Inorg. Chem. 3, 34-43.

33. Pick, M., Rabani, J., Yost, F. and Fridovich, I. (1974) J. Am. Chem. Soc. 96, 7329-7333.

MAGNETIC PROBES OF METAL BINDING SITES IN COPPER PROTEINS

JACK PEISACH

Departments of Molecular Pharmacology and Molecular Biology,
Albert Einstein College of Medicine of Yeshiva University,
Bronx, New York 10461

ABSTRACT

For paramagnetic copper proteins, EPR spectroscopy has been used to describe the chemical composition of the metal binding site by comparison of magnetic parameters with those for Cu(II) complexes of known composition. For the blue copper proteins, this cannot be done because of the lack of suitable model compounds. By studying the linear electric field effect (LEFE) in EPR, we are able to demonstrate near tetrahedral geometry by comparisons with model compounds. Our data are also in accord with the view that there is a charge transfer between cysteinyl sulfur and Cu(II). Other magnetic experiments based on electron nuclear interactions leads us to assign imidazole as a ligand to copper in both blue and non-blue copper proteins. In the case of cytochrome c oxidase, evidence is presented that the EPR heretofore assigned to Cu(II) should be assigned to a free radical.

There are three different types of cupric copper found in proteins (Table I). These have been classified on the basis of their optical and magnetic properties [1,2]. The first, designated Type 1 and often called "blue copper" has a molar extinction coefficient near 600 nm that can vary from 3000-6000. For Type 2 copper, often referred to as "non-blue copper," extinction coefficients can vary from 100 to 400, much the same as found for simple cupric model compounds and peptide complexes.

TABLE I

PHYSICAL PROPERTIES OF Cu(II)-PROTEINS

TYPE	ϵ_M ($\lambda_{max} \sim 600nm$)	$A_\|$ (mK)
I	3000 - 5000	3.5-9
II	100 - 400	13-20
III	—*	none

* $\lambda_{max} \sim 325\text{-}345nm$, $\epsilon_M > 3000$

Type 1 and Type 2 Cu(II) also differ magnetically. The nuclear hyperfine coupling constant, $A_\|$, may vary from 14-20 mK for Type 2 copper. It is never larger than about 9 mK for Type 1 copper. In Fig. 1 we indicate what is meant by the hyperfine coupling constant. Due to the interaction of the unpaired electron of Cu(II) with its nuclear spin ($I = 3/2$), the $g_\|$ feature of the axial EPR spectrum, here shown for Cu(II)-aquo, is split into four features. The separation of the middle one, ΔH, expressed in gauss, is related to $A_\|$, expressed in millikaisers (mK) by:

$$A_\| = 0.0467\ g_\| \Delta H$$

where $g_\|$ is determined from the magnetic field H, half way between the second and third hyperfine lines by the expression:

$$g = \frac{\nu}{1.4\ H}$$

Here, ν is the microwave frequency in MHz of the EPR spectrometer employed for the measurement.

Type 3 copper is considered to be a binuclear cupric pair that is antiferromagnetically coupled so that an EPR spectrum is never seen. The optical characteristic of this site consists of an optical absorp-

286

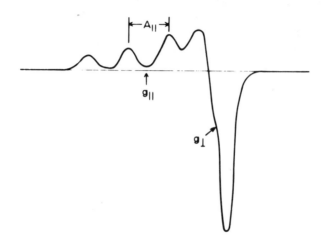

Fig. 1. X-band EPR spectrum of hydrated Cu(II) taken at 77°K. The magnetic parameters $A_\parallel$, $g_\parallel$ and $g_\perp$ can be determined directly from the spectrum.

tion between 325 and 345 nm with a molar extinction coefficient greater than 3000.

The various types of copper sites can be distributed in copper proteins in different ways (Table II). For example, the low molecular weight blue copper proteins azurin, stellacyanin and plastocyanin contains only Type 1 copper, one each per molecule. Examples of copper proteins that contain Type 2 copper include dopamine-β-hydroxylase, galactose oxidase and superoxide dismutase. Neurospora tyrosinase contains a single Type 3 site. For the blue copper oxidase, ceruloplasmin, laccase and ascorbate oxidase, all three types of sites are to be found.

For the various metal binding sites in copper proteins, our information regarding structure is rather limited. Little is known about the Type 3 site other than its capacity to accept two electrons in a redox reaction. More is known, however, about the Type 1 and 2 sites.

287

TABLE II

DIFFERENT TYPES OF COPPER FOUND IN COPPER (II)-PROTEINS

PROTEIN	TYPE		
	I	II	III
Azurin	+		
Stellacyanin	+		
Plastocyanin	+		
Monoamine oxidase		+	
Galactose oxidase		+	
Dopamine-β-hydroxylase		+	
Tyrosinase			+
Laccase	+	+	+
Ascorbate oxidase	+	+	+
Ceruloplasmin	+	+	+
Cytochrome c oxidase	?		

In an attempt to learn something about the metal ligands in copper proteins, we attempted to compare the EPR properties of these molecules with copper complexes of known structure[3]. A compilation of our findings is given in Fig. 2. Here values of $g_\parallel$ and $A_\parallel$ for a series of copper proteins containing non-blue, Type 2 copper and blue, Type 1 copper are shown. Within the arbitrary regions in the figure set off with dotted lines fall the EPR parameters for a series of copper model compounds where the close ligands to the metal ion are all oxygen (4O), all nitrogen (4N), both nitrogen and oxygen (2N2O), all sulfur (4S) and both sulfur and nitrogen (2N2S). The EPR parameters for Type 2 copper all fall within a region for copper complexes having nitrogen and/or oxygen ligands. None fall in the region for sulfur ligation. This suggests that Type 2 copper binding sites consist of nitrogen and/or oxygen, but not sulfur. A Type 2 copper protein, superoxide dismutase, which is the only copper protein studied to date by X-ray crystallography, has been shown to have a copper binding site consisting of four imidazole nitrogen ligands[4].

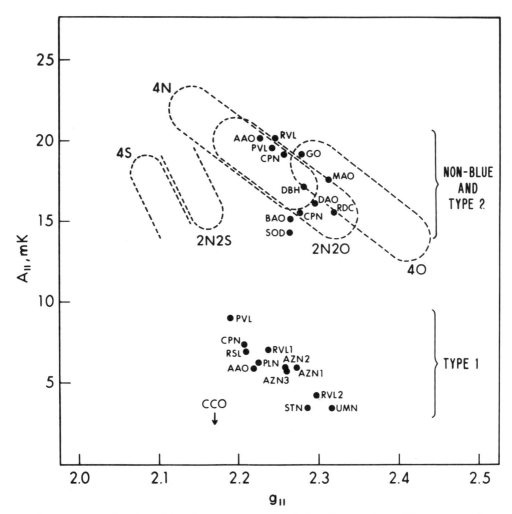

Fig. 2. The relationship between $g_{\parallel}$ and $A_{\parallel}$ for naturally occurring copper proteins. The dashed lines set off regions in the plot for the EPR parameters of model compounds which are all oxygen ligated (4O), all nitrogen ligated (4N), all sulfur ligated (4S) and with different ligand combinations. The abbreviations are: AAO, ascorbate oxidase, PVL, <u>Polyporous versicolor</u> laccase, RVL, <u>Rhus vernicifera</u> laccase, CPN, ceruloplasmin, GO, galactose oxidase, DBH, dopamine-β-hydroxylase, MAO, monoamine oxidase, DAO, diamine oxidase, BAO, benzylamine oxidase, SOD, superoxide dismutase, RDN, ribulose diphosphate carboxylase, PLN, plastocyanin, AZN, azurin, STN, stellacyanin, UMN, umecyanin, CCO, cytochrome <u>c</u> oxidase. (After Ref. 3).

289

For Type 1 copper, no conclusions can be drawn from this analysis. No well characterized model compounds have been prepared that possess the narrow hyperfine in the EPR that is associated with Type 1 copper. Thus, EPR alone cannot be used to define the structure.

A rather anomalous EPR spectrum that has been ascribed to copper is observed for cytochrome c oxidase[5]. Here, the hyperfine structure associated with Cu(II) is absent in the spectrum while the g values ($g_\parallel$ = 2.17 and $g_\perp$ = 2.03) are closer to those of a free radical, near g = 2, than those observed for any copper protein or simple copper model compound. Furthermore, the ratio of $(g_\perp-2)/(g_\parallel-2)$ is also smaller than for any copper protein or model compound indicating an appreciable transfer of unpaired spin from copper to a ligand. A possible electronic description for cytochrome c oxidase is one where the copper is formally cuprous and the bound ligand a free radical, such as cysteinyl sulfur.

Chemical evidence[6,7] has suggested the presence of cysteinyl sulfur ligation to Type 1 copper as well. Early experiments showed that holoplastocyanin could not be reconstituted from the apoprotein with Cu(II) if a mercurial were previously bound. Removal of the mercurial permitted reconstitution, suggesting that cysteinyl sulfur was a metal ligand. More recently, analysis of the optical spectrum of stellacyanin and its Co(II) substituted form[8] lead to the same conclusion. In this case it was suggested that the intense blue color of the protein could be ascribed to an RS^- ⟶ Cu(II) charge transfer. Type 1 copper differs from cytochrome c oxidase in that the presence of a nuclear hyperfine pattern in the EPR of the former clearly demonstrates that the oxidation state of the copper is formally cupric.

In more recent attempts to explain the unusual optical and magnetic properties of Type 1 copper sites, various model compounds have been prepared. The most recent reported[9], is one in which the Cu(II) ion is ligated to three imidazole nitrogens and a mercaptide sulfur. This model has the intense color of Type 1 copper, yet its EPR properties are not unusual, that is $A_\parallel$ is not smaller than 9 mK. Thus, mercaptide ligation alone cannot be responsible for the unusual magnetic properties of Type 1 copper, although one might think that the optical properties are explained.

In 1967, it was suggested[10] that the unusual physical properties of stellacyanin and other blue copper proteins are due, among other things, to a geometry of ligands that is greatly distorted away from the square planar orientation often presented for Cu(II). Although conventional EPR experiments have proven useful for identifying local site symmetry as isotropic, axial or rhombic, they cannot distinguish between a centrosymmetric arrangement of ligands, such as that occurring in the familiar square planar model of a copper complex, and a non-centrosymmetric arrangement such as that which would occur in a tetrahedral complex. In order to study odd symmetries, one may resort to experiments in which an electrostatic field is used in conjunction with the magnetic field[11-13]. By applying the electric field in selected directions, it is possible to detect imbalances in the coordination pattern of the complex and obtain information about the odd component of the ligand crystal field potential. These experiments are complementary to conventional EPR studies in which the even crystal field potentials are examined.

How are these experiments performed? For the sake of simplicity, let us consider a low spin heme complex, such as myoglobin azide where the ligands to the heme iron are azide and imidazole. This compound has a rhombic EPR spectrum with three g values. From single crystal studies[14], it has been demonstrated that g_{max}, the low field g value, is aligned roughly parallel to the heme normal while g_{mid} and g_{min} are aligned approximately in the plane of the porphyrin. If now the magnetic field H_o is set to g_{max}, then the application of an electric field in the same direction $(E_\parallel H_o)$ will shift the g tensor because there is an imbalance of crystal field in the direction of the heme normal since imidazole and azide nitrogens are different from one another. At g_{mid} and g_{min} this crystal field imbalance is not found and the shift in g will not be observed. If on the other hand, the electric field is applied perpendicular to the magnetic field, then the largest shifts of g will be observed away from g_{max} (For a more complete explanation of this problem, see Ref. 15, Fig. 4).

These shifts of g produced by electric fields can be extremely small, especially in non-crystalline materials, and conventional EPR spectroscopy cannot be employed for their study. It is possible, however, to employ a spin echo technique (see below). The shifts in g produced by electric field, here termed the linear electric field effect (LEFE), can be related to a shift in the spin precession frequency of a paramagnetic center which in turn shifts the magnetization responsible for generating the spin echo. In this way, the spin echo amplitude is reduced. We have arbitrarily chosen for means of comparison, electric fields that reduce the amplitude of the spin echo by 50%. The shift parameter, σ, then is related to the magnitude of that electric field that will reduce the spin echo by this amount.

Returning now to myoglobin azide we see that the largest LEFE (Fig. 3) is obtained when the electric and magnetic fields are aligned along the heme normal $(E_\| H_0)$, that is at the g_{max} setting. When both fields are perpendicular to each other, then the largest LEFE is observed at another g setting. This type of LEFE behavior is characteristic of low spin ferric heme complexes where there is an axis of asymmetry[16].

For copper complexes such as the one where the close ligands to the metal are all H_2O, a completely different LEFE behavior is observed (Fig. 4a)[17]. Here, the LEFE is reduced to a minimum at the ends of the EPR absorption envelope, except at the very high field end of the spectrum. Very little change is observed when the magnetic and electric field are oriented at different angles to one

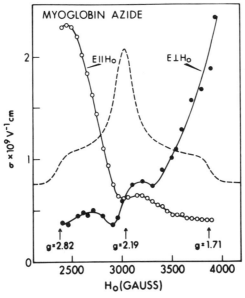

Fig. 3. Shift parameter σ as a function of applied magnetic field for myoglobin azide. Shifts were determined with the electric field aligned parallel to $(E_\| H_0)$ or perpendicular to $(E_\perp H_0)$ the applied magnetic field. The dotted line is a computed EPR absorption spectrum with the same g values as myoglobin azide. (After Ref. 16).

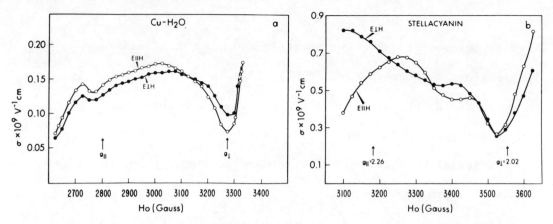

Fig. 4. Linear electric field effect as a function of applied magnetic field for a) hydrated Cu(II) and b) the blue copper protein,
stellacyanin. The shift parameter σ is described in the text. The
presence of an electric field induced shift of g demonstrates that
both complexes are non-centrosymmetric. The rise of σ at the low
field and of the EPR spectrum at the $E_\perp H_o$ setting for stellacyanin
is characteristic of all blue copper sites and is indicative of a
$RS^- \longrightarrow Cu(II)$ charge transfer. (After Refs. 17,18).

another. The fact that an LEFE is observed at all for Cu(II) in

frozen solution is rather surprising since this result is contrary

to the view, based on X-ray studies of crystalline copper complexes,

that the ligands are disposed around the metal in a distorted octa-

hedral environment which is characterized by the centrosymmetric point

group D4h. This cannot give rise to an LEFE. However, an LEFE has

been obtained and we must look for modification of the tetragonal

geometry which could account for this result.

The LEFE results that were obtained for hydrated Cu(II) have been

interpreted[17] as being characteristic of tetrahedral distortion of

the square planar configuration for the metal site. A similar study

has been performed for copper complexes with various N and O

ligands[18]. As a means of comparison, we list the shift parameters

for a series of these compounds taking maximal values in the middle

of the LEFE curve when $E_\parallel H_o$ (Table III). For the most part, the

LEFE behavior is approximately the same. There are some notable

294

TABLE III

MAGNITUDE OF LINEAR ELECTRIC FIELD EFFECT IN EPR FOR VARIOUS COPPER COMPLEXES

Copper Complex	Ligand Atoms	σ
uroporphyrin	4N	0.04
H_2O	4O	0.17
NH_3	4N	0.19
bis(glycine), pH 8.2	2N, 2O	0.14
gly-gly, pH 3.0	1N, 3O	0.16
bis(ethylenediamine)	4N	0.18
3-ethoxy-2-ketobutyraldehyde bis(thiosemicarbazone)	2N, 2S	0.14
o-phenanthroline dichloride	2N, 2Cl	0.65
stellacyanin	-	0.85
azurin	-	0.70
laccase (Type 1 site)	-	1.25

The shift parameter σ in units of $10^9 V^{-1}$ cm is taken as the maximum in the $E_\parallel H_0$ curve of the linear electric field effect. (After Refs. 18,20)

examples, however. One of these is the o-phenanthroline dichloride complex of Cu(II). Here, it has been demonstrated from X-ray crystallographic and single crystal EPR studies that the copper site is nearly tetrahedral[19]. The LEFE is four times larger than for the aquo complex. On the other hand, where the ligands to the Cu(II) are constrained to a square planar configuration, as in the Cu(II) uroporphyrin monomer, the maximum LEFE is four times smaller than for the Cu aquo complex and sixteen times smaller than for the phenanthroline dichloride complex. These results taken together suggest that all Cu(II) complexes in frozen solution tend to adopt a weakly tetrahedral coordination rather than the square coordination often assumed for copper.

The LEFE for blue copper in Type 1 sites (Fig. 4b) is considerably larger than for simple Cu(II) complexes[18]. For stellacyanin, the maximal shift is comparable to that for Cu(II)-o-phenan Cl_2. For the Type 1 site of laccase[20], it is larger still. These results demonstrate that Type 1 copper is tetrahedrally coordinated and this coordination is characteristic of the site. This coordination per se cannot give rise to the intense blue color since many tetrahedral Cu(II) model compounds do not share this property. However, the presence of the mercaptide sulfur could result in a charge transfer and thus produce the intense color. Charge transfer would also be associated with an odd ligand field component and it should therefore be detectable in an LEFE experiment. The rise of the LEFE shift parameters at the low field end of the EPR spectrum when $E_{\perp}H_0$ suggests that there is indeed a strong crystal field component, probably from a $RS^- \longrightarrow Cu(II)$ charge transfer, oriented in a direction approximately perpendicular to the $g_{\parallel}$ axis. One might then, summarize by saying that the unusual physical properties of Type 1 copper are attributable both to the tetrahedral geometry and to the $RS^- \longrightarrow Cu(II)$ charge transfer.

What else can we say about the chemical constitution of the Type 1 copper site? There has been a plethora of evidence based on NMR comparisons of apo and holo azurin and plastocyanin[21-23] showing that at least one copper ligand is a histidine imidazole. We too have made such an assignment but based on a different type of experiment that involves the use of electron spin echoes.

A spin echo experiment differs from a conventional continuous wave

296

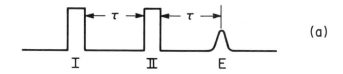

(a)

(b)

Fig. 5. (a) Two pulsed electron spin-echo sequence consisting of
microwave pulses (I and II), and spin echo signal (E). The time,
τ, is the period between applied pulses and between the second
applied pulse and the spin echo. In the experiments described here
the microwave pulses were each 20 ns. long and the power level
0.1-1.0 kW. (b) Superposition of a number of electron spin-echo
signals illustrating meaning of the envelope modulation function.
As τ is increased the amplitude of the echo signal varies, giving
rise to the modulation patterns shown in Figs. 6,7 and 9. (After
Ref. 26).

EPR experiment in that the microwave frequency is imparted to the

frozen paramagnetic sample in two high powered pulses which, in our

experiments, are each about 20 nsec in duration. These are indicated

as I and II in Fig. 5. If the application of pulses I and II are

separated in time by τ, then the spin echo E is generated at time τ

after pulse II. The echo amplitude does not however decay mono-

tonically with τ, as one might expect for a normal relaxation pro-

cess, but undergoes a series of oscillations. Theoretical calcula-

tions show that the frequencies which thus "modulate" the echo decay

are due to the superhyperfine splitting of the electron spin levels

which is caused by coupling with nearby nuclei, i.e. they are the

"ENDOR" frequencies of the microwave resonance transition[24],[25].

The modulation pattern can be obtained experimentally by increasing

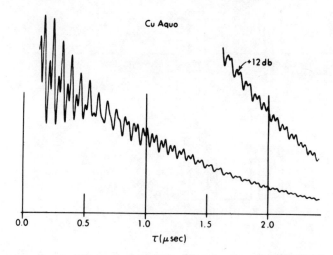

Fig. 6. Nuclear modulation pattern for hydrated Cu(II) demonstrating coupling between the paramagnetic electron and surrounding protons. (After Ref. 26).

the period τ between pulse I and pulse II and then plotting the amplitude of the spin echo observed at the same τ after the second pulse, as shown in Fig. 5b. (In practice this is done with an X Y recorder and suitable electronic circuitry[26]).

Fig. 6 shows a modulation pattern obtained in this way for a frozen solution of hydrated Cu(II). The ~90 nsec period observable at the beginning of the trace corresponds to the water proton precession period in the applied magnetic field. The shorter period of about 45 nsec corresponds to the sum of the ENDOR frequencies that are also characteristic of protons. For deuterons, either bound specifically to Cu(II) or other paramagnetic metal ligand, the frequency is quite different and one is able to demonstrate proton-deuteron exchanges near a paramagnetic center using this technique[27]. The intensity of the pattern can be related quantitatively to the number of interacting nuclei and their distance from the paramagnetic center[28].

The nuclear modulation pattern for stellacyanin (Fig. 7a) is

298

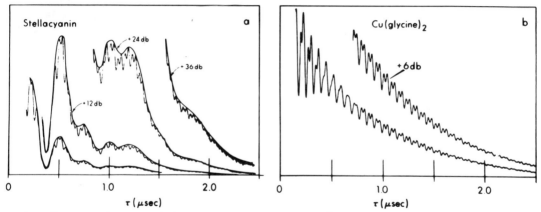

Fig. 7. Nuclear modulation patterns for a) stellacyanin and b) Cu(II)-(bisglycine). In both traces, the high frequency pattern is due to proton. In a, the low frequency pattern is due to imidazole ^{14}N. Directly coupled ^{14}N as in b does not contribute. (After Ref. 26).

completely different in shape. Upon careful examination, one can still observe the pattern ascribed to protons (Fig. 6), but this no longer predominates the trace. The major features, which are emphasized by the solid line drawn through the data, consist of a low frequency complex pattern. We have been able to ascribe this pattern to imidazole nitrogen. Almost an identical pattern[26] is obtained for Cu(II)-bovine serum albumin where it has been demonstrated[29] that the ligands to copper consist of one amino nitrogen, two peptide nitrogens and one nitrogen from a histidine imidazole. By comparison with a peptide complex (Fig. 7b), here Cu(II)-glygly at pH 8 where the imidazole is absent but the amino and peptide nitrogens are still present[30], one no longer observes the slow period as is found for stellacyanin and for bovine serum albumin. Fig. 7b points out the fact that directly coordinated nitrogen to Cu(II) does not contribute to the modulation pattern and for the example given, only the characteristic proton pattern is observed.

Other evidence that the dominating features in the modulation pattern for stellacyanin arises from imidazole nitrogen is obtained from experiments with model compounds. For example, a qualitatively similar pattern, though with greater modulation depth, is seen for Cu(II)-(imid)$_4$ but not for Cu(II)-(guanidine)$_4$[36]. In addition, a model compound consisting of Cu(II)-diethylenetriamine and imidazole (Fig. 8) has a modulation pattern[20] having strong resemblance

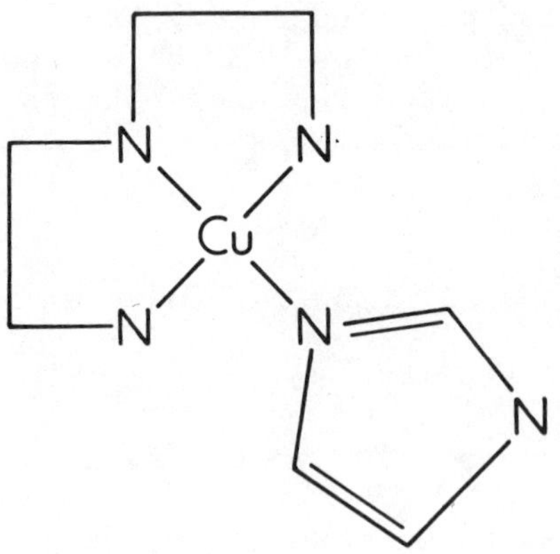

Fig. 8. Chemical structure of Cu(II)-diethylenetriamine imidazole, a chemical model in which the single imidazole gives rise to the nuclear modulation pattern in Fig. 7a.

(Fig. 9d) to that for stellacyanin (Fig. 7a). Furthermore, the same complex prepared with ^{15}N imidazole has a radically different modulation pattern[31]. It should be pointed out that Cu(II)-diethylenetriamine in the absence of either ^{14}N- or ^{15}N-imidazole has a modulation pattern qualitatively not very different from that for Cu(II)-glygly (Fig. 7b).

We have been able to extend our study to other copper proteins and have been able to demonstrate imidazole ligation in these cases

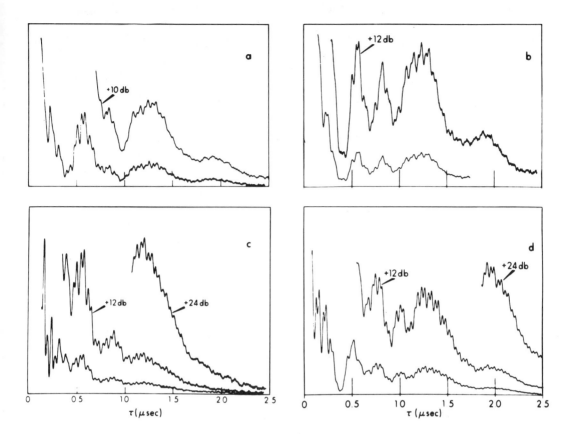

Fig. 9. Nuclear modulation patterns for a) Type 2 Cu(II) of laccase
b) Type 2 Cu(II) of ceruloplasmin, c) Type 1 copper of laccase and
d) the model shown in Fig. 8. (After Ref. 20).

as well. For example, R. vernicifera laccase and porcine ceruloplasmin each contain two forms of paramagnetic Cu(II) (Table II).
Due to the large difference in nuclear hyperfine interactions of the
two paramagnetic types of copper in these proteins (Fig. 1 and
Table I), one ca.. easily examine the nuclear modulation effect of
Type 2 copper by studying the spin echo amplitude in the region of
the EPR where the spectra of the two paramagnetic forms do not overlap. The result of such a study is shown in Fig. 9a and b. As can
be seen, the characteristic imidazole nitrogen pattern is present.

In another study[20], the modulation pattern for Type 1 copper of laccase was examined in a preparation where the Type 2 copper had been reversibly removed[32]. Here too, a pattern characteristic of ligated imidazole was seen (Fig. 9c).

Imidazole ligation to Cu(II) is not limited to blue copper proteins but its presence has also been demonstrated for Cu-transferrin, Cu-carbonic anhydrase and galactose oxidase[32]. Two noteworthy cases where this pattern is not seen for copper containing proteins is for bovine cytochrome c oxidase and for the mercaptoethanol derivative of Neurospora crassa tyrosinase. The latter protein contains a single Type 3 site[34] and in the native state has no EPR. However, the addition of mercaptoethanol converts the protein to a paramagnetic form, the EPR of which[35] strongly resembles that heretofore ascribed to the Cu(II) in cytochrome c oxidase[5]. For both paramagnetic proteins the nuclear hyperfine pattern cannot be resolved and the EPR resembles that of an axial free radical. The nuclear modulation patterns for these proteins is shown in Fig. 10. Noteworthy, in both traces is the absence of the pattern which we have ascribed to imidazole nitrogen and which is present for all Cu(II) proteins we have studied. The weak, low frequency patterns resemble that observed for Cu(II)-hexagly where some distant dipolar interaction with nitrogen produces a small perturbation in the modulation pattern (See Ref. 26, Fig. 7). Furthermore, the intensity of the pattern ascribed to protons is much weaker than for most Cu(II) complexes. This finding suggests that the unpaired electrons in both these proteins are relatively isolated from close lying protons, a condition not observed for copper proteins in general. These findings taken together with the lack of nuclear hyperfine pattern in the

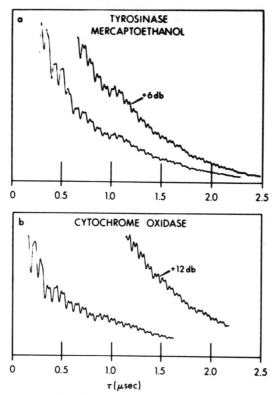

Fig. 10. Nuclear modulation pattern for a) cytochrome c oxidase and for the mercaptoethanol derivative of b) N. crassa tyrosinase. In a, the nuclear modulation was studied in the region of the EPR heretofore ascribed to Cu(II) and which we ascribe to a sulfur free radical.

EPR of these proteins lead us to conclude that the paramagnetic species in both proteins is not Cu(II). Our earlier suggestion[3] that the EPR of cytochrome oxidase is that of a sulfur radical seems reasonable based on the uniqueness of the nuclear modulation pattern. The best electronic description then, is one where the unpaired electron resides primarily on the sulfur radical and not on the copper. If we wish to include copper in the paramagnetic site, one might say that in cytochrome c oxidase, a Cu(I) $\longrightarrow$ ·SR charge transfer is present as compared to the charge transfer as $RS^- \longrightarrow$ Cu(II) in Type 1 copper site. Although it has not been demonstrated that in the

303

oxidase the copper is at all involved in the paramagnetic site, it is interesting to speculate that both the metal ion and sulfur are present.

ACKNOWLEDGEMENTS

This work was supported by U. S. Public Health Service Research Grant HL-13399 from the Heart and Lung Institute, and by National Cancer Institute Contract NO1-CP-55606, and as such is Communication No. 373 from the Joan and Lester Avnet Institute of Molecular Biology.

REFERENCES

1. Malkin, R. and Malmström, B. G. (1970) Adv. Enzymol. 33, 177-244.

2. Fee, J. A. (1975) Structure and Bonding 23, 1-60.

3. Peisach, J. and Blumberg, W. E. (1974) Arch. Biochem. Biophys. 165, 691-708.

4. Richardson, J. S., Thomas, K. A., Rubin, B. H. and Richardson, D. C. (1975) Proc. Natl. Acad. Sci. (U.S.A.) 72, 1349-1353.

5. Beinert, H., Griffiths, O. E., Wharton, D. C. and Sands, R. H. (1962) J. Biol. Chem. 237, 2337-2346.

6. Katoh, S. and Takamiya, A. (1964) J. Biochem. (Tokyo) 55, 378-387.

7. Finazzi-Agró, A., Rotilio, G., Avigliano, L., Guerrier, P., Boffi, V. and Mondovi, B. (1970) Biochemistry 9, 2009-2014.

8. McMillin, D. R., Holwerda, R. A. and Gray, H. B. (1974) Proc. Natl. Acad. Sci. (U.S.A.) 71, 1339-1341.

9. Thompson, J. S., Marks, T. J. and Ibers, J. A. (1977) Proc. Natl. Acad. Sci. (U.S.A.) 74, 3114-3118.

10. Peisach, J., Levine, W. G. and Blumberg, W. E. (1967) J. Biol. Chem. 242, 2847-2858.

11. Royce, E. B. and Bloembergen, N. (1963) Phys. Rev. 131, 1912-1923.

12. Roitsin, A. B. (1971) Usp. Fiz. Nauk 105, 677-705.

13. Mims, W. B. (1976) The Linear Electric Field Effect in Paramagnetic Resonance, Clarendon Press, Oxford.

14. Helcké, G. A., Ingram, D. J. E. and Slade, E. F. (1968) Proc. Royal Soc. (London) B169, 275-279.

15. Mims, W. B. and Peisach, J. (1974) Biochemistry 13, 3346-3349.

16. Mims, W. B. and Peisach, J. (1976) J. Chem. Phys. 64, 1074-1091.

17. Peisach, J. and Mims, W. B. (1976) Chem. Phys. Letters, 37, 307-310.

18. Peisach, J. and Mims, W. B. (1978) Eur. J. Biochem., in press.

19. Kokoszka, G. F., Reiman, C. W. and Allen, H. C. Jr. (1967) J. Phys. Chem. 71, 121-126.

20. Mondoví, B., Graziani, M. T., Mims, W. B., Oltzik, R. and Peisach, J. (1977) Biochemistry 16, 4198-4202.

21. Markley, J. L., Ulrich, E. L., Berg, S. P. and Krogman, D. W. (1975) Biochemistry 14, 4428-4433.

22. Hill, H. A. O., Leer, J. C., Smith, B. E., Storm, C. B. and Ambley, R. P. (1976) Biochem. Biophys. Res. Commun., 70, 331-338.

23. Ugurbil, K., Norton, R. F., Allerhand, A. and Bersohn, R. (1977) Biochemistry 16, 886-894.

24. Geschwind, S. (1972) Electron Paramagnetic Resonance, Plenum Press, New York, Chapter 4.

25. Salikhov, V. M., Semenov, A. G. and Tsvetkov, Y. D. (1976) Electron Spin Echoes and Their Application, Science Press, Novosibirsk.

26. Mims, W. B. and Peisach, J. (1976) Biochemistry 17, 3863-3868.

27. Peisach, J., Orme-Johnson, N. R., Mims, W. B. and Orme-Johnson, W. H. (1977) J. Biol. Chem. 252, 5643-5650.

28. Mims, W. B., Peisach, J. and Davis, J. L. (1977) J. Chem. Phys. 66, 5536-5550.

29. Peters, T. and Blumenstock, F. A. (1967) J. Biol. Chem. 242, 1574-1578.

30. Freeman, H. C. (1966) in The Biochemistry of Copper, J. Peisach, P. Aisen and W. E. Blumberg, eds., New York, N. Y. Academic Press, pp. 77-113.

31. Peisach, J. and Mims, W. B. to be published.

32. Graziani, M. T., Morpurgo, L., Rotilio, G. and Mondovi, B. (1976) FEBS Letters, 70, 87-90.

33. Kosman, D. J., Peisach, J. and Mims, W. B. to be published.

34. Lerch, K. (1976) FEBS Letters, 69, 157-160.

35. Deinum, J., Lerch, K. and Reinhammer, B. (1976) FEBS Letters, 76, 161-164.

INDEX

pyridine-nucleotide, 94-98, 105-106
succinate, 143-53
thiamine, 45-54
trimethylamine, 127-41
Deoxycholate
formate dehydrogenase and,156
glutathione-insulin dehydrogenase and, 34-35, 37
glutathione-insulin transhydrogenase and, 34-35
nitrate reductase and, 157
Deuterium isotope effect
aldehyde exchange and, 70-73
bacterial luciferase and, 69-77
carbon-hydrogen band scission and, 70, 71, 73
pyridine coenzymes and, 81
stopped-flow kinetics and, 69, 70-71, 73-74
temperature and, 73
Dimethylsuberimidate, 251, 254-55, 257-58
Dioxygen
reduction by laccase, 263-70
superoxide dimutase and, 277-83
Dismutation
catalysed, 274-77
iron-superoxide, 274, 277-83
spontaneous, 273-74
Distal sulfur, disulfide reductase, 21
Disulfide bond
in glutathione-insulin transhydrogenase, 30-32, 40
in glutathione reductase, 20-21
in oxytocin, 30
in vasopressin, 30
Dithionite, 4, 128-30, 135
Dopamine-B-hydroxylase, 287-89

Electrode potential, clostridial hydrogenase, 121, 123
Electron density map, glutathione reductase, 17
Electron paramagnetic resonance (epr) studies
of ascorbate oxidase, 289
of azurin, 289
of ceruloplasmin, 289
of clostridial hydrogenase, 121-22
of copper proteins, 286-91, 301, 302
of cytochrome c oxidase, 222-30, 239-40, 245, 285, 290
of cytochrome c peroxidase, 216
of dopamine-B-hydroxylase, 289
of FeMo cofactor, 165-66, 171
of galactose oxidase, 289
of laccase, 264-70, 289, 296

of myoglobin azide, 291-93
of nitrate reductase, 206-209
of nitrogenase, 166-67, 174, 177, 178
of QH_2:cytochrome c oxidoreductase, 181, 182-84, 187-88
of stellacyanin, 289
of sulfite reductase, 206-209
of trimethylamine dehydrogenase, 129-35
of tyrosinase, 289
Electron transfer
covalent catalysis and, 3, 6-7, 15
cytochrome c and, 251
in disulfide reductase, 21
in flavoproteins, 4, 6, 14-15
pH dependence of, 5-6, 9
pyridine nucleotides and, 3, 10-12
in superoxide dismutase, 275
Escherichia coli, nitrate reductase complex of
cytochrome b_1 in, 158-61
formate dehydrogenase and, 155, 156, 159-61
genetic control of, 161-62
nitrate reductase and, 155, 157-61
regulation of, 159-61

Facultative bacteria, nitrate respiration in, 155, 159
FAD-binding domain, glutathione reductase, 20
FeMo cofactor
epr spectra of, 165, 169, 171
Mössbauer spectra of, 165, 169-70
Fe protein, nitrogenase, 165, 166, 173
ATP activation of, 176, 179
reduction of, 174, 175-76
Ferrocytochrome c
cytochrome c oxidase and, 221, 239, 243-44
cytochrome c peroxidase and, 216-18
Ferryl heme iron, cytochrome c peroxidase, 217
Flavoproteins
active site cysteine residues and, 3-6
glupo sequence of, 19, 20
N-oxidase and, 194
Fluorescence
of cytochrome c peroxidase, 215, 216,219
in liver alcohol dehydrogenase, 108, 110-13
Fluoride inhibition, 277, 280-81
Formate dehydrogenase, 155, 156, 159-61
Free radical
cytochrome c oxidase ligation and, 290
cytochrome c peroxidase and, 217

Galactose oxidase
   epr study of, 289
   imidazole ligation and, 302
   type 2 copper in, 287, 288
Glucose-6-phosphate dehydrogenase
   catalytic properties of, 25
   homogeneity of, 23, 24-25
   molecular associations of, 23,
     25-27
   physical properties of, 23-28
   purification of, 23, 24
   subunit composition of, 23,
     25-27
Glupo sequence, flavoprotein, 19,
   20
Glutamate dehydrogenase
   ADP interaction with, 95-96, 99,
     106
   calorametric measurements of,
     94, 99-100, 101
   carbinolamine complex and,
     103-105
   coenzyme binding and, 94-95
   evolutionary aspects of, 105-106
   hydrogen exchange and, 96-97
   iminiglutamate complex and, 103-
     105
   reaction energetics of, 99-105
   reaction kinetics of, 97-99
   solvent isotope effect and, 98
Glutathione-insulin transhydro-
     genase
   amino acid composition of, 39
   disulfide bond probe and, 30-32
   mechanism of, 29-35
   structure of, 35, 37-40
Glutathione reductase
   active site structure, 17-20
   catalytic properties of, 20, 25
   crystal structure of, 17-21
   cysteine residue generation
     and, 3-7
   disulfide bridge of, 20-21
   homogeneity of, 23, 24-25
   molecular associations of, 23,
     25-26, 26-27
   physical properties of, 23-28
   purification of, 23, 24
   structural domains of, 20
   subunit composition of, 23,
     25-27
   thiol reactivity and, 12-15
Glycollate
   adduct structure of, 57-67
   lactate oxidase and, 55, 56, 58
   stereodensity of, 60-62
Glycollyl-N(5)-flavin adduct
   lactate oxidase reaction
     mechanism and, 55, 56, 58-59
   structure of, 57-62, 64-66

Hormone inactivation, 29, 30

Horseradish peroxidase, 216-18
Hydrogen peroxide substrate, 215
Hydrogenase, Clostridium pasteurianum
   electrode potential of, 121, 123
   epr study of, 121-22
   function of, 122-24
   inhibition of, 124
   methyl viologen, 122-23
   optical absorption spectra of,
     120-21
   preparation of, 119
   tetrameric iron-sulfur center of,
     120
Hydroperoxidases, 215-16

Imidazole ligation, 299-302
   of carbonic anhydrase, 302
   of ceruloplasmin, 301
   of cytochrome oxidase, 232
   of galactose oxidase, 302
   of laccase, 301
   of superoxide dismutase, 289
   of transferrin, 302
Imidoesters, 254
Iminoglutarate complex, 103-105
Inhibition, enzyme
   of clostridial hydrogenase, 124
   of glutathione-insulin trans-
     hydrogenase, 34-36
   of lactate oxidase, 55, 58
   of nitrite reductase, 206-207
   of nitrogenase, 165, 167, 178
   of succinate dehydrogenase, 143-44
     148-53
   of sulfite reductase, 206-207
Insulin
   degradation of, 29, 32-34
   disulfide bond cleavage in, 30-32
   glutathione-insulin transhydro-
     genase and, 29-32
Interface domain, glutathione re-
     ductase, 20
Iron-sulfur clusters (centers)
   clostridial hydrogenase, 120
   extrusion of, 120
   formate dehydrogenase, 156
   line shape changes and, 181, 183
   nitrate reductase, 158, 207-210
   nitrogenase, 165, 169, 173, 175-76
   oxidoreductase, 181-88
   Rieske, 181, 183
   succinate dehydrogenase, 144
   sulfite reductase, 207-10
   trimethylamine dehydrogenase, 127-
     130
Iron superoxide dismutase
   azide inhibition of, 277, 281-82
   dioxygen dismutation catalysis by,
     277-83
   fluoride inhibition of, 277, 280-8
Isoalloxazine ring, 20

D